INTERNATIONAL UNION OF CRYSTALLOGRAPHY
TEXTS ON CRYSTALLOGRAPHY

This volume forms part of a series of books sponsored by the International Union of Crystallography (IUCr) and published by Oxford University Press. There are three IUCr series: IUCr Monographs on Crystallography, which are in-depth expositions of specialized topics in crystallography; IUCr Texts on Crystallography, which are more general works intended to make crystallographic insights available to a wider audience than the community of crystallographers themselves; and IUCr Crystallographic Symposia, which are essentially the edited proceedings of workshops or similar meetings supported by the IUCr.

IUCr Monographs on Crystallography

1 *Accurate molecular structures: Their determination and importance*
  A. Domenicano and I. Hargittai, *editors*
2 *P. P. Ewald and his dynamical theory of X-ray diffraction*
  D. W. J. Cruickshank, H. J. Juretscke, and N. Kato, *editors*
3 *Electron diffraction techniques, Volume 1*
  J. M. Cowley, *editor*
4 *Electron diffraction techniques, Volume 2*
  J. M. Cowley, *editor*
5 *The Rietveld method*
  R. A. Young, *editor*

IUCr Texts on Crystallography

1 *The solid state: From superconductors to superalloys*
  A. Guinier and R. Jullien, *translated by* W. J. Duffin
2 *Fundamentals of crystallography*
  C. Giacovazzo, *editor*
3 *The basics of crystallography and diffraction*
  C. Hammond
4 *X-ray charge densities and chemical bonding*
  P. Coppens

IUCr Crystallographic Symposia

1 *Patterson and Pattersons: Fifty years of the Patterson function*
  J. P. Glusker, B. K. Patterson, and M. Rossi, *editors*
2 *Molecular structure: Chemical reactivity and biological activity*
  J. J. Stezowski, J. Huang, and M. Shao, *editors*
3 *Crystallographic computing 4: Techniques and new technologies*
  N. W. Isaacs and M. R. Taylor, *editors*
4 *Organic crystal chemistry*
  J. Garbarczyk and D. W. Jones, *editors*
5 *Crystallographic computing 5: From chemistry to biology*
  D. Moras, A. D. Podjarny, and J. C. Thierry, *editors*
6 *Crystallographic computing 6: A window on modern crystallography*
  H. D. Flack, L. Parkanyi, and K. Simon, *editors*
7 *Correlations, transformations, and interactions in organic crystal chemistry*
  D. W. Jones and A. Katrusiak, *editors*

X-RAY CHARGE DENSITIES AND CHEMICAL BONDING

Philip Coppens

Chemistry Department
State University of New York at Buffalo

International Union of Crystallography
Oxford University Press
1997

Oxford University Press

Oxford   New York
Athens   Auckland   Bangkok   Bogota   Bombay   Buenos Aires
Calcutta   Cape Town   Dar es Salaam   Delhi   Florence   Hong Kong
Istanbul   Karachi   Kuala Lumpur   Madras   Madrid   Melbourne
Mexico City   Nairobi   Paris   Singapore   Taipei   Tokyo   Toronto

and associated companies in
Berlin   Ibadan

Library of Congress Cataloging-in-Publication Data
Coppens, Philip.
X-ray charge densities and chemical bonding / Philip Coppens.
p.   cm. — (International Union of Crystallography texts on
crystallography ; 4)
Includes bibliographic references and index.
ISBN 0-19-509823-4
1. X-ray crystallography. 2. Chemical bonds. 3. Electron
distribution. I. Title. II. Series
QD945.C597   1997
548'.3—dc21   96-48736

1 3 5 7 9 8 6 4 2

Printed in the United States of America
on acid-free paper

Preface

It seems to me that the experimental study of scattered radiation, in particular from light atoms, should get more attention, since along this way it should be possible to determine the arrangement of the electrons in the atoms

P. Debye (1915)

An accurate set of nuclear coordinates and a detailed map of the electron density can be obtained, by X-ray diffraction, only jointly and simultaneously, never separately or independently

F. L. Hirshfeld (1992)

The ability to measure the experimental charge distribution in crystals from the intensities of the scattered X-rays was realized almost immediately after the discovery of X-ray diffraction. Notwithstanding this early recognition, the technical developments of the 1960s and beyond, which occurred in diffractometry, automation of data collection, low-temperature techniques, and computers, were needed to achieve a breakthrough in the method. The accurate crystallographic methods developed during the past decades led not only to a much better precision in atomic coordinates, but also to crucial information on the charge distribution in crystals. This experimentally obtained distribution can be compared directly with theoretical results, and can be used to derive other physical properties, such as electrostatic moments, the electrostatic potential, and lattice energies, which are accessible by spectroscopic and thermodynamic measurements. This broad interface with other physical sciences is one of the most appealing aspects of the field.

The aim of this volume is to provide the background necessary for interpretation of the results of accurate crystallographic methods, and to present the concepts to a wider community of nonspecialized scientists. Though a number of excellent conference proceedings exist, there is no single text summarizing the subject. This text is not meant as a comprehensive review of the existing literature; experimental results are presented as an illustration of principles and methods. While this leaves out a great many valuable studies available in the literature, the restraint is unavoidable if the text is to be of manageable size.

Experimental methods have not been covered, as they are still developing rapidly. New developments, in particular the advent of third-generation synchrotron sources combined with parallel data collection methods and new interpretative software packages, are bound to further enhance the scope of the field. It is hoped that this text will serve to stimulate its continuing development.

I am grateful to the many colleagues who have commented on the contents of this volume, and especially to Dr. Zhengwei Su who played a key role in the development of the formalisms described in chapters 7–9 and appendixes D, E, G, and H, and who has pointed out many errors and omissions in earlier versions of the text. Others, including D. Feil, T. R. Furlani, C. Lecomte, V. Petricek, and S. Price provided additional valuable suggestions and corrections. I would like to thank S. Priore-Fensore for her outstanding dedication, which made the completion of the manuscript possible, and I. Novozhilova for assistance in the final stages of this project.

Philip Coppens
Buffalo, NY
November 1996

Contents

X-RAY CHARGE
DENSITIES AND
CHEMICAL BONDING

1

Scattering of X-rays and Neutrons

1.1 Outline of this Chapter

This chapter starts with a discussion of the classical treatment of X-ray scattering, followed by a brief overview of the quantum-mechanical theory in the first Born approximation. The scattering of a periodic arrangement is derived by considering the crystal as a convolution of the unit cell contents and a periodic lattice. The atomic description of the charge density, which is the basis for structure analysis, is introduced. The origin of resonance anomalous scattering is discussed. While its effect must be accounted for before charge densities can be derived from the X-ray scattering amplitudes, resonance scattering itself can give invaluable information on the electronic states of the resonating atoms. The final section of this chapter deals with the scattering of neutrons by atomic nuclei. Nuclear neutron scattering is independent of the distribution of the electrons, and can provide atomic positions and thermal amplitudes unbiased by the bonding effects which are the subject of this book.

1.2 Introduction to the Theory of X-ray Scattering

1.2.1 Classical Treatment of X-ray Scattering

In the classical theory of scattering (Cohen-Tannoudji et al. 1977, James 1982), atoms are considered to scatter as dipole oscillators with definite natural frequencies. They undergo harmonic vibrations in the electromagnetic field, and emit radiation as a result of the oscillations.

The equation of motion for a single harmonic oscillators of mass m, and force constant k_f ($k_f = m\omega_s^2$), is (following Newton) given by

$$m\ddot{x} + k_f x = 0 \tag{1.1}$$

where $\ddot{x}$ is the second time derivative of the displacement x. When the oscillator is a particle with charge $-e$, exposed to an oscillating electric field $E = E_0 e^{i\omega t}$, one obtains

$$\ddot{x} + \omega_s^2 x = \frac{eE_0}{m} e^{i\omega t} \tag{1.2}$$

When the oscillation is damped, with damping factor k, and damping $k\dot{x}$ proportional to the speed of motion $\dot{x}$, one obtains

$$\ddot{x} + k\dot{x} + \omega_s^2 x = \frac{eE_0}{m} e^{i\omega t} \tag{1.3}$$

The corresponding time-dependent value of the damped oscillating dipole equals

$$\mathbf{M} = e\mathbf{x} = -\frac{e^2}{m} \frac{\mathbf{E}_0 e^{i\omega t}}{\omega^2 - \omega_s^2 - ik\omega} \tag{1.4}$$

as can be verified by direct substitution of Eq. (1.4) into Eq. (1.3).

The oscillating dipole is a source of electromagnetic radiation of the same frequency, polarized in the direction of the oscillations. At large distances, the wave is spherical. According to the electromagnetic theory, the resulting electric vector at a point in the equatorial plane of the dipole is $\omega^2/|\mathbf{r}|c^2$ times the moment of the dipole at time $t - |\mathbf{r}|/c$. The amplitude of the spherically scattered wave at unit distance in the equatorial plane is therefore

$$A = -\frac{e^2}{mc^2} \frac{\omega^2 E_0}{\omega^2 - \omega_s^2 - ik\omega} \tag{1.5a}$$

The scattering of the free electron for $E_0 = 1$ is obtained from this expression by setting both the force constant ω_s and the damping factor k equal to zero, which gives

$$A_{\text{free electron}} = -e^2/mc^2 \tag{1.5b}$$

The quantity e^2/mc^2 is the *scattering amplitude of the classical electron*, denoted by the symbol r_e, and generally used as the unit of electron scattering. Its numerical value equals $2.818 \cdot 10^{-13}$ cm $= 2.818$ fermi.[1]

For an assembly of several free point electrons, interference occurs between radiation scattered by different centers. If the incident and diffracted beams are defined by two unit vectors, $\mathbf{s}_0$ and $\mathbf{s}$, respectively (Fig. 1.1), the phase difference of the radiation scattered by two points, separated by the vector $\mathbf{r}$, equals $2\pi\mathbf{S}\cdot\mathbf{r}$, where $\mathbf{S}$ is the scattering vector, equal to $(\mathbf{s} - \mathbf{s}_0)/\lambda$. Vector $\mathbf{S}$ bisects $\mathbf{s}$ and $\mathbf{s}_0$, and has the length $2\sin\theta/\lambda$. In the physics literature, an alternative notion is commonly used. The incident and diffracted beams are defined by the vectors $\mathbf{k}$

---

[1] If physical constants in SI units are used in the evaluation of e^2/mc^2, the result has to be divided by the factor 4π times the permittivity of free space $4\pi\varepsilon_0 = 1.112\,626\,5\cdot10^{-10}$ C^2 N^{-1} m^{-2}.

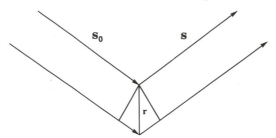

FIG. 1.1 Geometry of scattering

and $\mathbf{k}_0$, of length $2\pi/\lambda$, such that the scattering vector $\mathbf{K} = \mathbf{k} - \mathbf{k}_0$ has a magnitude $|K| + 4\pi \sin \theta/\lambda$.

For the scattering by two point electrons at distance $\mathbf{r}$, using the rules for addition of coherent waves, we get per unit incident amplitude E_0

$$A = (1 + e^{2\pi i \mathbf{S} \cdot \mathbf{r}})\frac{e^2}{mc^2} \tag{1.6}$$

in which the negative sign of Eq. (1.5b) has been factored out.

For a continuous electron distribution $\rho(\mathbf{r})$, the summation over waves of different phase must be replaced by an integration leading to a diffraction amplitude:

$$A(\mathbf{S}) = \int \rho(\mathbf{r}) \exp(2\pi i \mathbf{S} \cdot \mathbf{r}) \, d\mathbf{r} \tag{1.7}$$

Thus, for $\omega \gg \omega_s$ and $\omega^2 \gg ik\omega$, the amplitude of scattering is the Fourier transform of the electron density. This important result is confirmed by the first-order quantum-mechanical treatment discussed in the next section. When $\omega \cong \omega_s$, both the amplitude and the phase of the scattered radiation depend on the frequency.

1.2.2 Quantum-Mechanical Treatment: The First Born Approximation

The treatment of the interaction of a quantum-mechanical system with radiation is described in detail in the literature (Feil 1975; Cohen-Tannoudji et al. 1977; Blume 1985, 1994). Only an outline will be given here.

The Hamiltonian $\hat{H}$ of a system in a radiation field, in the absence of interaction with the field and a perturbation Hamiltonian H' describing the interaction with the field, is given by

$$\hat{H} = \hat{H}_{\text{system}} + \hat{H}_{\text{radiation}} + \hat{H}' \tag{1.8}$$

The interaction of a system with the electromagnetic radiation leads to a time-dependence of the wave function ψ, which follows from the time-dependent Schrödinger equation

$$H\psi = \frac{ih}{2\pi}\frac{d\psi}{dt} \tag{1.9}$$

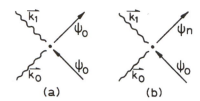

FIG. 1.2 Feynman diagrams describing (a) elastic scattering, (b) inelastic scattering. *Source*: Feil (1975).

The time-dependent wave function is described by

$$\psi = c_m(t)\psi_m(x, t) + c_n(t)\psi_n(x, t) \tag{1.10a}$$

in which the x values are the coordinates of the scattering particles, and the transition is from an initial state ψ_m to a final state ψ_n. The wave function ψ describes both the scattering assembly M and the radiation field; in shorthand notation,

$$|\psi_m\rangle \equiv |\psi_{M,0}, 1_{k_0}, 0_{k_1}\rangle \tag{1.10b}$$

for the initial-state wave function and an incident photon with propagation vector $k_0(|k| = 2\pi/\lambda)$.

In the *first Born approximation*, the interaction between the photons and the scattering system is weak and no excited states are involved in the elastic scattering process. Furthermore, there is no rescattering of the scattered wave, that is, the single-scattering approximation is valid. In the Feynman diagrams (Fig. 1.2), there is only one point of interaction for first-Born-approximation processes.

The interaction Hamiltonian contains the operator $\hat{A}$, corresponding to the vector potential A of the electromagnetic field.[2] Excluding magnetic scattering, the interaction Hamiltonian is given by

$$\hat{H}' = \frac{q^2}{2mc^2}(\hat{A} \cdot \hat{p}) + \frac{q^2\hat{A}^2}{2m} \tag{1.11}$$

in which q is the charge of the scattering particle, and m is its mass. Note that scattering by the heavier nuclei is orders of magnitude smaller than the scattering by the electrons. Vector $\hat{p}$ is the momentum operator equal to $(h/2\pi i)\nabla$.

The second term in Eq. (1.11) is the origin of the scattering in the first Born approximation. It leads to an amplitude for the scattering of photons with propagation vector k_0 into photons with vector k equal to

$$F = C\langle\psi_n| \exp(i\mathbf{K} \cdot \mathbf{r})|\psi_m\rangle \tag{1.12}$$

with $\mathbf{K} = \mathbf{k} - \mathbf{k}_0$. In this expression, ψ is the wave function describing the assembly only, and C is a proportionality constant which includes the polarization factor.[3] The positive sign of the exponent in Eq. (1.12) is the convention followed in crystallography (James 1982) and adopted in this text. To avoid confusion, it

---

[2] The electric field E and the magnetic field B are related to the vector potential by $E = -\partial A/\partial t$ and $B = \nabla \times A$. See, for example, Jackson (1977).

[3] Polarization is explicitly included in the treatment in section 1.3.2.

should be pointed out that in the physics literature the plane-wave expression $\exp(-i\mathbf{K}\cdot\mathbf{r})$ is used.

For independently scattering systems, without specific phase relation, we get for the total intensity in the scattered beam

$$I_{total} = C^2 \sum_n \langle \psi_n | \exp(i\mathbf{K}\cdot\mathbf{r}) | \psi_m \rangle^2 \qquad (1.13)$$

The integration is over the coordinates of all the electrons, and the wave function ψ describes all particles of the scattering system.

For elastic scattering, illustrated by Fig. 1.2(a), the initial and final states are the same, that is, $n = m$. For a system of electrons, represented by a many-electron wave function, we obtain in the approximation that the N electrons are scattering independently:

$$I_{coherent,\,elastic}(\mathbf{K}) = C^2 \left\langle \psi_0 \left| \sum_{i=1}^{N} \exp(i\mathbf{K}\cdot r_i) \right| \psi_0 \right\rangle^2 \qquad (1.14)$$

As the electrons are indistinguishable in the antisymmetrized wave function, the one-electron scattering can be obtained by integration over all coordinates but those of the jth electron. Summation over all equivalent electrons then leads to

$$I_{coherent,\,elastic}(\mathbf{K}) = \left| \int \rho(\mathbf{r}) \exp(i\mathbf{K}\cdot\mathbf{r}) \, d\mathbf{r} \right|^2 \qquad (1.15)$$

where $\rho(\mathbf{r})$ is the electron distribution. Or, for the scattering amplitude $A(\mathbf{K})$,

$$A(\mathbf{K}) = \int \rho(\mathbf{r}) \exp(i\mathbf{K}\cdot\mathbf{r}) \, d\mathbf{r} = \hat{F}[\rho(\mathbf{r})] \qquad (1.16)$$

where $\hat{F}$ is the Fourier transform operator. This result is equivalent to the classical expression (1.7). It is sometimes referred to as the *form-factor approximation*, and is based on the assumption that the particles scatter independently of each other. This assumption neglects binding effects among the electrons and between electrons and nucleons, which become important as very low and very high energies (Kissel et al. 1995).

1.2.3 Scattering by a Periodic Crystal

According to Eq. (1.16), the elastic coherent X-ray scattering amplitude is the Fourier transform of the electron density in the crystal. The crystal is a three-dimensional periodic function described by the convolution of the unit cell density and the periodic translation lattice. For an infinitely extended lattice,

$$\rho_{crystal}(\mathbf{r}) = \sum_n \sum_m \sum_p \rho_{unit\,cell}(\mathbf{r}) * \delta(\mathbf{r} - n\mathbf{a} - m\mathbf{b} - p\mathbf{c}) \qquad (1.17)$$

where n, m, and p are integers, and δ is the Dirac delta function. Equation (1.16) states that the scattering of the crystal is the Fourier transform of $\rho_{crystal}(\mathbf{r})$. According to the Fourier convolution theorem, the Fourier transform of a convolution of two functions is the product of the Fourier transform of the individual functions, or

$$A(\mathbf{S}) = \hat{F}\{\rho(\mathbf{r})\} = \sum_n \sum_m \sum_p \hat{F}\{\rho_{unit\,cell}(\mathbf{r})\} \hat{F}\{\delta(\mathbf{r} - n\mathbf{a} - m\mathbf{b} - p\mathbf{c})\} \qquad (1.18)$$

The Fourier transform of the direct space δ function is a δ function in reciprocal space, representing the reciprocal lattice. We thus obtain

$$A(\mathbf{S}) = \hat{F}\{\rho_{\text{unit cell}}(\mathbf{r})\} \sum_h \sum_k \sum_l \delta(\mathbf{S} - h\mathbf{a}^* - k\mathbf{b}^* - l\mathbf{c}^*) \tag{1.19}$$

It follows from Eq. (1.19) that a crystal scatters with an amplitude proportional to $\hat{F}\{\rho_{\text{unit cell}}(\mathbf{r})\}$ in directions defined by the scattering vectors

$$\mathbf{S} = \mathbf{H} = h\mathbf{a}^* + k\mathbf{b}^* + l\mathbf{c}^* \tag{1.20}$$

where the reciprocal axes $\mathbf{a}^*_{i, i=1, 3}$ are defined by

$$\mathbf{a}_i \cdot \mathbf{a}^*_j = \delta_{ij} \tag{1.21}$$

Thus, the scattering of a periodic lattice occurs in discrete directions. The larger the translation vectors defining the lattice, the smaller $\mathbf{a}^*_{i, i=1, 3}$, and the more closely spaced the diffracted beams. This inverse relationship is a characteristic property of the Fourier transform operation. The scattering vectors terminate at the points of the *reciprocal lattice* with basis vectors $\mathbf{a}^*_{i, i=1, 3}$, defined by Eq. (1.21).

The Fourier transform of the unit cell density $\hat{F}\{\rho_{\text{unit cell}}(\mathbf{r})\}$ is referred to as the *structure factor F*:

$$F(\mathbf{H}) = \hat{F}\{\rho_{\text{unit cell}}(\mathbf{r})\} = \int_{\text{unit cell}} \rho(\mathbf{r}) \exp(2\pi i \mathbf{H} \cdot \mathbf{r}) \, d\mathbf{r} \tag{1.22}$$

Expression (1.19) is valid for a crystal with a very large number of unit cells, for which particle-size broadening is negligible, as indicated by the infinitely sharp δ function. However, a finite lattice is the mathematical product of the infinite lattice used in the derivation of Eq. (1.19) and a three-dimensional step function describing the shape of the crystal. Using again the Fourier convolution theorem, which states that the Fourier transform of a product is the convolution of the Fourier transforms of the individual functions, shows that in the scattering expression (1.19), the periodic delta function is to be convoluted with the Fourier transform of a three-dimensional step function describing the crystal.

For simplicity, we treat the one-dimensional case. The step function $f(x)$, for a crystal of N unit cells in the $\mathbf{a}$ direction, is given by $f(x) = 1$ for $-Na/2 < x < Na/2$, and $f(x) = 0$ elsewhere. The Fourier transform of this function is

$$\hat{F}[f(x)] = \frac{\sin \pi N S a}{\pi S} \tag{1.23}$$

and is illustrated in Fig. 1.3. The value of $\hat{F}[f(x)]$ has its maximum equal to Na at $S = 0$; that is, for a three-dimensional crystal, it is proportional to the crystal's volume. The first zero of $\hat{F}[f(x)]$ occurs when $\pi N S a = \pm \pi$, or $NSa = \pm 1$. Since the diffraction maxima at $H = h\mathbf{a}^* + k\mathbf{b}^* + l\mathbf{c}^*$ are convoluted with Eq. (1.23), subsidiary maxima will occur. They become negligibly small for larger N. The net effect of the finite crystal size is then that each diffraction maximum is broadened. This is *particle-size broadening*, which dominates the width of the diffracted beams for very small particle sizes of the order of 1000 Å or less. The

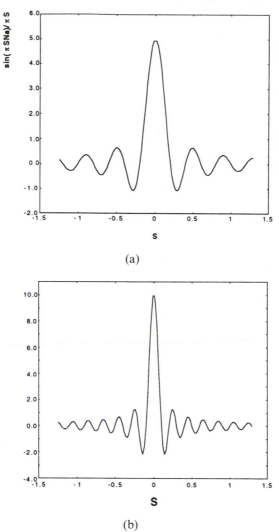

FIG. 1.3 (a) The shape transform of a one-dimensional crystal, and (b) the effect of a factor 2 increase in particle size.

well-known Scherrer equation for the particle-size broadening B,

$$B(2\theta) = \frac{0.94\lambda}{\cos\theta} \qquad (1.24)$$

follows from this theory (Warren 1967).

1.2.4 The Structure Factor Formalism in Terms of Atomic Densities

The structure factor F can be simplified by approximating the unit cell density distribution by a summation over atomic densities, each centered at the nuclear position $\mathbf{r}_j$. In mathematical terms, the unit cell density can then be formulated as a sum over the convolution of the atomic densities and delta functions centered at the nuclear positions:

$$\rho_{\text{unit cell}}(\mathbf{r}) = \sum \rho_j(\mathbf{r}) * \delta(\mathbf{r} - \mathbf{r}_j) \tag{1.25}$$

To obtain the corresponding structure factor expression, the Fourier convolution theorem is applied, or

$$F(\mathbf{H}) = \hat{F}\{\rho_{\text{unit cell}}(\mathbf{r})\}$$

$$= \int_{\text{unit cell}} \sum \rho_j(\mathbf{r}) * \delta(\mathbf{r} - \mathbf{r}_j) \exp(2\pi i \mathbf{H} \cdot \mathbf{r}) \, d\mathbf{r}$$

$$= \sum f_j(\mathbf{H}) \exp(2\pi i \mathbf{H} \cdot \mathbf{r}_j) \tag{1.26}$$

in which the *atomic scattering* factor $f_j(\mathbf{H})$ is the Fourier transform of the atomic density $\rho_j(\mathbf{r})$ at $\mathbf{H}$. In the approximation common in structure determination, the atomic densities are assumed to be spherically symmetric, with a radial dependence equal to that of the theoretical ground state atom. This is the independent-atom model, abbreviated as IAM.

The Fourier transform of the spherical atomic density is particularly simple. One can select $\mathbf{S}$ to lie along the z axis of the spherical polar coordinate system (Fig. 1.4), in which case $\mathbf{S} \cdot \mathbf{r} = Sr \cos \vartheta$. If $\rho_j(r)$ is the radial density function of the spherically symmetric atom,

$$f_j(S) = \int_{\text{atom}} \rho_j \exp 2\pi i \mathbf{S} \cdot \mathbf{r} \, d\mathbf{r}$$

$$= -\int_{\vartheta=0}^{\pi} \int_{\phi=0}^{2\pi} \int_{r=0}^{\infty} \rho_j(r) \exp(2\pi i Sr \cos \vartheta) r^2 \sin \vartheta \, dr \, d\vartheta \, d\phi \tag{1.27}$$

Performing the integration over ϕ gives

$$f_j(S) = 2\pi \int_{\vartheta=0}^{\pi} \int_{r=0}^{\infty} \rho_j(r) \exp(2\pi i Sr \cos \vartheta) r^2 \sin \vartheta \, dr \, d\vartheta$$

$$= -\frac{2\pi}{2\pi i Sr} \int_{\vartheta=0}^{\pi} \int_{r=0}^{\infty} \rho_j(r) \exp(2\pi i Sr \cos \vartheta) r^2 \, dr \, d(2\pi i Sr \cos \vartheta)$$

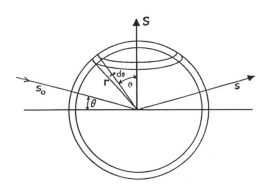

FIG. 1.4 Integration of the scattering over the density of a spherical atom.

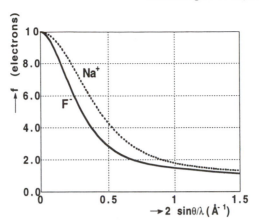

FIG. 1.5 Spherical atom scattering factors for the isoelectronic F^- and Na^+ ions.

Finally, integration over ϑ gives the desired result,

$$f_j(S) = \int_0^\infty 4\pi r^2 \rho_j(r) \frac{\sin 2\pi Sr}{2\pi Sr} dr \equiv \int_0^\infty 4\pi r^2 \rho_j(r) j_0 \, dr \equiv \langle j_0 \rangle \qquad (1.28)$$

where $4\pi r^2 \rho_j(r)$ is the probability that the electron is found in the shell bounded by spheres with radii r and $r + dr$. The term j_0 is known as a zero-order spherical Bessel function (see, e.g., Arfken 1970), and the integral in Eq. (1.27), labeled $\langle j_0 \rangle$, is referred to as the Fourier–Bessel transform of the atomic density. For an atomic density expressed as a sum over exponential functions, Eq. (1.28) can be expressed in closed form as discussed in chapter 3. The appropriate expressions are listed in appendix G.

Scattering factors for the isoelectronic F^- and Na^+ ions are shown in Fig. 1.5. The inverse relationship between direct space and reciprocal space is evident, as the more compact Na^+ ion has the more "diffuse" scattering factor. For the same reason, the scattering of the atomic valence electrons is concentrated in the low-order region of reciprocal space, while the core electron scattering persists to high values of $\sin\theta/\lambda$. This is the basis for the *high-order refinement method*, in which only high-order data are used to reduce the bias in the structural parameters due to deviations from the IAM. The high-order refinement method is further discussed in chapter 3.

1.3 Resonance Scattering of X-rays

1.3.1 Classical Treatment

The classical scattering amplitude of an electron was derived in section 1.2.1 as

$$A = -\frac{e^2}{mc^2} \frac{\omega^2 E_0}{\omega^2 - \omega_s^2 - ik\omega} \qquad (1.5a)$$

The amplitude A can be separated into real and imaginary parts by multiplication of both the numerator and the denominator by $\omega_s^2 - \omega^2 - ik\omega$. For unit value of E_0, in units of $-e/mc^2$, the result is

$$A_0 = \frac{\omega^2(\omega^2 - \omega_s^2)}{(\omega^2 - \omega_s^2)^2 + k^2\omega^2} - \frac{ik\omega^3}{(\omega^2 - \omega_s^2)^2 + k^2\omega^2} \tag{1.29}$$

The anomalous contribution to the real part of the scattering amplitude can be separated by subtraction of the classical Thompson scattering, $-e^2/mc^2$, from the first term of Eq. (1.29) to give

$$A_0(\text{anomalous, real}) = \frac{\omega^2(\omega_s^2 - k^2) - \omega_s^4}{(\omega^2 - \omega_s^2)^2 + k^2\omega^2} \approx \frac{\omega_s^2}{\omega^2 - \omega_s^2}, \quad \text{for small } k \tag{1.30}$$

According to Eqs. (1.29) and (1.30), the real part of the resonance scattering amplitude of an oscillator is negative below the absorption edge, where $\omega < \omega_s$, and positive above the edge, while the imaginary part of the amplitude is negative everywhere (i.e., has a sign opposite to that of the classical electron scattering). This conclusion is confirmed by more advanced theory and by experiment, except in the vicinity of the absorption edges, where the behavior is more complicated than can be accounted for by the simple classical theory.

Including resonance effects, the atomic scattering factor for a many-electron atom is written as

$$f(S, \omega) = f_0(S) + f'(S, \omega) + if''(S, \omega) \tag{1.31}$$

where the last two terms describe the real and imaginary parts of the resonance contribution, respectively. In the classical theory, the electron scattering for each electron shell is multiplied by the *oscillator strength* $g(s)$ of the shell. Separation of the real and imaginary parts in units of scattering of the classical electron, leads to

$$f_0 + f' = \sum_s g(s) \frac{\omega^2(\omega^2 - \omega_s^2)}{(\omega^2 - \omega_s^2)^2 + k^2\omega^2} \tag{1.32}$$

and

$$f'' = -\sum_s g(s) \frac{k\omega^3}{(\omega^2 - \omega_s^2)^2 + k^2\omega^2} \tag{1.33}$$

These equations neglect the angle dependence of the atomic scattering and thus assume that the atomic dimensions are small relative to the wavelength of the photons. This is a good approximation for the dimensions of the atomic shells and the resonance wavelengths of the corresponding absorption edges.

For values of $\omega \gg \omega_s$, that is, for negligible effect of the resonance, the real contribution $f_0 + f' = \sum_s g(s)$, and $f'' \to 0$. Thus, the real contribution becomes equal to the oscillator strength. As discussed by James (1982), there is a close correspondence between the classical oscillator strengths and the matrix element of quantum-mechanical scattering theory.

1.3.2 Quantum-Mechanical Treatment: The Second Born Approximation

In the vicinity of the atomic absorption edges, the participation of free and bound excited states in the scattering process can no longer be ignored. The first term in the interaction Hamiltonian of Eq. (1.11) leads, in second-order perturbation theory, to a resonance scattering contribution (in units of classical electron scattering) equal to (Gerward et al. 1979, Blume 1994)[4]

$$
f'(\omega) + if''(\omega) = -\frac{1}{m}\sum_n \left[\frac{\langle \psi_0 | e_1 \cdot p\, e^{-ik\cdot r} | \psi_n\rangle\langle\psi_n| e_0 \cdot p\, e^{ik_0\cdot r}|\psi_0\rangle}{E_n - E_0 - \dfrac{h}{2\pi}\omega - i\Gamma/2} \right.
$$

$$
\left. + \frac{\langle \psi_0 | e_1 \cdot p\, e^{ik\cdot r} | \psi_n\rangle\langle\psi_n| e_0 \cdot p\, e^{-ik_0\cdot r}|\psi_0\rangle}{E_n - E_0 + \dfrac{h}{2\pi}\omega - i\Gamma/2} \right] \qquad (1.34a)
$$

where p is the electron momentum operator, and m is the electron mass, $|\psi_0\rangle$ the wave function of the initial (and final) state including the photon, and e_0 and e_1 are the polarization vectors of the incident and reflected beams, respectively. The summation is over all unoccupied orbitals, including the free states in the continuum. For the latter, the summation must be replaced by an integration.

The two terms in Eq. (1.34a) include processes in which the initial photon k_0 has been annihilated first, and those in which the final photon k has first been created. In the quantum-mechanical description of the first type of process, the photon k_0 is absorbed and then, in a very small time interval, the photon k is emitted through a stimulated emission process. This process and three-beam multiple scattering are illustrated by the Feynman diagrams in Fig. 1.6.

The first term in Eq. (1.34a) represents the resonance scattering. It becomes large when E_{photon} $(=\hbar\omega) \approx E_n - E_0$. In comparison, the second term is small and is usually neglected. The imaginary term in the denominator contains Γ, the inverse lifetime (related to linewidth) of the intermediate state $|\psi_n\rangle$.

For a system of independently scattering electrons, the resonance scattering amplitude, ignoring the second term, becomes

$$
f'(\omega) + if''(\omega) = -\frac{1}{m}\sum_n \left[\frac{\langle \psi_0 | e_1 \sum_i e^{-ik\cdot r}\, p_i | \psi_n\rangle\langle\psi_n| e_0 \sum_i e^{ik_0\cdot r}\, p_i|\psi_0\rangle}{E_n - E_0 - \dfrac{h}{2\pi}\omega - i\Gamma/2} \right]
$$

$$
(1.34b)
$$

where the sum in the numerator is over all the electrons.

The terms in the summation for which $E_n - E_0 = h\omega/2\pi$ correspond to an absorption of radiation and give rise to the imaginary part of the dispersion f''.

---

[4] To be consistent with the physics literature, in this section the incident photon wave function is defined as $\exp(ik_0\cdot r)$, rather than as $\exp(-ik_0\cdot r)$.

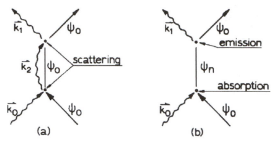

FIG. 1.6 Feynman diagrams for (a) multiple scattering, and (b) resonance scattering. *Source*: Feil (1975).

The separation between the real and imaginary components is achieved by use of the expression

$$\lim_{(\varepsilon \to 0)} \frac{1}{x + i\varepsilon} = P\frac{1}{x} - i\pi\delta(x) \tag{1.35}$$

where P indicates that the term is excluded when $x = 0$. The real part is then

$$f'(\omega) = -\frac{1}{m} P \sum_n \left[\frac{\langle \psi_0 | e_1 \sum_i e^{-i\mathbf{k}\cdot\mathbf{r}} \mathbf{p}_i | \psi_n \rangle \langle \psi_n | e_0 \sum_i e^{i\mathbf{k}_0\cdot\mathbf{r}} \mathbf{p}_i | \psi_0 \rangle}{E_n - E_0 - \frac{h}{2\pi}\omega} \right] \tag{1.36}$$

where P represents the principal part of the summation in the sense that all terms with zero denominator are left out, while the imaginary part f'' is given by

$$f''(\omega) = -\frac{\pi}{m} \sum_n \delta\left(E_n - E_0 - \frac{h}{2\pi\omega}\right) \left[\langle \psi_0 | e_1 \sum_i e^{-i\mathbf{k}\cdot\mathbf{r}} \mathbf{p}_i | \psi_n \rangle \langle \psi_n | e_0 \sum_i e^{i\mathbf{k}_0\cdot\mathbf{r}} \mathbf{p}_i | \psi_0 \rangle \right] \tag{1.37}$$

Using

$$\hat{p}_x = -\frac{ih}{2\pi} \frac{d}{dx} \tag{1.38}$$

for the components of the momentum operator $\mathbf{p}$, and

$$-\frac{h^2}{4\pi m} \frac{d}{dx} = [\hat{H}_0, x] = \hat{H}_0 x - x\hat{H}_0 \tag{1.39}$$

the term defined by the numerator in Eq. (1.36) and the expression in square brackets in Eq. (1.37) can be reduced to

$$-\frac{m^2}{\hbar^2} (E_0 - E_n) e_1 \cdot e_0 \left[\langle \psi_0 \rangle \sum_i e^{i\mathbf{k}\cdot\mathbf{r}} \mathbf{r}_i | \psi_n \rangle \langle \psi_n | \sum_i e^{i\mathbf{k}_0\cdot\mathbf{r}} \mathbf{r}_i | \psi_0 \rangle \right] \tag{1.40}$$

provided the operators between the vertical lines in Eqs. (1.36) and (1.37) are Hermitian (Feil 1992). The factor $e_1 \cdot e_0$ is the polarization vector routinely applied in the reduction of X-ray intensities to structure factor amplitudes.

As is evident from Eqs. (1.36) and (1.37), and from the classical treatment as well, the effect of resonance on the intensity of X-ray scattering is pronounced when $E_n \approx E_0$, that is, in the vicinity of the absorption edges. Even for data collected at other wavelengths, it is necessary to correct the structure factors for anomalous scattering before the electron density can be calculated by the Fourier inversions of Eqs. (1.22) and (1.26), as further discussed in chapter 5. The anomalous scattering factors needed for this purpose are available in the literature (*International Tables for X-ray Crystallography* 1974, Kissel and Pratt 1990).

The position of the absorption edge, and its fine structure, give information on the ionization energy of the resonating electrons, and on the nature of spectroscopic transitions from the resonating shell to unoccupied bound states of the scattering entity (Pickering et al. 1993, Sorensen et al. 1994). The ionization energy I is, in turn, related to the binding energy ε of the electrons and to the electrostatic potential Φ at the atomic nucleus, which is one of the physical quantities that can be derived from the charge distribution (see chapter 8 for a discussion of the relation between I, ε, and Φ).

1.3.3 The Power Series Expansion of the Scattering Operator

Scattering operators of the form $\exp(i\mathbf{k}\cdot\mathbf{r})$, as occur in Eqs. (1.34)–(1.40), may be developed in a power series:

$$\exp(i\mathbf{k}\cdot\mathbf{r}) = 1 + i\mathbf{k}\cdot\mathbf{r} - (\mathbf{k}\cdot\mathbf{r})^2/2 + \cdots \tag{1.41}$$

The approximation in which only the leading term in the expansion is retained is referred to as the *dipolar approximation*. The dipolar approximation will be more closely obeyed for small values of $k = 2\pi/\lambda$, that is, for longer wavelengths and, in particular, for visible light. The higher-order terms are also smaller when the scattering object is compact relative to the wavelength used, that is, if either the initial state or the final state has a compact core-type wave function.

In the dipolar approximation, the matrix elements in Eq. (1.40) become equal to $\langle \psi_n | \mathbf{e}\cdot\mathbf{r} | \psi_k \rangle$, for a polarization direction defined by the unit vector $\mathbf{e}$. The anomalous scattering effects are then independent of the magnitude of $\mathbf{K} = \mathbf{k} - \mathbf{k}_0$. Factoring out the polarization effects implicit in $\mathbf{e}_1 \cdot \mathbf{e}_0$, the angular dependence of the scattering factor is described by the products $r_j r_k$. In other words, the scattering amplitude is a second-order tensor property when either the ground state or the excited stated wave function is not spherically symmetric.

Retention of additional terms of the expansion Eq. (1.41) in Eq. (1.40) gives rise to a more complex angular dependence. Templeton has formulated the total resonance scattering in terms of a tensor equation, which, including terms up to fourth order, is given by (Templeton 1994)

$$f' + if'' = e_j e'_k S^{jk} + i e_j e'_k (k'_m - k_m) T^{jkm} + e_j e'_k k'_m k_n U^{jkmn}$$

$$+ e_j e'_k (k'_m k'_n + k_m k_n) V^{jkmn} + \cdots$$

$$\equiv Q_{22} + i Q_{24} + Q_{44} + Q_{28} + \cdots \tag{1.42}$$

where the primed quantities refer to the scattered beam, and summation over repeated indices is assumed. The imaginary Q_{24} term is a mixed dipolar-quadrupolar term, which is absent for atoms at centrosymmetric sites. A second fourth-rank contribution is due to the dipolar–octapolar term Q_{28}. Experimental evidence for the Q_{24} term has been obtained from intensity measurements on potassium chromate (Templeton and Templeton 1993). The existence of the Q_{44} term was demonstrated in a study of the forbidden (111) reflection in hematite (α-Fe$_2$O$_3$) (Finkelstein et al. 1992).

1.3.4 The Optical Theorem and the Relation Between f'' and f'

The scattering of the X-ray beam in the forward direction adds out-of-phase components to the propagating beam, as the classical electron scattering has a negative sign. This implies that the X-ray refractive index n differs from unity. The reduction in n leads to *total reflection* at very small angles, which is applied in the design of X-ray mirrors.

Since the scattering amplitude has both real and imaginary components, we may write

$$n = 1 - \delta = 1 - \alpha - i\beta \qquad (1.43)$$

Here, α, the real component of δ, is related to the atomic scattering factors f by the expression

$$\alpha = (\lambda^2 e^2/2\pi mc^2) \sum N_i(f_i + f'_i) \qquad (1.44)$$

where the sum is over all atom types, each with number density N_i, and f_i and f'_i are expressed in units of the classical electron scattering. An equivalent formula is obtained by substitution of $\lambda = 2\pi c/\omega$, where ω is the angular frequency

$$\alpha = (2\pi e^2/m\omega^2) \sum N_i(f_i + f'_i) \qquad (1.45)$$

The complex term β in Eq. (1.41) depends on f'' in a manner analogous to that expressed by Eq. (1.44):

$$\beta = (\lambda^2 e^2/2\pi mc^2) \sum N_i f''_i \qquad (1.46)$$

Since β and f'' represent an absorption of the propagating beam, they are related to the linear absorption coefficient $\mu(\omega)$. This relation is called the *optical theorem*

$$\mu(\omega) = (4\pi e^2/m\omega c) \sum N_i f''_i(\omega) = (2\lambda e^2/mc^2) \sum N_i f''_i(\lambda) \qquad (1.47)$$

At energies just above an absorption edge, the contribution of the strongly absorbing atom is dominant. The absorption of other atoms is often small and in first approximation is independent of the energy. Expression (1.47) can be used to obtain $f''(\lambda)$ for the dominantly resonating atom, over a large energy range, from the experimental absorption curve. Substitution of the classical scattering amplitude e^2/mc^2, equal to $0.2818 \cdot 10^{-14}$ m, in Eq. (1.47) gives a numerical relation between $\mu(\lambda)$ and $f''(\lambda)$:

$$\mu(\lambda) = 0.5636 \cdot 10^{-14} \lambda N f''(\lambda) \qquad (1.48)$$

in which μ is in m^{-1}, λ in m, and N, the number density, in m^{-3}. Or, in more common units:

$$\mu(\lambda)(\text{cm}^{-1}) = 0.5636 \cdot 10^{4} \lambda(\text{Å}) N(\text{Å}^{-3}) f''(\lambda) \tag{1.49}$$

The expression for the atomic cross section σ follows from $\sigma = \mu/N$. The mass absorption coefficient μ/ρ of the element is given by $\mu/\rho = A\sigma/M$, where ρ is the density, and A and M are Avogadro's number and the atomic weight, respectively. This gives, analogously to Eq. (1.49),

$$\mu(\lambda)/\rho(\text{cm}^{2}\,\text{g}^{-1}) = 3.3940 \cdot 10^{3}\,\lambda(\text{Å}) f''(\lambda)/M \tag{1.50}$$

The two anomalous components of the scattering factor, f'' and f', are interrelated through the Kramers–Kronig transforms, which have the form

$$f'(\omega_0) = \frac{2}{\pi} \int_0^\infty \frac{\omega}{\omega_0^2 - \omega^2} f''(\omega)\, d\omega \tag{1.51a}$$

or, equivalently,

$$f'(E_0) = \frac{2}{\pi} \int_0^\infty \frac{E}{E_0^2 - E^2} f''(E)\, dE \tag{1.51b}$$

and for the inverse transformation

$$f''(\omega_0) = -\frac{2\omega_0}{\pi} \int_0^\infty \frac{f'(\omega)}{\omega_0^2 - \omega^2}\, d\omega \tag{1.52}$$

Since both expressions have a singularity at $\omega = \omega_0$, special mathematical techniques are used for precise evaluation. A symbol P is often inserted in front of the integral to indicate that it must be evaluated as the Cauchy principal value, excluding the infinite contribution at $\omega_0 = \omega$. The proper mathematical procedure to integrate across the singularity has been discussed by Hoyt et al. (1984).

The integration in the Kramers–Kronig expression is over an infinite range. Theoretical values of f'' in regions remote from the absorption edge are therefore required for the application of the transform. They can be calculated as described by Cromer and Liberman (1970). The fitting of the absorption curve to the calculated values for the dominant scatterer provides a scale for f'', and eliminates the slowly varying contributions by other atoms to Eq. (1.47) (see, e.g., Hendrickson et al. 1988). As a result of errors in the extrapolation procedure and the absorption curves, an uncertainty of 0.2–0.3 electrons in the height of the whole curve appears to be present. This should be compared with the total changes in f', which may have magnitudes of 6–10 electrons or larger.

A typical relation between f' and f'' near an absorption edge is illustrated in Fig. 1.7. It shows that f' has a minimum when f'' has increased by about half its total variation, that is, halfway up the slope of the edge.

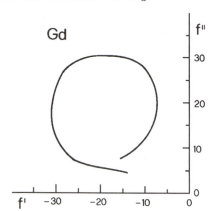

FIG. 1.7 Anomalous scattering terms f' and f'' for gadolinium near the L_3 edge. *Source:* Giacovazzo (1992).

1.4 Neutron Scattering

1.4.1 Properties of Neutrons

The wavelength of the neutron is related to its mass m $(=1.675 \cdot 10^{-27}\,\text{kg})$ and velocity v by the de Broglie relation

$$\lambda = \frac{h}{mv} \tag{1.53}$$

The velocity distribution in a neutron gas at equilibrium is subject to the laws of the kinetic theory of gases. The neutron velocities at equilibrium obey the Maxwell distribution

$$f(v) = 4\pi \left(\frac{m}{2\pi k_B T}\right)^{3/2} v^2 \exp\left(-\tfrac{1}{2}mv^2/k_B T\right) \tag{1.54}$$

where k_B is Boltzmann's constant.

From Eq. (1.54), the rms value of the velocity of the neutrons is equal to

$$v = \left(\frac{3k_B T}{m}\right)^{1/2} \tag{1.55}$$

which gives, for the average kinetic energy,

$$\langle E \rangle = 1/2mv^2 = (3/2)k_B T \tag{1.56}$$

Substituting Eq. (1.56) into Eq. (1.53) gives, for the average neutron wavelength,

$$\lambda_{\text{average}} = \frac{h}{\sqrt{3mk_B T}} \quad \text{or} \quad \lambda_{\text{average}}(\text{Å}) = 25.15\,\frac{1}{\sqrt{T\,(\text{K})}} \tag{1.57}$$

As neutrons from research reactors or spallation sources are brought to an equilibrium temperature by collisions with a moderator, the temperature T in Eq.

(1.57) is the temperature of the moderator. Neutrons moderated at a few hundred degrees Celsius have wavelengths comparable to interatomic distances, and can thus be used in diffraction. The rms energy of neutrons moderated at 400 K, is, according to Eq. (1.56), equal to $8.3 \cdot 10^{-21}$ J, or ≈ 0.05 eV. This is comparable to the spacing of the energy levels of phonons in a crystal (see chapter 2), and the basis for phonon spectroscopy based on inelastic neutron scattering. On the other hand, the energy of 1 Å photons is higher by several orders of magnitude, as can be easily verified from $E_{photon} = hc/\lambda$. A photon scattered inelastically by energy exchange with a phonon therefore loses (or gains) an extremely small fraction of its energy, which can only be detected with supreme energy resolution. In the case of neutrons, the change is easily measurable by energy analysis of the diffracted beam.

1.4.2 The Neutron Scattering Length

The principal contribution to neutron scattering is due to the interaction between the atomic nuclei and the neutrons (Bacon 1962, Squires 1978). Since the atomic nuclei have dimensions of 10^{-13} cm (1 fermi), and nuclear forces have about this range, the nucleus acts as a point scatterer for neutrons with $\lambda \approx 1$ Å. For practical purposes, the neutron scattering amplitude, or *scattering length b*, is therefore independent of $\sin \theta/\lambda$, unlike the X-ray form factor. Furthermore, since the scattering length is a property of the nucleus, it is different for different isotopes of the same element. For some isotopes with energy-level spacings close to the energy of thermal neutrons, scattering is a resonance phenomenon, and wavelength dependent, in analogy to anomalous X-ray scattering. Examples are ^{103}Rh, ^{113}Cd, and ^{157}Gd. For most nuclei, however, the scattering length is wavelength independent.

The value of the scattering length depends also on the spin state of the compound nucleus consisting of the neutron and the scattering nucleus. As the neutron has a spin 1/2, every nucleus with nonzero spin has two values of the scattering length. None of these values can be predicted by nuclear theory, so experimentally determined values must be used. The random occurrence of different scatterers on identical crystallographic sites is equivalent to structural disorder, and leads to an incoherent component of the scattering. If the relative frequency of occurrence of the isotopes in specific spin states is f_i, the average nuclear scattering length $\bar{b}$ will be given by

$$\bar{b} = \sum f_i \bar{b}_i \qquad (1.58)$$

Some values for $\bar{b}$ are listed in Table 1.1. It is the amplitude of scattering to be used in the structure factor expression for the amplitude of the Bragg reflections. In analogy to Eq. (1.26),

$$F(\mathbf{H}) = \sum_j \bar{b}_j \exp\left(2\pi i \mathbf{H} \cdot \mathbf{r}_j\right) \qquad (1.59)$$

The incoherent scattering is given by the difference between the total scattering and the coherent scattering. For a monatomic crystal with one atom per unit cell,

$$I_{incoherent} = 4\pi \sum \left(\overline{b^2} - \bar{b}^2\right) \qquad (1.60)$$

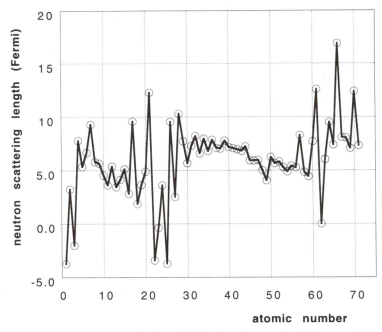

FIG. 1.8 Coherent neutron scattering lengths for the elements in their natural abundance, in units of Fermi (1 fermi = 10^{-13} cm).

The spin-incoherent scattering is especially pronounced for the hydrogen atom. Atom scattering lengths for the elements in their natural abundance are shown in Fig. 1.8, while some coherent and incoherent scattering lengths are listed in Table 1.1. It is of interest that the scattering lengths are of the same order of magnitude as the scattering amplitude of the free electron, which equals $2.818 \cdot 10^{-13}$ cm, or 2.818 fermi. The incoherent scattering length is zero for zero-spin nuclei with even numbers of protons and neutrons. As a consequence of the large incoherent values for ^{1}H, it is important in accurate work to substitute deuterium for hydrogen whenever possible, in order to reduce the incoherent background in the diffraction pattern.

The scattering lengths discussed so far refer to a fixed nucleus. If the nucleus is free to vibrate, it will recoil under the impact of the neutron. In that case the effective mass is that of the compound nucleus, consisting of the neutron and the scattering nucleus. This means that the neutron mass m must be replaced by the reduced mass of the compound nucleus $\mu = mM/(M + m)$, where M is the mass of the scattering atom. As a result, the scattering length of the free atom is related to that of the bound atom by

$$b_{\text{free}} = \frac{\mu}{m} b \qquad (1.61)$$

The difference is usually small, except for the very light elements. For ^{1}H, for example, the free scattering length is only half that of the bound proton.

TABLE 1.1 Values of Coherent and Incoherent Scattering
Lengths in Fermi (1 fermi $= 10^{-13}$ cm) for a Number of Nuclei

Nucleus	Abundance (%)	I	b_{coh}	b_{inc}
^1_1H	99.985	1/2	−3.74	25.22
^2_1H	0.015	1	6.67	4.03
$^{12}_6C$	98.90	0	6.65	0
$^{13}_6C$	1.10	1/2	6.19	0.52
$^{50}_{24}Cr$	4.35	0	−4.50	0
$^{52}_{24}Cr$	83.79	0	4.92	0
$^{53}_{24}Cr$	9.50	3/2	−4.20	6.86
$^{54}_{24}Cr$	2.36	0	4.55	0
$^{55}_{25}Mn$	100	5/2	−3.73	1.79
$^{234}_{92}U$	0.005	0	12.4	0
$^{235}_{92}U$	0.72	7/2	10.5	1.3
$^{238}_{92}U$	99.275	0	8.4	0

Source: International Tables for Crystallography, Vol. C (1992).

The great advantage of elastic neutron scattering over X-ray scattering is that it gives information directly on the nuclear positions, without being influenced by the details of the charge density distribution. The combination of the two techniques can be used advantageously to separate the effects of thermal vibrations from the effects of chemical bonding on the X-ray scattering, as discussed further in chapter 5.

2

The Effect of Thermal Vibrations on the Intensities of the Diffracted Beams

The atoms in a crystal are vibrating with amplitudes determined by the force constants of the crystal's normal modes. This motion can never be frozen out because of the persistence of zero-point motion, and it has important consequences for the scattering intensities.

Since X-ray scattering (and, to a lesser extent, neutron scattering) is a very fast process, taking place on a time scale of 10^{-18} s, the photon–matter interaction time is much shorter than the period of a lattice vibration, which is of the order $1/v$, or $\approx 10^{-13}$ s. Thus, the recorded X-ray scattering pattern is the sum over the scattering of a large number of instantaneous states of the crystal. To an extremely good approximation, the scattering averaged over the instantaneous distributions is equivalent to the scattering of the time-averaged distribution of the scattering matter (Stewart and Feil 1980). The structure factor expression for coherent elastic Bragg scattering of X-rays may therefore be written in terms of $\langle \rho(\mathbf{r}) \rangle$, the thermally averaged electron density:

$$F(\mathbf{H}) = \int_{\text{unit cell}} \langle \rho(\mathbf{r}) \rangle \exp\left(2\pi i \mathbf{H} \cdot \mathbf{r}\right) d\mathbf{r} \qquad (2.1)$$

The smearing of the electron density due to thermal vibrations reduces the intensity of the diffracted beams, except in the forward $|S| = 0$ direction, for which all electrons scatter in phase, independent of their distribution. The reduction of the intensity of the Bragg peaks can be understood in terms of the diffraction pattern of a more diffuse electron distribution being more compact, due to the inverse relation between crystal and scattering space, discussed in chapter 1.

The reduction in intensity due to thermal motion is accompanied by an increase in the incoherent elastic scattering, ensuring conservation of energy. In this respect, thermal motion is much like disorder, with the Bragg intensities representing the average distribution, and the deviations from the average appearing as a continuous, though not uniform, background, generally referred to as *thermal diffuse scattering* or TDS.

2.1 The Normal Modes of a Crystal

2.1.1 Phonons, Internal and External modes

A crystal with n atoms per unit cell has $3nN$ degrees of freedom, N being the number of unit cells in the crystal. Thus, subtracting the translations and rotations of the crystal as a whole, there are $3nN - 6$ ($\approx 3nN$) *normal modes*. Since the displacements of atoms in different cells are correlated, the normal modes are waves, or *phonons*, extending over the crystal, with force constants Φ, obtained from a sum over the interactions between atoms in all unit cells, and wavevector $\mathbf{q}$.

For a small change in magnitude of $\mathbf{q}$, the change in frequency ω is small, and ω is a continuous function of $|q|$ ($=2\pi/\lambda$). The dependence of ω on $|q|$ is referred to as the *dispersion relation*. The number of *phonon branches* with continuously varying ω equals $3n$, but some of these may be degenerate due to the symmetry of the crystal.

For a molecular crystal, the description can be simplified considerably by differentiating between internal and external modes. If there are M molecules in the cell, each with n_M atoms, the number of *external* translational phonon branches will be $3M$, as will the number of external rotational branches. When the molecules are linear, only $2M$ external rotational modes exist. For each molecule, there are $3n_M - 6$ ($3n_M - 5$ for a linear molecule) internal modes, the wavelength of which is independent of $\mathbf{q}$. Summing all modes gives a total number of $N\{M(3n_M - 6) + 6M\} = 3nN$, as required, because each of the modes that have been constructed is a combination of the displacements of the individual atoms.

In the harmonic approximation, a mode is described by a single force constant k_f, equal to the ratio between the magnitudes of the displacement and the restoring force (Hooke's law). For molecular crystals, the force constants of the internal modes are usually much larger than those of the external modes. Frequencies of the former are typically in the $500-3000$ cm^{-1} range, similar in value to those of the isolated gas-phase molecule. Phonon frequencies tend to be very much lower, in the $20-200$ cm^{-1} range. The variation of the phonon frequency with the wavelength of the phonon is pronounced for *acoustical* modes, for which all particles (molecules for a molecular crystal, and atoms otherwise) in the unit cell move in phase. If the unit cell contains more than one particle, *optical* modes exist, for which two particles in the same unit cell have opposite phase. The dispersion curves for the rock-salt structure are illustrated in Fig. 2.1, and for the rigid body vibrational modes of the molecular crystals anthracene and naphthalene in Fig. 2.2.

2.1.2 The Frequency of the Normal Modes

The theory of lattice vibrations was developed by Born and von Kármán (1912, 1913), and is described in detail in a monograph by Born and Huang (1954). The equations of motion of a vibrating assembly of atoms j, each of mass $m(j)$, are given by

$$m(j)\ddot{\mathbf{u}}(j, t) = -\sum_{j'} \Phi_{\alpha\alpha'}(j, j')\mathbf{u}(j', t) \tag{2.2a}$$

in which $\ddot{\mathbf{u}}$ is the vector of the second time-derivatives of the displacements $\mathbf{u}$ and Φ is a 3×3 force constant matrix, with elements

$$\left(\frac{\partial^2 V}{\partial u_\alpha \, \partial u_{\alpha'}} \right)_0$$

where V is the potential giving rise to the forces. Thus, $-\Phi_{\alpha\alpha'}(j, j')$ is the force exerted in the α direction on atom j when atom j' is given a small unit displacement in the α' direction. The summation in Eq. (2.2a) is over all atoms in the crystal, but, in general, the largest contributions originate from nearest-neighbor interactions.

The number of independent elements of Φ may be restricted by symmetry. In the face-centered cubic structure, for example, the force constant matrix for two atoms 1/2 1/2 0 apart is given by (Willis and Pryor 1975)

$$\Phi = \begin{pmatrix} \alpha & \gamma & 0 \\ \gamma & \alpha & 0 \\ 0 & 0 & \beta \end{pmatrix} \tag{2.3}$$

in which the zero elements correspond to cases in which one of the two directions α and α' is perpendicular to the internuclear vector, and the second has a component along this vector.

For harmonic oscillations the time-dependent displacements u of atom j are related to the amplitudes of vibration $\mathbf{U}(j)$ by

$$\mathbf{u}(j, t) = \mathbf{U}(j)e^{-i\omega t} \tag{2.4}$$

in which ω is the angular frequency, equal to $2\pi v$.

The amplitude $\mathbf{U}(j)$ is complex when the phase at $t = 0$ is taken into account, that is,

$$\mathbf{U}(j) = |\mathbf{U}(j)|e^{i\phi_j} \tag{2.5}$$

Substituion of Eq. (2.4) into Eq. (2.2a) gives

$$m(j)\omega^2\mathbf{U}(j) = \sum_{j'} \Phi(jj')\mathbf{U}(j) \tag{2.6}$$

The phonons are not stationary modes, but traveling waves extending through the whole crystal. The momentum of a phonon can be assigned as equal to $\hbar\mathbf{q}$, in analogy with the momentum of a photon, though it is not strictly defined, as the phonon can be described equivalently in an extended Brillouin zone (see Fig. 2.1), corresponding to a different value of the wavevector $\mathbf{q}$.

The displacements of a particle j at $\mathbf{r}$ in unit cell l, subject to a phonon wave with wave vector $\mathbf{q}$, obey the equation

$$\mathbf{u}(jl, t) = |\mathbf{U}(j\,|\,\mathbf{q})| \exp\left[i(\mathbf{q}\cdot\mathbf{r}(jl) - \omega(\mathbf{q})t + \phi(j, \mathbf{q}))\right] \tag{2.7a}$$

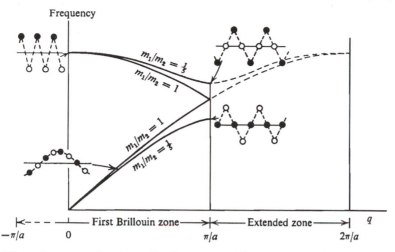

FIG. 2.1 Dispersion curves for the rock-salt structure. The terms m_1 and m_2 are the masses of the two ions in the rock-salt structure. The relative directions of displacement of the two types of atoms at a number of points along the branches are indicated. *Source:* Willis and Pryor (1975). Reprinted with the permission of Cambridge University Press.

Using Eq. (2.5), this becomes

$$\mathbf{u}(jl, t) = \mathbf{U}(j \,|\, \mathbf{q}) \exp \left[i(\mathbf{q} \cdot \mathbf{r}(jl) - \omega(\mathbf{q})t) \right] \qquad (2.7b)$$

Note that atoms in different unit cells have the same *amplitude* of vibration for a particular phonon; they differ in the phase of the traveling wave because of the $\mathbf{q} \cdot \mathbf{r}(jl)$ term.

The equation of motion (2.2a) can now be rewritten as the sum over the atoms in a unit cell and the sum over all unit cells:

$$\mathbf{m}(j)\ddot{\mathbf{u}}(jl, t) = -\sum_{l'} \sum_{j'} \Phi\begin{pmatrix} j & j' \\ l & l' \end{pmatrix} \mathbf{u}(j'l', t) \qquad (2.2b)$$

As before, the element $-\Phi_{\alpha\alpha'}\begin{pmatrix} j & j' \\ l & l' \end{pmatrix}$ is the force in the α direction exerted on atom (jl) when atom $(j'l')$ is given a small unit displacement in the α' direction; the index j runs from 1 to n, the number of atoms in the unit cell, and l runs over all unit cells.

It is clear from Eq. (2.2b) that the frequency ω in Eq. (2.7) is a function of $\mathbf{q}$, because $\mathbf{q}$ governs the relative displacement of two interacting atoms. The $\omega(\mathbf{q})$ dependence on $\mathbf{q}$ (the dispersion relationships) is illustrated in Fig. 2.1 for the rock-salt structure. It can be shown that all normal modes can be represented in the first Brillouin zone, which extends from 0 to π/a in the a direction of the rock-salt structure, or, more generally, is bounded by faces located halfway between the reciprocal lattice points in the space defined by[1] $\mathbf{a}_i \cdot \mathbf{a}_j = 2\pi\delta_{ij}$. The

[1] Note that compared with Eq. (1.21) a factor 2π occurs on the right-hand side of this equation.

number of normal modes within the first Brillouin zone along each of the branches in Fig. 2.1 is equal to the number of unit cells in the crystal along the a direction. Substitution of Eq. (2.7b) into Eq. (2.2b) leads to the matrix equation

$$\omega^2 \mathbf{U}_m = \mathbf{D}\mathbf{U}_m \qquad (2.8)$$

where $\mathbf{U}_m$ is a column matrix of the *mass-adjusted* displacement coordinates

$$\mathbf{U}_m = \mathbf{m}^{1/2}\mathbf{U} \qquad (2.9)$$

in which $\mathbf{m}$ is a $3n \times 3n$ diagonal matrix obtained by repeating the masses three times along the diagonal (for the three displacements of each atom). Term $\mathbf{D}$ is the mass-adjusted *dynamical matrix* of the crystal, an element of which is given by

$$D_{\alpha\alpha'}(jj' | \mathbf{q}) = m(j)^{-1/2}m(j')^{-1/2} \sum_{l'} \Phi_{\alpha\alpha'}\begin{pmatrix} j & j' \\ 0 & l' \end{pmatrix} \exp \left[i\mathbf{q}\cdot\{\mathbf{r}(j'l') - \mathbf{r}(j0)\} \right] \qquad (2.10)$$

The elements of $\mathbf{D}$ represent the sum over all unit cells of the interaction between a pair of atoms. $\mathbf{D}$ has $3n \times 3n$ elements for a specific $\mathbf{q}$ and j, though the numerical value of the elements will rapidly decrease as pairs of atoms at greater distances are considered. Its eigenvectors, labeled $e_\alpha(j | k\mathbf{q})$, where k is the *branch index*, represent the directions and relative size of the displacements of the atoms for each of the normal modes of the crystal. Eigenvector $e_\alpha(j | k\mathbf{q})$ is a column matrix with three rows for each of the n atoms in the unit cell. Because the dynamical matrix is Hermitian, the eigenvectors obey the orthonormality condition

$$\sum_{\alpha j} e_\alpha^*(j | k\mathbf{q})e_\alpha(j | k'\mathbf{q}) = \delta_{kk'} \qquad (2.11)$$

that is, when summed over all three directions α and all atoms, the product is equal to 1 if $k = k'$, and zero otherwise. The eigenvectors are, in general, complex, as a phase vector is included, except for an atom or molecule located on a center of symmetry, for which the phase is symmetry restricted to be zero or π.

The mean-square displacements of each of the atoms in the crystal, which affect the the X-ray scattering amplitudes, are obtained by summation over the displacements due to all normal modes, each of which is a function of $e_\alpha(j | k\mathbf{q})$, as further discussed in section 2.3. The eigenvalues of $\mathbf{D}$ are the frequencies of the normal modes.

For a molecular crystal, the internal modes tend to be $\mathbf{q}$ independent and thus appear as horizontal lines in Fig. 2.1; n is then equal to the number of molecules M in the cell, leading to a considerable simplification. The resulting dynamical matrix has $6M \times 6M$ elements, considering both translational and rotational motions, and atom–atom potential functions may be used for its evaluation. Dispersion curves obtained in this manner for anthracene and naphthalene, are illustrated in Fig. 2.2.

The relation between the crystallographic temperature factors and the eigenvectors $\mathbf{e}$ of $\mathbf{D}$ is discussed in section 2.3.

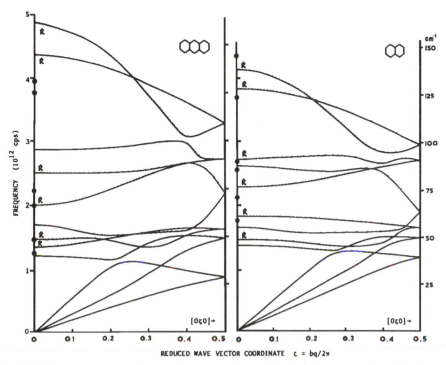

FIG. 2.2 Dispersion curves for anthracene and naphthalene for external-mode phonons traveling in the *b* direction. In each crystal there are two molecules in the unit cell, leading to 12 external phonon branches. The librational modes are labeled R. *Source*: Pawley (1967).

2.2 The Effect of Thermal Vibrations on the Bragg Intensities

2.2.1 The Born–Oppenheimer Approximation

Since electrons are much lighter than nuclei, they move much faster, and may be assumed to adjust instantaneously to a change in nuclear configuration. This is the background for the *Born–Oppenheimer approximation*, which allows separation of the nuclear and electronic energies. In the Born–Oppenheimer approximation, the electronic energy and the electron distribution are functions of the instantaneous nuclear coordinates, the dependence on the nuclear coordinates resulting from the electrostatic attractions between the electrons and the nuclei. The time-averaged electron density $\langle \rho(\mathbf{r}) \rangle$ is then the weighted average of the electron density for each of the nuclear configurations which occur along the nuclear vibration path, the weights being determined by the nuclear probability distribution $P(\mathbf{u}_1, \ldots, \mathbf{u}_N)$, where $\mathbf{u}_1, \ldots, \mathbf{u}_N$ are the displacement coordinates.

Thus, if $\rho(\mathbf{r}, \mathbf{u}_1, \ldots, \mathbf{u}_N)$ is the electron density at $\mathbf{r}$ corresponding to the nuclear geometry $\mathbf{u}_1, \ldots, \mathbf{u}_N$, the time-averaged electron density is

$$\langle \rho(\mathbf{r}) \rangle = \int \rho(\mathbf{r}, \mathbf{u}_1, \ldots, \mathbf{u}_N) P(\mathbf{r}, \mathbf{u}_1, \ldots, \mathbf{u}_N) \, d\mathbf{u}_1 \ldots d\mathbf{u}_N \qquad (2.12)$$

When the electrons can be assigned to specific nuclei, and follow these nuclei perfectly, the density for such a rigid group can be written as

$$\langle \rho_{\text{rigid group}}(\mathbf{r}) \rangle = \int \rho_{\text{static}}(\mathbf{r} - \mathbf{u}) P(\mathbf{u}) \, d\mathbf{u} = \rho_{\text{static}}(\mathbf{r} - \mathbf{u}) * P(\mathbf{u}) \qquad (2.13)$$

The label of rigid group as used here may apply to atoms or rigidly connected groups of atoms. In the former case, we obtain

$$\langle \rho_{\text{atom}}(\mathbf{r}) \rangle = \rho_{\text{atom, static}}(\mathbf{r}) * P(\mathbf{u}) \qquad (2.14)$$

Expression (2.14), referred to as the *convolution approximation*, is widely applied in crystallographic work.

2.2.2 The Harmonic Temperature Factor

According to the Fourier convolution theorem, further discussed in section 5.1.3, the Fourier transform of the convolution in expression (2.14) is the product of the Fourier transforms of the individual functions, or

$$\langle f(\mathbf{S}) \rangle = \hat{F} \langle \rho_{\text{atom}}(\mathbf{r}) \rangle = f(\mathbf{S}) T(\mathbf{S}) \qquad (2.15)$$

Thus, the *temperature factor* $T(\mathbf{S})$ is the Fourier transform of the probability distribution $P(\mathbf{u})$: $T(S) = \hat{F}\{P(\mathbf{u})\}$. In the common case that the rigidly vibrating groups are considered to be the individual atoms, $T(\mathbf{S})$ is the Fourier transform of the atomic probability distribution.

For a harmonic oscillator, the probability distribution averaged over all populated energy levels is a Gaussian function, centered at the equilibrium position. For the classical harmonic oscillator, this follows directly from the expression of a Boltzmann distribution in a quadratic potential. The result for the quantum-mechanical harmonic oscillator, referred to as Bloch's theorem, is less obvious, as a population-weighted average over all discrete levels must be evaluated (see, e.g., Prince 1982).

For an isotropic potential, the three-dimensional probability distribution is given by[2]

$$P(u) = (2\pi \langle u^2 \rangle)^{-3/2} \exp \{ -u^2/2\langle u^2 \rangle \} \qquad (2.16)$$

where $\langle u^2 \rangle$ is the (isotropic) mean-square displacement. The corresponding temperature factor follows from

$$T(S) = \hat{F}\{P(u)\} = (2\pi \langle u^2 \rangle)^{-3/2} \int e^{-u^2/2\langle u^2 \rangle} e^{2\pi i \mathbf{S} \cdot \mathbf{r}} \, d\mathbf{r} \qquad (2.17)$$

The integral in Eq. (2.17) is the product of the integrals over each of the Cartesian

---

[2] In practice, the probability distribution often includes static disorders in the crystal. The temperature parameter B in such cases is more properly described as a mean-square displacement parameter.

coordinates, $I = I_x I_y I_z$. For the integration over the x coordinate,

$$I_x = \int_{-\infty}^{\infty} e^{-x^2/2\langle u^2 \rangle}\, e^{2\pi i S_x x}\, dx = \int_{-\infty}^{\infty} e^{-x^2/2\langle u^2 \rangle + 2\pi i S_x x}\, dx$$

$$= (2\pi \langle u^2 \rangle)^{1/2}\, e^{-2\pi S_x^2 \langle u^2 \rangle} \tag{2.18}$$

as can be verified by completing the square in the exponent of the second integral by multiplication with the x-independent term $e^{2\pi^2 S_x^2 \langle u^2 \rangle}$. Substitution into Eq. (2.17) gives, for the isotropic harmonic temperature factor,

$$T(S) = \exp\left(-2\pi^2 S^2 \langle u^2 \rangle\right) \tag{2.19}$$

or, equivalently,

$$T(S) = \exp\left(-B \sin^2\theta/\lambda^2\right) \qquad \text{with } B = 8\pi^2 \langle u^2 \rangle \tag{2.20}$$

It is noted that both the probability distribution of Eq. (2.16) and the temperature factor of Eq. (2.19) are Gaussian functions, but with inversely related mean-square deviations. Analogous to the relation between direct and reciprocal space, the Fourier transform of a diffuse atom is a compact function in scattering space, and vice versa.

A trivariate normal distribution describes the probability distribution for *anisotropic* harmonic motion in three-dimensional space. In tensor notation (see appendix A for the notation, and appendix B for the treatment of symmetry and symmetry restrictions of tensor elements), with j and k ($= 1, 3$) indicating the axial directions,

$$P(\mathbf{u}) = \frac{|\sigma^{-1}|^{1/2}}{(2\pi)^{3/2}} \exp\left\{-\tfrac{1}{2}\sigma_{jk}^{-1}(u^j u^k)\right\} \tag{2.21a}$$

where u^j are the contravariant displacement coordinates with respect to the covariant axes x_j, σ is the matrix with elements $\langle u^j u^k \rangle$, and $|\sigma^{-1}|$ is the determinant of the inverse of σ. As is common in tensor notation, summation over repeated indices has been assumed. The corresponding equation in matrix notation is

$$P(\mathbf{u}) = \frac{|\sigma^{-1}|^{1/2}}{(2\pi)^{3/2}} \exp\left\{-\tfrac{1}{2}(\mathbf{u})^T \sigma^{-1}(\mathbf{u})\right\} \tag{2.21b}$$

where the superscript T indicates the transpose.

The anisotropic temperature factor will be the Fourier transform of $P(\mathbf{u})$, given by

$$T(\mathbf{S}) = \exp\left\{-2\pi^2 \sigma^{jk} h_j h_k\right\} \tag{2.22a}$$

or, in matrix notation,

$$T(\mathbf{S}) = \exp\left\{-2\pi^2 \mathbf{S}^T \sigma \mathbf{S}\right\} \tag{2.22b}$$

With the change of variable $\beta^{jk} = 2\pi^2 \sigma^{jk}$, Eq. (2.22) is, for a reflection at $\mathbf{S} = \mathbf{H}$,

$$T(\mathbf{H}) = \exp\left\{-\beta^{jk} h_j h_k\right\} \tag{2.23}$$

In the atomic model, the tensor β^{jk} describes the anisotropic motion of an

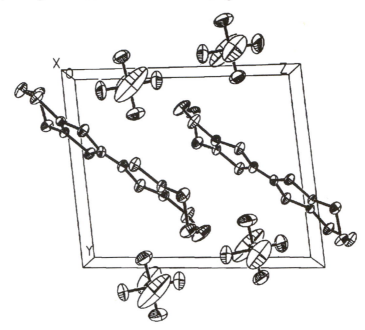

FIG. 2.3 Equal probability ellipsoids for δ-(BEDT-TTF)PF$_6$ [BEDT-TTF = bis(ethylene-dithiotetrathiofulvalene)] *Source*: Bu et al. (1992).

atom. Since $\beta^{jk} = 2\pi^2\sigma^{jk} = 2\pi^2\langle u^j u^k \rangle$, then

$$\beta^{jk} = 2\pi^2\langle u^j u^k \rangle \tag{2.24}$$

It is clear from Eq. (2.24) that β^{jk} is a symmetric tensor, which must have positive principal components in order to be physically meaningful.

We are often interested in the rms thermal displacements in Å. They correspond to the contravariant components U^{jk} along covariant axes of unit length, rather than along the non-unit length **a**, **b**, **c** axes. The rms displacements are obtained from

$$\beta^{jk} = 2\pi^2 U^{jk}|a^j||a^k| \tag{2.25a}$$

or

$$U^{jk} = \frac{\beta^{jk}}{2\pi^2|a^j||a^k|} \tag{2.25b}$$

The tensor U^{jk} may be represented by its *equal probability surface*, which, according to Eq. (2.21) is given by

$$u^j\sigma_{jk}^{-1}u^k = c^2 \tag{2.26}$$

where, as in Eq. (2.21), σ_{jk}^{-1} is the inverse of σ^{jk}, and c is a constant. For $c = 1.5382$, the volume of the ellipsoid is equal to one half, that is, the probability that the atom is inside the ellipsoid is 50% (appendix C). This ellipsoid is referred to as the 50% probability ellipsoid. An example is given in Fig. 2.3.

The mean-square displacement of an atom in a direction defined by the unit vector $\hat{v}$ is given by

$$\langle u^2 \rangle = \sigma^{ij} \hat{v}_i \hat{v}_j \qquad (2.27)$$

The surface defined by σ^{ij} is not an ellipsoid, but is, in general, peanut shaped (Hummel et al. 1990).

2.2.3 Beyond the Harmonic Approximation

The probability distribution of Eq. (2.21) was derived assuming rectilinear motion in a harmonic potential. The true potential in a crystal is often more complex, especially in the upper parts of the potential surface, which are of importance at higher temperatures.

Three more general distributions and the corresponding temperature factors are discussed in the following.

2.2.3.1 The Gram–Charlier Expansion

The three-dimensional Gram–Charlier expansion, first applied to thermal motion analysis by Johnson and Levy (1974), is a statistical expansion in terms of the zero and higher derivatives of a normal distribution (Kendal and Stuart 1958). If D_j is the operator d/du^j, the expansion is defined by

$$P(\mathbf{u}) = \left[1 - c^j D_j + \frac{1}{2!} c^{jk} D_j D_k - \frac{1}{3!} c^{jkl} D_j D_k D_l + \cdots \right.$$

$$\left. + (-1)^r \frac{c^{\alpha_1} \dots c^{\alpha_r}}{r!} D_{\alpha_1} \dots D_{\alpha_r} \right] P_0(\mathbf{u}) \qquad (2.28)$$

The leading term $P_0(\mathbf{u})$ is the harmonic probability distribution, $\alpha_i = 1, 2,$ or 3, and $D_{\alpha_1} \dots D_{\alpha_r}$ is the rth partial derivative operator $\partial^r/(\partial u^{\alpha_1} \dots \partial u^{\alpha_r})$. The Einstein convention of summation over repeated indices is implied.

For a distribution expanded around the equilibrium position, the first derivative is zero, and may be omitted, while the second derivatives are redundant as they merely modify the harmonic distribution. Since $P_0(\mathbf{u})$ is a Gaussian distribution, Eq. (2.28) can be simplified by use of the Tchebycheff–Hermite polynomials, often referred to simply as Hermite polynomials,[3] $H_{\alpha_1 \dots \alpha_2}$, related to the derivatives of the three-dimensional Gaussian probability distribution by

$$H_{\alpha_1 \dots \alpha_r}(\mathbf{u}) \exp\left(-\tfrac{1}{2} p_{jk} x^j x^k\right) = (-1)^r D_{\alpha_1} \dots D_{\alpha_r} \exp\left(-\tfrac{1}{2} p_{jk} x^j x^k\right) \qquad (2.29)$$

The result is

$$P(\mathbf{u}) = \left[1 + \frac{1}{3!} c^{jkl} H_{jkl}(\mathbf{u}) + \frac{1}{4!} c^{jklm} H_{jklm}(\mathbf{u}) \right.$$

$$\left. + \frac{1}{5!} c^{jklmn} H_{jklmn}(\mathbf{u}) + \frac{1}{6!} c^{jklmnp} H_{jklmnp}(\mathbf{u}) + \cdots \right] P_0(\mathbf{u}) \qquad (2.30a)$$

---

[3] The polynomials defined here are different from the Hermite polynomials which occur in the solutions of the Schrödinger equation for the harmonic oscillator.

TABLE 2.1 Low-Order Hermite Polynomials

$$H(\mathbf{u}) = 1$$
$$H_j(\mathbf{u}) = w_j$$
$$H_{jk}(\mathbf{u}) = w_j w_k - p_{jk}$$
$$H_{jkl}(\mathbf{u}) = w_j w_k w_l - (w_j p_{kl} + w_k p_{lj} + w_l p_{jk}) = w_j w_k w_l - 3w(jp_{kl})$$
$$H_{jklm}(\mathbf{u}) = w_j w_k w_l w_m - 6w(jw_k p_{lm}) + 3p_j(kp_{lm})$$
$$H_{jklmn}(\mathbf{u}) = w_j w_k w_l w_m w_n - 10w(lw_m w_n p_{jk}) + 15w(np_{jk}p_{lm})$$
$$H_{jklmnp}(\mathbf{u}) = w_j w_k w_l w_m w_n w_p + 45w(jw_k p_{lm}p_{np}) - 15w(jw_k w_l w_m p_{jk}) - 15p_j(kp_{lm}p_{np})$$

Note: $w_j \equiv p_{jk}x^k$, where the Gaussian function is defined as $\exp(-\frac{1}{2}p_{jk}x^j x^k)$. Brackets indicate averaging of the term over all permutations of the bracketed indices which produce distinct terms, noting that $p_{jk} = p_{kj}$ and $w_j w_k = w_k w_j$.
Source: Johnson (1969), Johnson and Levy (1974), Zucker and Schulz (1982).

Here, the permutations of $j, k, l, \ldots$ include all combinations which produce different terms. The multivariate Hermite polynomials are listed in Table 2.1 for orders ≤ 6. Like the spherical harmonics, the Hermite polynomials form an orthogonal set of functions (Kendal and Stuart 1958, p. 156).

The coefficients c in this probability distribution are referred to as the *quasimoments* of the distribution. Because of the orthogonality of Hermite polynomials, the quasimoments of a function are obtained by integration of the product of the function and the related Hermite polynomial over all space. For the one-dimensional case,

$$c^j = \frac{1}{j!} \int f(x) H_j(x)\, dx \tag{2.30b}$$

Thus, the quasimoments are directly related to the moments μ of a distribution defined by $\mu^{ijk \cdots} = \int_{-\infty}^{+\infty} f(\mathbf{x}) x^i x^j x^k \ldots d\mathbf{x}$. The relations between the two sets follow by substituting the expressions for the Hermite polynomials into Eq. (2.30b). They can also be derived by writing $f(\mathbf{x})$ as an expansion, both in terms of its moments [see Eq. (2.33)], and in terms of the quasimoments [as in Eq. (2.30a)], and equating equivalent terms.

The Gram–Charlier temperature factor is the Fourier transform of Eq. (2.30), which is given by

$$T(\mathbf{H}) = [1 - \tfrac{4}{3}\pi^3 i c^{jkl} h_j h_k h_l + \tfrac{2}{3}\pi^4 c^{jklm} h_j h_k h_l h_m + \tfrac{4}{15}\pi^5 i c^{jklmn} h_j h_k h_l h_m h_n$$
$$- \tfrac{4}{45}\pi^6 c^{jklmnp} h_j h_k h_l h_m h_n h_p + \cdots] T_0(\mathbf{H}), \tag{2.31}$$

where $T_0(\mathbf{H})$ is the harmonic temperature factor.

As Eq. (2.31) shows, the Gram–Charlier temperature factor is a power-series expansion about the harmonic temperature factor, with real even terms, and imaginary odd terms. This is an expected result, as the even-order Hermite polynomials in the probability distribution of Eq. (2.30) are symmetric, and the odd-order polynomials are antisymmetric with respect to the center of the distribution.

2.2.3.2 The Cumulant Expansion

A second statistical expansion that may be used to describe the atomic probability distribution is due to Edgeworth (Kendal and Stuart 1958, Johnson 1969). It expresses a distribution in terms of its *cumulants* κ. If D is the differential operator, $P(\mathbf{u})$ is described as

$$P(\mathbf{u}) = \left[\exp \left(\kappa^j D_j + \frac{1}{2!} \kappa^{jk} D_j D_k - \frac{1}{3!} \kappa^{jkl} D_j D_k D_l + \frac{1}{4!} \kappa^{jklm} D_j D_k D_l D_m - \cdots \right) \right] P_0(\mathbf{u})$$

(2.32)

where $P_0(\mathbf{u})$ is the harmonic probability distribution. Though the differential operator in Eq. (2.32) occurs in the exponent, use of a Taylor expansion leads to an expression of a more common form.

A cumulant of rank s is a symmetric tensor with $(s^2 + 3s + 2)/2$ unique elements for a three-dimensional distribution. Like the moments μ and the quasimoments c, the cumulants are descriptors of the distribution. For a one-dimensional distribution, the relations between the cumulants and the moments are defined by equating the two expansions:

$$\exp \left\{ \kappa_1 x + \frac{\kappa_1 x^2}{2!} + \cdots + \frac{\kappa_r x^r}{r!} + \cdots \right\} = 1 + \mu_1 x + \frac{\mu_1 x^2}{2!} + \cdots + \frac{\mu_r x^r}{r!}$$

(2.33)

As is evident from Eq. (2.18), the Fourier transform of an exponential is an exponential. Fourier transform of Eq. (2.32), omitting, as in Eq. (2.30), the first- and second-order terms, gives (Kendal and Stuart 1958)

$$T(\mathbf{H}) = \exp \left[\frac{(2\pi i)^3}{3!} \kappa^{jkl} h_j h_k h_l + \frac{(2\pi i)^4}{4!} \kappa^{jklm} h_j h_k h_l h_m + \cdots \right] T_0(\mathbf{H})$$

$$= \exp \left[-\tfrac{4}{3}\pi^3 i \kappa^{jkl} h_j h_k h_l + \tfrac{2}{3}\pi^4 \kappa^{jklm} h_j h_k h_l h_m + \cdots \right] T_0(\mathbf{H})$$

(2.34)

Compared with the Gram–Charlier temperature factor of Eq. (2.31), the entire series now occurs in the exponent, so, in the cumulant formalism, terms are added to the exponent of the harmonic temperature factor $T_0(\mathbf{H}) = \exp \{ -\beta^{jk} h_j h_k \}$.

Application of the Taylor expansion $\exp(i\mathbf{H}) = \sum (i\mathbf{H})^N/N!$ to Eq. (2.34) shows that the two expressions are identical if all terms up to infinity are included. The Taylor expansion of Eq. (2.34) is

$$T(\mathbf{H}) = \left[1 + \frac{(2\pi i)^3}{3!} \kappa^{jkl} h_j h_k h_l + \frac{(2\pi i)^4}{4!} \kappa^{jklm} h_j h_k h_l h_m + \cdots \right.$$

$$+ \frac{(2\pi i)^6}{6!} \left\{ \kappa^{jklmnp} + \frac{6!}{2!(3!)^2} \kappa^{jkl} \cdot \kappa^{mnp} \right\} h_j h_k h_l h_m h_n h_p$$

$$\left. + \text{ higher-order terms} \right] T_0(\mathbf{H})$$

(2.35)

This formulation, referred to as the *Edgeworth approximation* (Zucker and

Schulz 1982), corresponds to a probability distribution which is the Taylor expansion of Eq. (2.32), and similar to the Gram–Charlier distribution of Eq. (2.30):

$$P(\mathbf{u}) = P_0(\mathbf{u}) \left[1 + \frac{1}{3!} \kappa^{jkl} H_{jkl}(\mathbf{u}) + \frac{1}{4!} \kappa^{jklm} H_{jklm}(\mathbf{u}) + \cdots \right.$$

$$\left. + \frac{1}{6!} \{ \kappa^{jklmnp} + 10\kappa^{jkl}\kappa^{mnp} \} H_{jklmnp} + \text{higher-order terms} \right] \quad (2.36)$$

Relations between the cumulants κ^{jkl} and the quasimoments c^{jkl} follow from comparison of Eqs. (2.36) and (2.30):

$$c^{jkl} = \kappa^{jkl}$$

$$c^{jklm} = \kappa^{jklm}$$

$$c^{jklmn} = \kappa^{jklmn} \quad\quad (2.37)$$

$$c^{jklmnp} = \kappa^{jklmnp} + 10\kappa^{jkl}\kappa^{mnp}$$

The result shows that the sixth- and higher-order cumulants and quasimoments differ. The third-order cumulant κ^{jkl} contributes not only to the coefficient of H^{jkl}, but also to higher-order terms of the probability distribution function. The situation is analogous for cumulants of higher orders. It follows that for a finite truncation of the temperature factor defined by Eq. (2.34), the probability distribution cannot be represented by a finite number of quasimoments. This is a serious difficulty when a probability distribution is to be derived from an experimental temperature factor of the cumulant type. A second complication in the use of the cumulant expansion, pointed out by Scheringer (1985), is that the probability function always has some physically unrealistic negative regions. In certain cases of large anharmonicity, the Fourier transform of Eq. (2.34) may, in fact, not exist at all. As a result of such considerations, the Gram–Charlier expression is generally preferred over the cumulant expansion, because its truncation is equivalent in real and reciprocal space, and it does not lead to negative regions in the probability distribution (Kuhs 1983, 1992; Scheringer 1985).

2.2.3.3 The One-Particle Potential (OPP) Model

Unlike the Gram–Charlier and cumulant formalisms, the OPP model has a physical rather than a statistical basis. It assumes that each atom vibrates in a potential well $V(\mathbf{u})$, determined by the interaction with the other atoms in the crystal, without any correlation between vibrations of adjacent atoms.

In the classical high-temperature limit, $k_B T \gg h\nu$, where k_B is the Boltzmann constant, and $h\nu$ is the spacing of the quantum-mechanical harmonic oscillator energy levels. If this condition is fulfilled, the energy levels may be considered as continuous, and Boltzmann statistics apply. The corresponding distribution is

$$P(\mathbf{u}) = N \exp\{-V(\mathbf{u})/k_B T\} \quad (2.38)$$

with N, the normalization constant, defined by $\int P(\mathbf{u})\, d\mathbf{u} = 1$.

The potential function may be expanded in terms of increasing order of products of the contravariant displacement coordinates (Dawson 1967, Willis 1969)

$$V = V_0 + \alpha_j u^j + \beta_{jk} u^j u^k + \gamma_{jkl} u^j u^k u^l + \delta_{jklm} u^j u^k u^l u^m + \cdots \quad (2.39)$$

As in the probability distributions of Eqs. (2.30) and (2.32), the first derivatives vanish at the equilibrium position, so $\alpha_j = 0$. The constant V_0 term does not affect the probability distribution, and may also be omitted. Substitution of Eq. (2.39) in Eq. (2.38), and use of the approximation $\exp(-\Delta) = 1 - \Delta$ for the higher-order terms, leads to

$$P(\mathbf{u}) = N \exp\{-\beta'_{jk} u^j u^k\}\{1 - \gamma'_{jkl} u^j u^k u^l - \delta'_{jklm} u^j u^k u^l u^m - \cdots\} \quad (2.40)$$

in which $\beta' = \beta/(k_B T)$, etc. In the description, the higher-order terms appear as corrections to the harmonic temperature factor. We note that the normalization factor N depends on the level of truncation of the series.

The Fourier transform of the OPP distribution, in a general coordinate system, is (Johnson 1970, Scheringer 1985)

$$T(\mathbf{H}) = T_0(\mathbf{H})[1 - \tfrac{4}{3}\pi^3 i \gamma'_{jkl} G^{jkl}(\mathbf{H}) + \tfrac{2}{3}\pi^4 \delta'_{jklm} G^{jklm}(\mathbf{H})$$
$$+ \tfrac{4}{15}\pi^5 i \varepsilon'_{jklmn} G^{jklmn}(\mathbf{H}) - \tfrac{4}{45}\pi^6 \varphi'_{jklmnp} G^{jklmnp}(\mathbf{H}) \ldots] \quad (2.41)$$

where T_0 is the harmonic temperature factor, and G represents the Hermite polynomials in reciprocal space.

The harmonic term $T_0(\mathbf{H})$ of Eq. (2.41) is equal to

$$T_0(\mathbf{H}) = \exp -\{\pi^2 k_B T(\beta^{-1})^{ij} h_i h_j\} \quad (2.42)$$

Comparison with Eq. (2.23) shows that the classical OPP model predicts the elements of the anisotropic harmonic temperature parameter to be proportional to the absolute temperature.

Expressions simpler than Eq. (2.41) are obtained if the OPP temperature factor is expanded in the coordinate system which diagonalizes β_{jk}. In that case, the Hermite polynomials become products of the displacement coordinates u^j (Coppens 1980, Tanaka and Marumo 1983). The first terms in the expansion are given by

$$T(\mathbf{H}) = \exp(-\pi h_j^2 k_B T/\beta_{jj})\left\{ 1 - i\gamma_{jjj}\left[k_B T\left(\frac{3\pi h_j}{2\beta_{jj}^2}\right) - (k_B T)^2\left(\frac{\pi h_j}{\beta_{jj}}\right)^3 \right] \right.$$
$$+ i\gamma_{jjk}\left[k_B T\frac{\pi h_k}{2\beta_{jj}\beta_{kk}} - (k_B T)^2\frac{\pi^3 h_j^2 h_k}{\beta_{jj}^2 \beta_{kk}} \right]$$
$$\left. - i\gamma_{jkl}(k_B T)^2\frac{\pi^3 h_j h_k h_l}{\beta_{jj}\beta_{kk}\beta_{ll}} \right\} + \cdots \quad (2.43)$$

This result indicates that, according to the OPP model, the higher-order anharmonic terms have a stronger temperature dependence than the leading harmonic terms. The quartic terms, not specifically included in Eq. (2.43), have an even larger temperature dependence proportional to T^3. Thus, the effect of anharmonicity may be effectively reduced by cooling when the harmonic temperature factor is the leading term in the expansion. This is an important result for charge density analysis, especially for heavier atoms, for which anharmonicity affecting all electron shells may mask the bonding effects of the valence-electron distribution.

2.2.3.4 Application to Diamond-Type Structures

Dawson and coworkers pioneered the application of the OPP model to diamond-type structures (Dawson 1967, Dawson et al. 1967). In the diamond-type structure, common to diamond, silicon, and germanium, the atoms are located at 1/8, 1/8, 1/8, at the center-of-symmetry related position at $-1/8, -1/8, -1/8$, and repeated in a face-centered arrangement. The tetrahedral symmetry of the atomic sites greatly limits the allowed coefficients in the expansion of Eq. (2.39). With x, y, z expressed relative to the nuclear position, the potential is given by

$$V = V_0 + \beta(x^2 + y^2 + z^2) + \gamma xyz + \cdots + \tag{2.44}$$

with a corresponding OPP temperature factor, obtained by omitting the symmetry-forbidden terms from Eq. (2.43),

$$T(\mathbf{H}) = \exp\left\{-\pi(h^2 + k^2 + l^2)k_B T/\beta\right\}\left\{1 - i\gamma(k_B T)^2 \frac{\pi^3 hkl}{\beta^3}\right\} \tag{2.45}$$

When the atomic scattering factor is real (as it is when bonding effects on the charge density are neglected), and resonance scattering has been corrected for, the harmonic structure factor expression is equal to

$$F(\mathbf{H}) = 8f \cos\left\{2\pi(h + k + l)/8\right\}T_0 \tag{2.46}$$

where T_0 represents the harmonic temperature factor, defined by the exponential factor in Eq. (2.45). Thus, in this approximation, $F(\mathbf{H})$ equals zero for reflections with $h + k + l = 4n + 2$, such as (222), (442), and (662). However, the atoms vibrate more strongly into the void opposite the tetrahedrally arranged bonds, while the displacement in opposite directions into the bonds is more constrained. As a result, γ in Eqs. (2.44) and (2.45) has a significant negative value, leading to nonzero intensity due to anharmonicity for reflections with $h + k + l = 4n + 2$ and large values of the product hkl. The corresponding intensity increases with temperature, because the anharmonic term is proportional to T^2. This effect, and its relation to the scattering of the covalent bonding density, is further discussed in chapter 11.

2.2.4 Comparison of the Anharmonic Formalisms

The OPP formalism, though based on the assumption of independent motion, has the advantage of assigning a physical meaning to the terms in the expansion. By equating the OPP terms to the corresponding ones in the statistical expansions, the quasimoments and cumulants can be related to the parameters of the potential model, and their temperature dependence can be predicted.

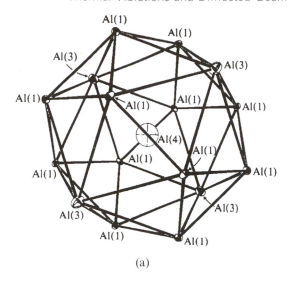

(a)

FIG. 2.4(a) The coordination of Al(4) in $VAl_{10.42}$. Thermal ellipsoids (100 K) are at the 50% probability level. Four nearest-neighbor Al(3) atoms are located 3.1330 (4) Å from Al(4) in the [111] directions, while 12 Al(1) atoms are at 3.1484 (5) Å, arranged in groups of three, capping each face of the tetrahedron formed by the Al(3) atoms. *Source*: Kontio and Stevens (1982).

For a cubic site, relations between the cumulants and the coefficients of the OPP model have been derived by Kontio and Stevens (1982), and applied to the Al(4) atom in the alloy $VAl_{10.42}$. The coordination of Al(4) is illustrated in Fig. 2.4(a), while the potential along [111], derived from the thermal parameter refinement, is shown in Fig. 2.4(b). It is clear from these figures that higher than third-order terms contribute to the potential, because the deviation from the harmonic curve is not exactly antisymmetric with respect to the equilibrium configuration. The potential appears steeper at the higher temperature, which is opposite to what is expected on the basis of the thermal expansion of the solid.

2.2.5 Quantum-Statistical Treatments

The classical model predicts thermal motion to vanish at very low temperatures, in contradiction to the zero-point vibrations which follow from the quantum-mechanical treatment of oscillators. For temperatures at which $hv \approx k_B T$, the spacing of the discrete energy levels cannot be neglected, so the classical model is no longer valid.

For a *harmonic oscillator* with frequency v, the potential energy equals $1/2 k_f \langle u^2 \rangle$, where k_f is the force constant. Using the virial theorem, which states that $E_{kin} = E_{pot} = E_{tot}/2$,

$$E_{tot} = k_f \langle u^2 \rangle \tag{2.47}$$

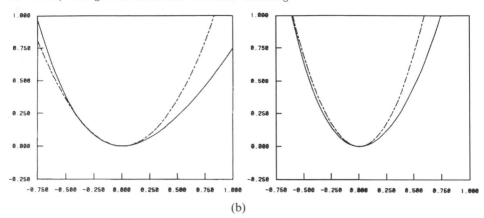

(b)

FIG. 2.4(b) The experimental potential along the [111] direction for the Al(4) atom in the alloy VAl$_{10.42}$, obtained with the one-particle potential temperature factor. Left: 100 K; right: room temperature. The broken lines represent the harmonic components. Differences between the classical and quantum-statistical results at both temperatures were found to be extremely small. *Source*: Kontio and Stevens (1982).

According to statistical mechanical theory, the total energy of the oscillator is a function of the partition function z, through the relation

$$E_{tot} = k_B T^2 \frac{\partial \ln z}{\partial T} \tag{2.48}$$

where T is the absolute temperature. The partition function z for the harmonic oscillator is obtained by the summation over all levels $E = vhv$, where v is the vibrational quantum number, and E is counted from the lowest ($v = 0$) level, at $hv/2$. The well-known result is

$$z = \frac{1}{1 - e^{-hv/k_B T}} \tag{2.49}$$

Substitution into Eq. (2.48), and adding the zero-point energy $hv/2$, gives

$$E_{tot} = hv \left[\frac{1}{2} + \frac{1}{e^{hv/k_B T} - 1} \right] \tag{2.50}$$

Combining this result with Eq. (2.47), and substitution of $k = 4\pi^2 m v$, where m is the reduced mass of the harmonic oscillator, gives the quantum-statistical temperature dependence of $\langle u^2 \rangle$ as

$$\langle u \rangle^2 = \frac{h}{4\pi^2 m v} \left[\frac{1}{2} + \frac{1}{e^{hv/k_B T} - 1} \right] \tag{2.51a}$$

or, equivalently,

$$\langle u \rangle^2 = \frac{h}{8\pi^2 m v} \coth \left(\frac{hv}{2k_B T} \right) \tag{2.51b}$$

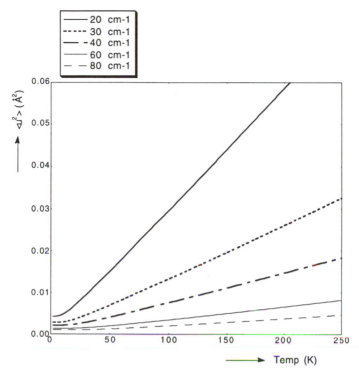

FIG. 2.5. Temperature dependence of the mean-square (ms) displacement of a quantum-mechanical harmonic oscillator with a mass of 200 daltons for a number of frequencies. The linearity of the ms displacement in the high-temperature region is evident.

This expression is illustrated in Fig. 2.5 for a number of frequencies. For large T, with $e^{\Delta} \approx 1 + \Delta$, this reduces to

$$\langle u^2 \rangle = \frac{k_B T}{h\nu} \tag{2.52}$$

Thus, in the high-temperature limit, the mean-square displacement of the harmonic oscillator, and therefore the temperature factor B, is *proportional* to the temperature, and *inversely* proportional to the frequency of the oscillator, in agreement with Eq. (2.43). At very low temperatures, the second term in Eq. (2.51a) becomes negligible. The mean-square amplitude of vibrations is then a constant, as required by quantum-mechanical theory, and evident in Fig. 2.5.

Quantum-statistical expressions such as Eq. (2.51) can be generalized for anharmonic motion, as shown by Mair (1980).

2.2.5.1 Numerical Application of Expression (2.51)

Substitution of values for the physical constants yields the numerical expression

$$\langle u \rangle^2 (\text{Å}^2) = \frac{33.69}{m(\text{dalton})\nu(\text{cm}^{-1})} \left[\frac{1}{2} + \frac{1}{e^{\nu(\text{cm}^{-1})/0.695T} - 1} \right] \tag{2.51c}$$

The temperature dependence according to this expression is illustrated in Fig. 2.5 for a number of frequencies, and a mass of 200 daltons. As the mean-square displacement scales with the inverse mass of the oscillator, the onset of the linear region is independent of mass for a given frequency.

As at very low temperatures, for very high frequencies the thermal motion becomes temperature independent. For example, for the C—H stretching mode, with frequency in the 2700–3300 cm^{-1} range, the exponential in the denominator of Eq. (2.51) is very large for common temperatures and the second term in the square brackets is negligible. Using for m the reduced mass of the oscillator (0.9231 dalton for diatomic C—H) gives, with $v = 3000$ cm^{-1}, a constant mean-square vibrational amplitude of 0.006 Å^2.

Vibrations in a real crystal are described by the lattice dynamical theory, discussed in section 2.1, rather than by the atomic oscillator model. Each harmonic phonon mode with branch index k and wavevector $\mathbf{q}$ then has, analogous to Eq. (2.50), an energy given by

$$E_{tot}(k\mathbf{q}) = hv(k\mathbf{q})\left[\frac{1}{2} + \frac{1}{e^{hv(k\mathbf{q})/k_B T} - 1}\right] \tag{2.53}$$

2.3 The Relation between the Atomic Temperature Factors and Lattice Dynamics

2.3.1 General Expression

As the oscillators of the OPP model vibrate independently of each other, the frequencies are dispersionless, that is, independent of a wavevector $\mathbf{q}$. For the internal modes of a molecular crystal, this tends to be a very good approximation. For the external modes, the dispersion can be pronounced, as shown in Figs. 2.1 and 2.2. In order to obtain the mean-square vibrational amplitudes for the latter, a summation over all phonon branches in the Brillouin zone must be performed.

The atomic displacements for the mode $(k\mathbf{q})$ are described by the $3n$ column matrix $\mathbf{U}_0$, which we used in Eq. (2.9). Matrix $\mathbf{U}$ is a function of the eigenvectors $\mathbf{e}(k\mathbf{q})$ of the mode, and its amplitude $A(k\mathbf{q})$:

$$\mathbf{U}(k\mathbf{q}) = m^{-1/2}U_m(k\mathbf{q}) = m^{-1/2}|A(k\mathbf{q})|\mathbf{e}(k\mathbf{q}) \tag{2.54}$$

The displacements are related to the energy of a mode by the expression

$$E(k\mathbf{q}) = N\omega^2(k\mathbf{q})|A(k\mathbf{q})|^2 \tag{2.55}$$

where N is the number of unit cells. This expression is analogous to Eq. (2.47), $E_j = k_f\langle u^2\rangle$, for the single harmonic oscillator, as can be verified by the substitution of $k_f = m\omega^2$ in the latter expression.

The displacement of an atom j, in unit cell l, at time t, is obtained by the summation over all $3nN$ normal modes, combined into $3n$ phonon branches,

$$\mathbf{u}(jl, t) = m(j)^{-1/2}\sum_{k\mathbf{q}}|A(k\mathbf{q})|\mathbf{e}(j \,|\, k\mathbf{q})\exp\left[i(\mathbf{q}\cdot\mathbf{r}(jl) - \omega(k\mathbf{q})t)\right] \tag{2.56}$$

or, using Eq. (2.55),

$$\mathbf{u}(j, t) = Nm(j)^{-1/2} \sum_{k\mathbf{q}} \left(\frac{E_k(\mathbf{q})}{\omega^2(k\mathbf{q})} \right)^{1/2} \mathbf{e}(j \,|\, k\mathbf{q}) \exp \left[i(\mathbf{q} \cdot \mathbf{r}(jl) - \omega(k\mathbf{q})t \right] \qquad (2.57)$$

The tensor $\langle \mathbf{U} \rangle_j$, describing the mean-square displacements of atom j, is the time average $\langle \mathbf{u}(j)\mathbf{u}(j)^T \rangle$, where $\mathbf{u}$ is the 3×1 column matrix of the displacements of atom j along the Cartesian axes, and T indicates the transpose. Since the normal modes are independent of each other, cross terms between modes disappear in the averaging. The result is

$$\langle \mathbf{U} \rangle_j = \frac{1}{Nm_j} \sum_{k\mathbf{q}} \frac{E_k(\mathbf{q})}{\omega^2(k\mathbf{q})} \mathbf{e}(j \,|\, k\mathbf{q})(\mathbf{e}^*(j \,|\, k\mathbf{q}))^T \qquad (2.58)$$

Application of Eq. (2.58) to calculate the temperature factors requires knowledge of the full frequency spectrum of the crystal throughout the Brillouin zone. Such information is only available for relatively simple crystal structures such as Al, Ni, KCl, and NaCl (Willis and Pryor 1975, p. 13ff.). Agreement between theory and experiment for such solids is often quite reasonable.

A considerable simplification is achieved when molecules can be treated as rigid bodies, as was done for naphthalene and anthracene (Fig. 2.2), the frequency spectra of which were derived using atom–atom potential functions. The mean-square displacements due to the internal modes can be calculated from the experimental infrared and Raman force constants, and added to the values obtained with Eq. (2.58). The rigid-body model for thermal vibrations is further discussed in section 2.3.3.

A very much simplified lattice-dynamical model is that of Debye. In the *Debye approximation*, discussed in the following section, a single phonon branch is assumed, with frequencies proportional to the magnitude of the wavevector $\mathbf{q}$.

2.3.2 The Debye Approximation

The Debye model assumes that there is a single acoustic branch, the frequency of which increases with constant slope (proportional to the average velocity of sound in the crystal) as q increases, up to the boundary of the Brillouin zone. The boundary is assumed to be of spherical shape, with a radius q_D determined by the total number of normal modes of the crystal. Thus,

$$\omega_k(\mathbf{q}) = v_s q \qquad (2.59)$$

The frequency v_D at the edge of the Brillouin zone is thus equal to $v_s q_D / 2\pi$. The *Debye temperature* Θ_D is defined as $hv_D/(k_B)$. As shown below, Θ_D is an inverse measure for the vibrational mean-square amplitudes of the atoms in a crystal at a given temperature.

As the normal modes are assumed to be uniformly distributed in reciprocal space, the frequency distribution $g(\omega)$ will be proportional to ω^2, that is,

$$g(\omega) = C\omega^2 \qquad (2.60)$$

in which the proportionality constant C follows from the normalization condition.

For a crystal with nN atoms,

$$3nN = \int g(\omega)\, d\omega \tag{2.61}$$

For a monatomic cubic crystal, the corresponding mean-square displacement is (Willis and Pryor 1975)

$$\langle u^2 \rangle = \frac{1}{3mN} \int_0^{\omega_D} \frac{E_\omega}{\omega^2} g(\omega)\, d\omega \tag{2.62}$$

which may be compared with Eq. (2.58). Using the vibrational partition function, this can be shown to be equal to

$$\langle u^2 \rangle = \frac{3\hbar^2 T}{mk_B\Theta_D^2} \left[\Phi\!\left(\frac{\Theta_D}{T}\right) + \frac{1}{4} \cdot \frac{\Theta_D}{T} \right] \tag{2.63}$$

with

$$\Phi(x) = \frac{1}{x} \int_0^x \frac{y}{e^y - 1}\, dy \approx 1 - \frac{x}{4} + \frac{x^2}{36} + \cdots \tag{2.64}$$

where $x = \Theta_D/T$.

In the high-temperature limit, $x \to 0$, and Eq. (2.63) becomes

$$\langle u^2 \rangle = \frac{3\hbar^2 T}{mk_B\Theta_D^2} = \frac{3k_B T}{4\pi^2 m v_D^2} \tag{2.65a}$$

The mean-square displacement is again proportional to T, as it is for the harmonic oscillator [Eq. (2.52)], but the slope of the $\langle u^2 \rangle$ versus T curve is quite different from the value of $k_B/(h\nu)$, predicted for the single harmonic oscillator [Expression (2.52)]. Expression (2.65) represents the temperature dependence of an assembly of harmonic oscillators, rather than that of a single oscillator.

In the low-temperature region, x becomes very large, and $\Phi(x)$ becomes equal to zero. We then obtain from Eq. (2.63),

$$\langle u^2 \rangle = \frac{3\hbar^2}{4mk_B\Theta_D} \tag{2.65b}$$

that is, the mean-square displacements again will be independent of the temperature.

In any crystal, the low-frequency acoustic modes dominate at low temperatures, so that the approximation that ω is proportional to q becomes increasingly valid as is evident from Fig. 2.2. In particular, the T dependence of the specific heat at very low temperatures is well predicted by the Debye approximation.

2.3.3 The Rigid-Body Model for Molecular Crystals

In molecular crystals, the separation between internal and external modes is of importance. Except for torsional oscillations in some types of molecules, the internal modes have much higher frequencies than the external modes. According to expressions such as Eqs. (2.51) and (2.58), the latter are then the dominant contributors to the atomic mean-square displacements. This is the basis for the *rigid-body approximation* developed by Cruickshank (1956), and expanded into a more general theory by Schomaker and Trueblood (1968). The following discussion is based on a treatment by Dunitz (1979).

The most general motions of a rigid body consist of rotations about three axes, coupled with translations parallel to each of the axes. Such motions correspond to screw rotations. A libration around a vector λ $(\lambda_1, \lambda_2, \lambda_3)$, with length corresponding to the magnitude of the rotation, results in a displacement $\delta\mathbf{r}$, such that

$$\delta\mathbf{r} = (\lambda \times \mathbf{r}) = \mathbf{D}\mathbf{r} \qquad (2.66)$$

with

$$\mathbf{D} = \begin{bmatrix} 0 & -\lambda_3 & \lambda_2 \\ \lambda_3 & 0 & -\lambda_1 \\ -\lambda_2 & \lambda_1 & 0 \end{bmatrix} \qquad (2.67)$$

or, in tensor notation, assuming summation over repeated indices,

$$\delta r_i = D_{ij}r_j = -\varepsilon_{ijk}\lambda_k r_j \qquad (2.68)$$

where the permutation operator ε_{ijk} equals $+1$ when i, j, k is a cyclic permutation of the indices 1, 2, 3; or -1 when the permutation is noncyclic; and zero when two or more indices are equal. For $i = 1$, for example, only the ε_{123} and ε_{132} terms are nonzero. Addition of a translational displacement gives

$$\delta r_i = D_{ij}r_j + t_i \qquad (2.69)$$

When a body undergoes vibrations, the displacements vary with time, so time averages must be taken to derive the mean-square displacements, as we did to obtain the lattice-dynamical expression of Eq. (2.58). If the librational and translational motions are independent, the cross products between the two terms in Eq. (2.69) average to zero, and the elements of the mean-square displacement tensor of atom n, U_{ij}^n, are given by

$$\begin{aligned}
U_{11}^n &= +L_{22}r_3^2 + L_{33}r_2^2 - 2L_{23}r_2r_3 + T_{11} \\
U_{22}^n &= +L_{33}r_1^2 + L_{11}r_3^2 - 2L_{13}r_1r_3 + T_{22} \\
U_{33}^n &= +L_{11}r_2^2 + L_{22}r_1^2 - 2L_{12}r_1r_2 + T_{33} \\
U_{12}^n &= -L_{33}r_1r_2 - L_{12}r_3^2 + L_{13}r_2r_3 + L_{23}r_1r_3 + T_{12} \\
U_{13}^n &= -L_{22}r_1r_3 + L_{12}r_2r_3 - L_{13}r_2^2 + L_{23}r_1r_2 + T_{13} \\
U_{23}^n &= -L_{11}r_2r_3 + L_{12}r_1r_3 - L_{13}r_1r_2 - L_{23}r_1^2 + T_{23}
\end{aligned} \qquad (2.70)$$

where the coefficients $L_{ij} = \langle \lambda_i \lambda_j \rangle$ and $T_{ij} = \langle t_i t_j \rangle$ are the elements of the 3×3 libration tensor $\mathbf{L}$ and the 3×3 translation tensor $\mathbf{T}$, respectively. Since pairs of terms such as $\langle t_i t_j \rangle$ and $\langle t_j t_i \rangle$ correspond to averages over the same two scalar quantities, the $\mathbf{T}$ and $\mathbf{L}$ tensors are symmetrical.

If a rotation axis is correctly oriented, but incorrectly positioned, an additional translation component, perpendicular to the rotation axes, is introduced. The rotation angle and the parallel component of the translation are invariant to the position of the axis, but the perpendicular component is not. This means that the **L** tensor is unaffected by any assumptions about the position of the libration axes, whereas the **T** tensor depends on the assumption made concerning the location of the axes.

The quadratic correlation between librational and translational motions can be allowed for by including in Eq. (2.70) cross terms of the type $\langle D_{ik} t_j \rangle$, or

$$U_{ij} = \langle D_{ik} D_{jl} \rangle r_k r_l + \langle D_{ik} t_j + D_{ji} t_i \rangle r_k + \langle t_i t_j \rangle = A_{ijkl} r_k r_l + B_{ijk} r_k + \langle t_i t_j \rangle \quad (2.71)$$

which leads to the explicit expressions, such as

$$U_{11} = \langle \delta r_1 \rangle^2 = \langle \lambda_3^2 \rangle r_2^2 + \langle \lambda_2^2 \rangle r_3^2 - 2 \langle \lambda_2 \lambda_3 \rangle r_2 r_3 - 2 \langle \lambda_2 t_1 \rangle r_2 - 2 \langle \lambda_2 t_1 \rangle r_3 + \langle t_1^2 \rangle$$

$$U_{12} = \langle \delta r_1 \delta r_2 \rangle = - \langle \lambda_3^2 \rangle r_1 r_2 + \langle \lambda_1 \lambda_3 \rangle r_2 r_3 + \langle \lambda_2 \lambda_3 \rangle r_1 r_3 - \langle \lambda_1 \lambda_2 \rangle r_3^2$$
$$+ \langle \lambda_3 t_1 \rangle r_1 - \langle \lambda_1 t_1 \rangle r_3 - \langle \lambda_3 t_2 \rangle r_2 + \langle \lambda_2 t_2 \rangle r_3 + \langle t_1 t_2 \rangle \quad (2.72)$$

The products of the type $\langle \lambda_i t_j \rangle$ are the components of an additional tensor, **S**, called the screw tensor, as the coupling between translations and rotations describes a screw-type motion. Unlike the tensors **T** and **L**, **S** is unsymmetrical, since $\langle \lambda_i t_j \rangle$ is different from $\langle \lambda_j t_i \rangle$. The terms involving elements of **S** may be grouped as (for U_{12})

$$\langle \lambda_3 t_1 \rangle r_1 - \langle \lambda_3 t_2 \rangle r_2 + (\langle \lambda_2 t_2 \rangle - \langle \lambda_1 t_1 \rangle) r_3$$

or

$$S_{31} r_1 - S_{32} r_2 + (S_{22} - S_{11}) r_3 \quad (2.73)$$

As the diagonal elements occur as differences in this expression, a constant may be added to each of the diagonal terms without changing the observational equations. In other words, the trace of **S** is indeterminate.

In terms of the **L**, **T**, and **S** tensors, Eq. (2.70) is generalized as

$$U_{ij} = G_{ijkl} L_{kl} + H_{ijkl} S_{kl} + T_{ij} \quad (2.74)$$

It is clear from Eqs (2.70) and (2.72) that the arrays G_{ijkl} and H_{ijkl} involve the atomic coordinates $(x, y, z) = (r_1, r_2, r_3)$. They are listed in Table 2.2. Equations (2.74) for each of the atoms in the rigid body form the observational equations, from which the elements of **T**, **L**, and **S** can be derived by a linear least-squares procedure. One of the diagonal elements of **S** must be fixed in advance, or some other suitable constraint applied, because of the indeterminacy of $Tr(\mathbf{S})$. It is common practice to set $Tr(\mathbf{S})$ equal to zero. There are thus eight elements of **S** to be determined, as well as the six each of **L** and **T**, for a total of 20 variables. A shift of origin leaves **L** invariant, but it intermixes **T** and **S**.

If the origin is located at a center of symmetry, for each atom at **r** with vibration tensor $\mathbf{U}^n$, there will be an equivalent atom at $-\mathbf{r}$ with the same vibration tensor. When the observational equations for these two atoms are added, the terms involving elements of **S** disappear since they are linear in the components of **r**. The other terms, involving elements of the **T** and **L** tensors, are simply doubled, like the $\mathbf{U}^n$ components.

The physical meaning of the **T** and **L** tensor elements is as follows. The

TABLE 2.2 The Arrays G_{ijkl} and H_{ijkl} to be Used in Observational Equations: $U_{ij} = G_{ijkl}L_{kl} + H_{ijkl}S_{kl} + T_{ij}$ [Expression (2.74)]

G_{ijkl} \\ ij	kl	11	22	33	23	31	12
	11	0	z^2	y^2	$-2yz$	0	0
	22	z^2	0	x^2	0	$-2xz$	0
	33	y^2	x^2	0	0	0	$-2xy$
	23	$-yz$	0	0	$-x^2$	xy	xz
	31	0	$-xz$	0	xy	$-y^2$	yz
	12	0	0	$-xy$	xz	yz	$-z^2$

H_{ijkl} \\ ij	kl	11	22	33	23	31	12	32	13	21
	11	0	0	0	0	$-2y$	0	0	0	$2z$
	22	0	0	0	0	0	$-2z$	$2x$	0	0
	33	0	0	0	$-2x$	0	0	0	$2y$	0
	23	0	$-x$	x	0	0	y	0	$-z$	0
	31	y	0	$-y$	z	0	0	0	0	$-x$
	12	$-z$	z	0	0	x	0	$-y$	0	0

quantity $T_{ij}l_il_j$ is the mean-square amplitude of translational vibration in the direction of the unit vector l with components l_1, l_2, l_3 along the Cartesian axes, and $L_{ij}l_il_j$ is the mean-square amplitude of libration about an axis in this direction. The quantity $S_{ij}l_il_j$ represents the mean correlation between libration about the axis l and translation parallel to this axis. This quantity, like $T_{ij}l_il_j$, depends on the choice of origin, although the sum of the two quantities is independent of the origin.

The nonsymmetrical tensor **S** can be written as the sum of a symmetric tensor with elements $(S_{ij}^S = (S_{ij} + S_{ji})/2$ and a skew-symmetric tensor with elements $S_{ij}^A = (S_{ij} - S_{ji})/2$. Expressed in terms of principal axes, **S**^{S} consists of three principal screw correlations $\langle \lambda_I t_I \rangle$. Positive and negative screw correlations correspond to opposite senses of helicity. Since an arbitrary constant may be added to all three correlation terms, only the differences between them can be determined from the data.

The skew-symmetric part **S**^{A} is equivalent to a vector $(\lambda \cdot \mathbf{t})/2$ with components $(\lambda \cdot \mathbf{t})_i/2 = (\lambda_j t_k - \lambda_k t_j)/2$, involving correlations between a libration and a perpendicular translation. The components of **S**^{A} can be reduced to zero, and **S** made symmetric, by a change of origin. It can be shown that the origin shift that symmetrizes **S** also minimizes the trace of **T**. In terms of the coordinate system based on the principal axes of **L**, the required origin shifts $\hat{\rho}_i$ are

$$\hat{\rho}_1 = \frac{\hat{S}_{23} - \hat{S}_{32}}{\hat{L}_{22} + \hat{L}_{33}} \qquad \hat{\rho}_2 = \frac{\hat{S}_{31} - \hat{S}_{13}}{\hat{L}_{11} + \hat{L}_{33}} \qquad \hat{\rho}_3 + \frac{\hat{S}_{12} - \hat{S}_{21}}{\hat{L}_{11} + \hat{L}_{22}} \qquad (2.75)$$

in which the carets above the letters indicate that the quantities are referred to the principal axis system.

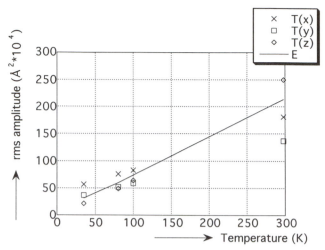

FIG. 2.6 Temperature dependence of the rigid-body translational mean-square amplitudes of quinolinic acid. The line represents the results from Eq. (2.51) with $v = 44 \text{ cm}^{-1}$. *Source*: Takusagawa and Koetzle (1979).

The description of the averaged motion can be simplified further by shifting to three generally nonintersecting libration axes, one each for each of the principal axes of **L**. Shifts of the $\mathbf{L}_1$ axis in the $\mathbf{L}_2$ and $\mathbf{L}_3$ directions by

$$^1\hat{\rho}_2 = -\hat{S}_{13}/\hat{L}_{11} \qquad \text{and} \qquad ^1\hat{\rho}_3 = \hat{S}_{12}/\hat{L}_{11} \qquad (2.76)$$

respectively, annihilate the S_{12} and S_{13} terms of the symmetrized **S** tensor and simultaneously effect a further reduction in $Tr(\mathbf{T})$ (the presuperscript denotes the axis that is shifted; the subscript denotes the direction of the shift component). Analogous equations for displacements of the $\mathbf{L}_2$ and $\mathbf{L}_3$ axes are obtained by permutation of the indices. If all three axes are appropriately displaced, only the diagonal terms of **S** remain. Referred to the principal axes of **L**, they represent screw correlations along these axes and are independent of origin shifts.

The elements of the reduced translation tensor **T** are:

$$^r T_{II} = \hat{T}_{II} - \sum_{K \neq I} (\hat{S}_{KI})^2/\hat{L}_{KK}$$

and

$$^r T_{IJ} = \hat{T}_{IJ} - \sum_K \hat{S}_{KI}\hat{S}_{KJ}/\hat{L}_{KK} \qquad J \neq I \qquad (2.77)$$

The resulting description of the average rigid-body motion is in terms of six independently distributed instantaneous motions—three screw librations about nonintersecting axes (with screw pitches given by $\hat{S}_{11}/\hat{L}_{11}$, etc.) and three translations. The parameter set consists of three libration and three translation amplitudes; six angles of orientation for the principal axes of **L** and **T**; six coordinates of axis displacement; and three screw pitches, one of which has to be chosen arbitrarily; again, for a total of 20 variables.

TABLE 2.3 Rigid-Bond Test for p-Nitropyridine-N-oxide at 30 K

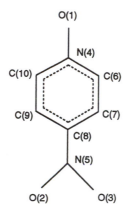

| | | Spherical-Atom Refinement $\sigma(u^2) \sim 0.0003$ Å^2 | | Charge-Density Model $\sigma(u^2) \sim 0.0002$ Å^2 | |
| | | $10^4 z_{A,B}^2$ (Å^2) | $10^4 z_{B,A}^2$ (Å^2) | $10^4 z_{A,B}^2$ (Å^2) | $10^4 z_{B,A}^2$ (Å^2) |
A	B				
N(4)	C(6)	60	48	52	52
C(6)	C(7)	66	65	56	53
C(7)	C(8)	55	66	52	58
C(8)	C(9)	57	53	51	45
C(9)	C(10)	62	69	55	57
C(10)	N(4)	64	70	64	65
N(4)	O(1)	46	28	41	38
C(8)	N(5)	40	56	42	47
N(5)	O(2)	72	51	63	56
N(5)	O(3)	66	42	58	50
rms discrepancy		14		5	

Source: Harel and Hirshfeld (1975).

Since diagonal elements of $\mathbf{S}$ enter into the expression for $^rT_{IJ}$, the indeterminacy of $Tr(\mathbf{S})$ introduces a corresponding indeterminacy in $^r\mathbf{T}$. The constraint $Tr(\mathbf{S}) = 0$ is unaffected by the various rotations and translations of the coordinate systems used in the course of the analysis.

2.2.3.1 The Multitemperature Study of Quinolinic Acid

The structure of quinolinic acid (2,3-pyridinedicarboxylic acid, $C_5H_3N(COOH)_2$) has been determined by neutron diffraction at four temperatures: 298, 100, 80, and

35 K (Takusagawa and Koetzle 1979). The rigid-body analysis indicates the largest translational tensor component to be along Z, perpendicular to the molecular plane at 298 K, but in the molecular plane along the X-axis, bisecting the two carboxylic groups at the other temperatures. The librations are largest around the two axes in the molecular plane at all temperatures. The **S** tensor components are small and never larger than 0.006 Å. In Fig. 2.6 temperature dependence of the translational mean-square amplitudes is plotted, and compared with the line calculated using expression (2.51) with $v = 44 \, cm^{-1}$.

2.3.4 The Rigid-Bond Test

The dominance of the external modes that underlies the success of the rigid-body model implies that bond-stretching vibrations give a minor contribution to the atomic vibrational amplitudes. The most important internal modes are, in fact, the torsional oscillations, and, to a lesser extent, the angle bending modes, which have lower frequencies than those of the stretching vibrations, but do not affect the relative amplitudes of bonded atoms A and B along the A—B bond. This is the basis for the *rigid-bond test* as a means to test the successful deconvolution of thermal and charge density effects in the refinement of X-ray data (Harel and Hirshfeld 1975, Hirshfeld 1976). The method requires the calculation of the vibration amplitudes in the direction of the atomic bonds, and is accomplished by using Eq. (2.27).

The use of more sophisticated scattering models, in which bonding effects on the charge density are taken into account, discussed in chapter 3, leads to a significant improvement in the results of the rigid-bond test. An example, based on a low-temperature analysis of *p*-nitropyridine-*N*-oxide, is given in Table 2.3.

3

Chemical Bonding and the X-ray Scattering Formalism

3.1 The Breakdown of the Independent-Atom Model

3.1.1 Qualitative Considerations

The assumption that the atomic electron density is well described by the spherically averaged density of the isolated atom has been the basis of X-ray structure analysis since its inception. The *independent-atom model* (IAM) is indeed a very good approximation for the heavier atoms, for which the valence shell is a minor part of the total density, but is much less successful for the lighter atoms. The lightest atom, hydrogen, has no inner shells of electrons, so that the effect of bonding is relatively pronounced. Because of the *overlap density* in covalent X—H bonds ($X = C, N, O$), the mean of the hydrogen electron distribution is significantly displaced inwards into the bond. When a spherical IAM hydrogen scattering factor is used in a least-squares adjustment of the atomic "position," the result will be biased because the centroid of the density associated with the H atom is shifted in the direction of the bond. The result is an apparent shortening of X—H bonds which is far beyond the precision of X-ray structure determination (Hanson et al. 1973). For sucrose, for example, the differences between X-ray and neutron bond lengths are 0.13 (1) Å averaged over 14 C—H bonds, and 0.18 (3) Å averaged over eight O—H bonds (Hanson et al. 1973). The observed discrepancy between X-ray results and spectroscopic values was first explained in terms of the electron distribution in the 1950s by Cochran (1956) and Tomii (1958).

That the bond density is also of significance for heavier atoms is evident from the occurrence of the spherical-atom forbidden (222) reflection of diamond and silicon, even at low temperatures where anharmonic thermal effects (see chapter 2) are negligible. The historical importance of the nonzero intensity of the diamond (222) reflection is illustrated by the following comment made by W. H. Bragg, in 1921:

Another point of interest is the existence of a small (222) reflection (in diamond). This has been looked for previously but without success. The structure of the diamond cannot be explained on the hypothesis that the field of force around the carbon atom is the same in all directions: or in other words, that the force between the two atoms can be expressed simply by a function of the distance between the centres. If this were so, the sphere, which would then represent the carbon atoms appropriately, would adopt the closed-packed arrangement. As a matter of fact, each atom is surrounded by four neighbours only. It is necessary, therefore, to suppose that the attachment of one atom to the next is due to some directed property, and the carbon atom has four such special directions: as indeed the tetra-valency of the atom might suggest. In that case the properties of the atom in diamond are based upon a tetrahedral not a spherical form. The tetrahedra point away from any (111) plane in case of half the atoms in diamond and towards it in case of the other half. Consecutive (111) sheets are not exactly of the same nature; and it might reasonably be expected that they would not entirely destroy each other's effects in the second order reflection from the tetrahedral plane. It is this effect which is now found to be quite distinct, though small.[1]

The IAM model further assumes the atoms in a crystal to be neutral. This assumption is contradicted by the fact that molecules have dipole and higher electrostatic moments, which can indeed be derived from the X-ray diffraction intensities, as further discussed in chapter 7. The molecular dipole moment results, in part, from the nonspherical distribution of the atomic densities, but a large component is due to *charge transfer* between atoms of different electronegativity. A population analysis of an extended basis-set SCF wave function of HF, for example, gives a net charge q of $+0.4$ electron units (e) on the H atom in HF; for CH_4 the value is $+0.12$ e (Szabo and Ostlund 1989).

The atomic dipole moment can be attributed to the preferential population of specific nonspherical atomic orbitals. In particular, this is the case for atoms with doubly-filled nonbonding *lone-pair* orbitals, such as the oxygen atoms in C—O—H and H—O—H, or oxygen in a terminal position as it is in the carbonyl group. An early demonstration of the bias introduced in X-ray positions of non-hydrogen atoms was the combined X-ray and neutron study of oxalic acid dihydrate (Coppens et al. 1969), which showed the X-ray positions of the oxygen atoms to be systematically displaced by small amounts into the direction of the lone pair density.

Some examples of such *asphericity shifts* are listed in Table 3.1. They indicate that bond lengths (and angles) from X-ray diffraction analysis using spherical scattering factors may be less reliable than implied by the quoted standard deviation, which are typically a few thousands of an Ångström. This is especially true when no high-angle data are included, as for most of the examples listed in Table 3.1. Since valence electron scattering is concentrated in the low-angle region, a low $\sin \theta/\lambda$ cut-off enhances the bias. But results from a data set including a large number of high-order reflections will be less affected by scattering-formalism inadequacies. This is the basis for the *high-order refinement* of X-ray data in

---

[1] It is interesting that, as recorded by James (1982), the original observation of the (222) reflection by W. H. Bragg (1921) was almost certainly due to multiple scattering, later discovered by Renninger.

TABLE 3.1 Some Discrepancies between X-ray and Neutron Positional Parameters

Compound	Nature of Shift (X-ray Relative to Neutron)	Magnitude (Å)	$(\sin\theta/\lambda)_{max}$ (Å^{-1})	Reference
Tetracyanoethylene oxide	$\overset{\uparrow}{O}$ ∕ \\ C C	0.013 (4)	0.81	Mathews and Stucky (1971), Mathews et al. (1971)
Tetracyanoethylene	$-C \rightarrow N$	0.008$_5$ (1$_5$)	0.94	Becker et al. (1973)
Oxalic acid	C—O ↗ \\ H	0.008 (2)	0.55	Coppens et al. (1969)
Sucrose	C—O ↗ \\ H	0.008 (2)	0·80	Brown and Levy (1973), Hanson et al. (1973)
	and $\overset{\uparrow}{O}$ ∕ \\ C C	0.007 (2)		
Ammonium oxalate H$_2$O	$\overset{\uparrow}{O}$ ∕ \\ H H	0.013 (3)	Not given	Taylor and Sabine (1972)
Cyanuric acid	$C{=}O \rightarrow$	0.005 (1)	0.80	Coppens and Vos (1971), Verschoor and Keulen (1971)
Sulfamic acid	$S{-}O \rightarrow$	0.0022 (6) 0.0015 (4)	0.65 1.23	Bats et al. (1977)

Source: Coppens (1978).

which low-order reflections are eliminated from the least-squares procedure (Jeffrey and Cruikshank 1953). The cut-off is often dictated by the need to have a sufficient excess of observations over parameters to be refined, but there is ample evidence that bonding effects are important to at least $\sin\theta/\lambda = 0.8$–0.9 Å^{-1}, as further discussed in chapter 5.

3.1.2 The Electron Density and the LCAO Formalism

What guidance for improving the scattering formalism can be obtained from theory? In the linear combination of atomic orbitals (LCAO) formalism, a molecular orbital (MO) is described as a combination of atomic *basis function* ϕ_μ:

$$\chi_i = \sum_\mu C_{i\mu}\phi_\mu \qquad (3.1)$$

To satisfy the exclusion principle, which requires the wave function to be antisymmetric with respect to the interchange of two electrons, the wave function is written in terms of Slater determinants, representing an antisymmetrized combination of occupied molecular orbitals. In the Hartree–Fock method, correlation between the instantaneous positions of electrons of opposite spins is neglected; each electron is assumed to be subject to the average potential of the other electrons. The *correlation energy* is defined as the difference between the exact nonrelativistic energy and the Hartree–Fock energy. A Hartree–Fock wave function consists of a single Slater determinant.

For a system of n electrons, the single-determinant wave function is

$$\psi = (n!)^{-1/2} \begin{vmatrix} \chi_1(1) & \cdots & \chi_n(1) \\ \vdots & & \vdots \\ \chi_1(n) & \cdots & \chi_n(n) \end{vmatrix} \qquad (3.2)$$

in which the number in brackets refers to the electron and the multiplier in front of the determinant is a normalization factor. In often-used shorthand notation,

$$\psi = |\chi_1(1)\chi_2(2)\cdots\chi_n(n)\rangle \qquad (3.3)$$

As implied by Eq. (3.2), the Slater determinant is a sum over products of molecular orbitals; in a different formulation:

$$\psi = \sum_i \sum_{j \geq i} \hat{P}_{ij}(\chi_1(1)\chi_2(2)\cdots\chi_n(n)) \qquad (3.4)$$

where the electron permutation operator $\hat{P}_{ij}$ is used, with eigenvalues $+1$ and -1 for even and odd permutations, respectively.

Interchange of two rows of the Slater determinant changes the sign of the wave function, which is therefore *antisymmetric* with respect to interchange of electrons. When two rows or columns are identical, the determinant is zero. The Slater determinant wave function therefore obeys the Pauli exclusion principle for fermions.

The wave function Ψ is a function of the **3n** space coordinates and n spin coordinates of the n electrons. The *one-electron density* $\rho(\mathbf{r})$ is obtained from the wave function by integration over all spin coordinates and the space coordinates of all but one electron:

$$\rho(\mathbf{r}) = \int |\Psi|^2 (\mathbf{r}_2, \mathbf{r}_3, \ldots, \mathbf{r}_n, s_1, s_2, \ldots, s_n)\, d\mathbf{r}_2 \cdots d\mathbf{r}_n, ds_1 \cdots ds_n \qquad (3.5)$$

Because the electrons are indistinguishable, this integration is independent of the choice of the electron for which integration is omitted. Since the MOs are orthonormal, cross products integrate to zero. This result is

$$\rho(\mathbf{r}) = \sum_i n_i \chi_i^2 \qquad (3.6)$$

where n_i equals 1 if the χ_i values are the spin orbitals describing both the spin and space characteristics (and obtained by omitting the integration over s_1 in Eq. 3.5), or either 1 or 2 if the spin part has been integrated. The electron density expression in terms of the atomic orbitals is obtained on substitution of the LCAO expression of Eq. (3.1), $\chi_i = \sum_\mu C_{i\mu} \phi_\mu$, in Eq. (3.7):

$$\rho(\mathbf{r}) = \sum_\mu \sum_\nu P_{\mu\nu} \phi_\mu(\mathbf{r}) \phi_\nu(\mathbf{r}) \tag{3.7}$$

in which **P**, with elements $P_{\mu\nu}$, is a *density matrix*. The $P_{\mu\nu}$ values are the populations of the orbital product density functions $\phi_\mu(\mathbf{r})\phi_\nu(\mathbf{r})$, and are given by

$$P_{\mu\nu} = \sum_i n_i C_{i\mu} C_{i\nu} \tag{3.8}$$

The wave function Ψ contains all information of the joint probability distribution of the electrons. For example, the *two-electron density* is obtained from the wave function by integration over the spin and space coordinates of all but *two* electrons. It describes the joint probability of finding electron 1 at $\mathbf{r}_1$ and electron 2 at $\mathbf{r}_2$. The two-electron density cannot be obtained from elastic Bragg scattering.

The single-Slater determinant includes correlation between the motion of two electrons with parallel spins that avoid each other because of the exclusion principle (Szabo and Ostlund 1989), but correlation between the motion of electrons with opposite spin is neglected. The wave function of Eq. (3.2) does not prevent the two electrons from being at the same point in space, which is physically impossible. The Slater determinant wave function is therefore described as *uncorrelated*.

In a more advanced treatment beyond the single-Slater determinant Hartree–Fock limit, additional determinants corresponding to excited-state configurations are added to the wave function. If the occupied spin orbitals are labeled a, b, c, etc., and the unoccupied orbitals r, s, t, etc., a single excited configuration representing excitation from orbital a to orbital r is given by

$$|\psi_a^r\rangle = |\chi_1 \chi_2 \cdots \chi_r \chi_b \cdots \chi_n\rangle \tag{3.9}$$

The *multiconfiguration* wave function is represented by the expression

$$|\psi\rangle = c_0 |\psi_0\rangle + \sum_{ra} c_a^r |\psi_a^r\rangle + \sum_{\substack{a<b \\ r<s}} c_{ab}^{rs} |\psi_{ab}^{rs}\rangle + \cdots \tag{3.10}$$

in which the coefficients c, like the LCAO coefficients of Eq. (3.1), are to be determined by energy minimization.

For a multi-Slater determinant wave function, orbitals which satisfy Eq. (3.6), and therefore Eq. (3.7), can still be defined. For these orbitals, referred to as the *natural spin orbitals*, the coefficients n_i are not necessarily integers, but have the boundaries $0 \leqslant n_i \leqslant 1$.

The summation of Eq. (3.7) contains one- and two-center terms for which ϕ_μ and ϕ_ν are centered on the same, and on different nuclei, respectively. The two-center terms represent the overlap density in a bond; they can only give a significant contribution to the density if $\phi_\mu(\mathbf{r})$ and $\phi_\nu(\mathbf{r})$ have an appreciable value in the same region of space, and are therefore not important for distant atoms.

We will give two simple examples, noting that in current work much larger sets of basis functions are used. The LCAO molecular spin orbitals (which describe both the electron position and the electron spin) for the H_2 molecule can be chosen as

$$\chi_1 = \sigma_g 1s\alpha = (1s_A + 1s_B)\alpha$$

$$\chi_2 = \sigma_g 1s\beta = (1s_A + 1s_B)\beta$$

$$\chi_3 = \sigma_u 1s\alpha = (1s_A - 1s_B)\alpha$$

$$\chi_4 = \sigma_u 1s\beta = (1s_A - 1s_B)\beta, \tag{3.11}$$

where the subscripts A and B label the two hydrogen atoms. The Slater determinant wave function for the configuration $(\sigma_g 1s)^2 = (\chi_1)(\chi_2)$ is given by

$$\psi = 2^{-1/2} \begin{vmatrix} \chi_1(1) & \chi_2(1) \\ \chi_1(2) & \chi_2(2) \end{vmatrix} = \frac{1}{\sqrt{2}} \{\sigma_g 1s(1)\sigma_g 1s(2)\}[\alpha(1)\beta(2) - \beta(1)\alpha(2)] \tag{3.12}$$

In this configuration, both electrons are in the same space orbital, but in different spin orbitals. If the electrons would have the same spin, Ψ would be zero, as required.

A second example is the *minimal-basis-set* (MBS) Hartree–Fock wave function for the diatomic molecule hydrogen fluoride, HF (Ransil 1960). The basis orbitals are six Slater-type (i.e., single exponential) functions, one for each inner and valence shell orbital of the two atoms. They are the $1s$ function on the hydrogen atom, and the $1s$, $2s$, $2p\sigma$, and two $2p\pi$ functions on the fluorine atom. The $2s_F$ function is an exponential function to which a term is added that introduces the radial node, and ensures orthogonality with the $1s$ function on fluorine. To indicate the orthogonality, it is labeled $2s_{\perp F}$. The orbital is described by

$$2s_{\perp F} = 2s_F - S_{1s2s} \tag{3.13}$$

where the overlap integral $S_{1s2s} = \int 1s_F 2s_F \, d\tau$.

A Hartree–Fock calculation leads to the following LCAOs, each occupied by two electrons:

$$1\sigma = 1.000(1s_F) + 0.012(2s_{\perp F}) + 0.002(2p\sigma_F) - 0.003(1s_H)$$

$$2\sigma = -0.018(1s_F) + 0.914(2s_{\perp F}) + 0.090(2p\sigma_F) - 0.154(1s_H)$$

$$3\sigma = -0.023(1s_F) - 0.411(2s_{\perp F}) + 0.711(2p\sigma_F) - 0.516(1s_H)$$

$$1\pi_{+1} = (2p\pi_{+1})_F$$

$$1\pi_{-1} = (2p\pi_{-1})_F \tag{3.14}$$

The first 1σ and the last two $2p\pi$ molecular orbitals are nonbonding orbitals, while the 2σ and 3σ orbitals are the bonding orbitals of the HF molecule.

The corresponding electron density is, according to Eq. (3.6), given by

$$\rho(\mathbf{r}) = 2.00(1s_F)^2 + 2.00(2s_{\perp F})^2 + 1.03(2p\sigma_F)^2 + 2.00(2p\pi_{+1})^2$$

$$+ 2.00(2p\pi_{-1})^2 + 0.29(1s_H)^2 - 0.80(2s_{\perp F})(2p\sigma_F)$$

$$+ 0.29(2s_{\perp F})(1s_H) + 1.52(2p\sigma_F)(1s_H) \tag{3.15}$$

in which terms with coefficients <0.02 have been omitted.

The density expression (3.15) contains several types of terms, which are to be considered in formulating an advanced X-ray scattering formalism:

(1) atom-centered, such as $(1s_F)^2$, $(2s_{\perp F})^2$, and $(2p\sigma_F)^2$.
(2) atom-centered cross terms, such as $(2s_{\perp F})(2p\sigma_F)$, which integrate to zero as the basis functions on the same atom are orthogonal, but correspond to a migration of the electron density from the negative to the positive lobe of the product function.
(3) two-center terms, such as $(2p\sigma_F)(1s_H)$, which represent the overlap density in the bond.

In a *Mulliken population analysis*, the electron density of the overlap terms is equally divided between the two atoms. Since the overlap integrals $S(2s_{\perp F})(1s_H)$ and $S(2p\sigma_F)(1s_H)$ are much smaller than 1, the hydrogen atom is positively charged.

In summary, the electron density for HF as described by Eq. (3.15) includes the effects of charge transfer between atoms, atomic orbital overlap, and preferential population of lone-pair orbitals, which are neglected in the independent-atom scattering formalism.

3.2 Improved Scattering Models

3.2.1 The Spherical Atom Kappa Formalism

A simple modification of the IAM model, referred to as the κ-*formalism*, makes it possible to allow for charge transfer between atoms. By separating the scattering of the valence electrons from that of the inner shells, it becomes possible to adjust the population and radial dependence of the valence shell. In practice, two charge-density variables, P_v, the valence shell population parameter, and κ, a parameter which allows expansion and contraction of the valence shell, are added to the conventional parameters of structure analysis (Coppens et al. 1979). For consistency, P_v and κ must be introduced simultaneously, as a change in the number of electrons affects the electron–electron repulsions, and therefore the radial dependence of the electron distribution (Coulson 1961).

In the κ-formalism, the atomic density is formulated as

$$\rho_{atom} = \rho_{core} + \rho'_{valence}(\kappa r) = \rho_{core} + P_v \kappa^3 \rho_{valence}(\kappa r) \qquad (3.16)$$

The parameter κ scales the radial coordinate r; when $\kappa > 1$, the same density is obtained at a smaller value of r, and the valence shell is therefore contracted. Conversely, for $\kappa < 1$, the valence shell is expanded. The κ^3 factor satisfies the normalization requirement

$$N 4\pi \int \rho_{valence}(\kappa r) r^2 \, dr = 1 \qquad (3.17)$$

As the corresponding unperturbed density is normalized to 1, that is, $4\pi \int \rho_{valence}(r) r^2 \, dr = 1$, N equals κ^3.

It is assumed in Eq. (3.16) that the inner or core electrons are not perturbed. There is abundant support for this approximation (Bentley and Stewart 1974), though very high resolution studies (Deutsch 1992) suggest that small devia-tions predicted by theory (Hirshfeld and Rzotkiewicz 1974) are accessible experi-mentally.

The scattering factor of the valence-density component in Eq. (3.16) is obtained by the Fourier transform

$$f'_{\text{valence}} = \int P_{\text{valence}} \kappa^3 \rho_{\text{valence}}(\kappa r) \exp\left(2\pi i \mathbf{S}\cdot\mathbf{r}\right) d\mathbf{r} \qquad (3.18)$$

By replacing $\mathbf{r}$ and $\mathbf{S}$ in the exponent by $\kappa\mathbf{r}$ and $\mathbf{S}/\kappa$, respectively, and writing $\kappa^3 d\mathbf{r} = 4\pi\kappa^2 r^2 d\kappa r$, one obtains

$$f'_{\text{valence}}(S) = f_{\text{valence}}(S/\kappa) \qquad (3.19)$$

We note that Eq. (3.19) again illustrates the inverse relation between direct and scattering space, a contraction of charge density corresponding to an expansion in scattering space, and vice versa. Equation (3.19) implies that the κ-modified scattering factor can be obtained directly from the unperturbed IAM scattering factors tabulated in the literature.

In summary, the κ-structure factor formalism is

$$F(\mathbf{H}) = \sum_j \left[\{P_{j,c}f_{j,\text{core}}(H) + P_{j,v}f_{j,\text{valence}}(H/\kappa)\} \exp\left(2\pi i \mathbf{H}\cdot\mathbf{r}_j\right) T_j(\mathbf{H})\right] \qquad (3.20)$$

where both scattering factors $f_{j,\text{core}}$ and $f_{j,\text{valence}}$ are normalized to one electron, and $P_{j,c}$ and $P_{j,v}$ are the core and valence electron populations, respectively.

3.2.2 Modified Spherical Scattering Factor for the Hydrogen Atom

Stewart, Davidson, and Simpson (SDS), in a seminal study published in 1965, addressed the bonding deformation of the hydrogen atom by fitting a flexible spherical H-atom form factor to the molecular scattering of a high-quality theoretical density of the hydrogen molecule. Though charge transfer between atoms is absent in this homonuclear diatomic molecule, the bonded atom is contracted considerably relative to the isolated H atom ($\kappa \approx 1.16$), and the centroid of its electron density is shifted by 0.07 Å into the bond, relative to the proton position. The fit to the density has a mean error of only 0.11% compared with 7.12% for the isolated $1s$ proton-centered density. The SDS form factor for hydrogen has become the standard form factor in X-ray structure analysis. An even better fit with a mean error of 0.013% is obtained for an H atom polarized into the bond.

It is not possible to determine κ for a hydrogen atom directly from experimental X-ray data, because its value correlates strongly with the temperature parameter due to the absence of unperturbed inner-shell electrons. The use of neutron temperature parameters provides an alternative. Combined analysis of X-ray and neutron data on glycylglycine and sulfamic acid suggests that for X—H ($X = C, N$) groups, the H atom is more contracted than for the H_2 molecule, with a κ value as large as 1.4 for both C—H and N—H bonds (Coppens et al. 1979).

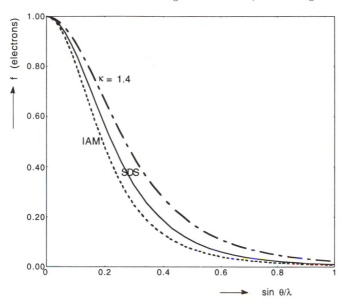

FIG. 3.1 Scattering factors for the hydrogen atom. IAM, free atom; SDS bonded atom as in H_2, according to Stewart et al. (1965); $\kappa = 1.4$, IAM density contracted with a kappa parameter equal to 1.4.

The $\kappa_H^{opt} = 1.4$ value seems large, and possibly corresponds to an overestimate of the contraction, as the analysis depends on the reliability of the neutron temperature factors. Nevertheless, it has been used successfully. There is no doubt that the contraction of the hydrogen-atom density must be taken into account in accurate structure analysis.

Several H-atom scattering factors are compared in Fig. 3.1.

3.2.3 Examples of Results Obtained with the κ-Formalism

Application of the kappa formalism leads to net charges in good agreement with accepted electronegativity concepts, and molecular dipole moments close to those from other experimental and theoretical methods (Table 3.2).

In agreement with theoretical prediction, the experimental analysis shows the more positive atoms to be contracted. This is explained by the decrease in electron–electron repulsions, or, in a somewhat different language, the decreased screening of the nuclear attraction forces by a smaller number of electrons. This contraction is incorporated in Slater's rules for approximate, single exponential (and therefore nodeless), hydrogen-like orbital functions (Slater 1932). For a $2p_x$ orbital of a second-row atom, for example, the orbital function is given by

$$\phi(2p_x) = N_{2p}x \exp\left(-cr/2a_0\right) \tag{3.21}$$

in which N is a normalization factor, x is the component of r along the x axis,

TABLE 3.2 Comparison of Dipole Moments from κ-Refinement of X-ray Data with Theoretical Values (D)

	X-Ray Data (κ-Refinement)	Theoretical	Other Experimental Values
Sulfamic acid (78 K)	9.6 (6)	9.33	10.2[a]
		12.2[b]	
		13.3[c]	
Formamide (90 K)	4.4 (5)	4.3	3.9[d]
	4.07		
Glycylglycine (82 K)	23.6 (1.3)		27.8[e]
p-Nitropyridine-N-oxide	0.4 (1.0)		0.69[f]
SCN^- in NH_4SCN	0.8 (4)		
SCN^- in NaSCN	1.3 (6)		

[a] In dimethyl sulfoxide.
[b] In N-methylpyrrolidine.
[c] In N,N-dimethylacetamide.
[d] Gas phase.
[e] In H_2O.
[f] In benzene.

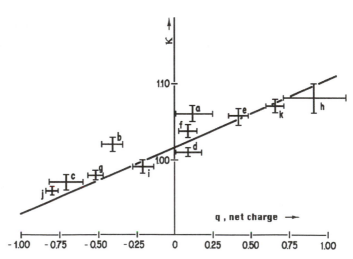

FIG. 3.2 Relation between κ and q (net charge) for N atoms in a number of structures: (a), (b) glycylglycine; (c) formamide; (d), (e) p-nitropyridine N-oxide; (f) sulfamic acid; (g), (h) NH_4SCN; (i) NaSCN; (j), (k) KN_3. Bars indicate estimated standard deviations. The full line is the relation predicted by Slater's recipe for atomic orbital exponents (Coppens et al. 1979). The line has been raised at $q = 0$ by a factor equal to the ratio of the Slater exponent (3.9 au^{-1}), and the energy-optimized exponent for the nitrogen $n = 2$ shell (3.84 au^{-1}) given by Clementi and Raimondi (1963).

and a_0 is the Bohr radius, equal to 0.5292 Å. According to the rules, c is equal to the nuclear charge Z, minus 0.35 for each other electron in the $n = 2$ shell, and minus 0.85 for each s electron in the $n = 1$ shell. For a neutral nitrogen atom, this gives $c = 7 - 2 \cdot 0.85 - 4 \cdot 0.35 = 3.9$. Since, for every additional valence electron, c is reduced by 0.35, the slope dc/dq (q being the net charge) is predicted to be 0.35/3.9, or about 9%.

The experimental κ/q relationship for a number of nitrogen-containing molecules is shown in Fig. 3.2. The nitrogen atoms included ranged from the terminal atoms in the azide ion in KN_3, with net charges of -0.080 (4) e; to the central atom of the azide ion, which is positive by 0.66 (6) charge units; and the N atom in the ammonium ion in NH_4SCN, which has a net charge of about 0.9 e. The observed slope of the curve is in remarkable agreement with Slater's recipe.

Results similar to those for the nitrogen compounds discussed here have been obtained by analysis of C, N, and O atoms in a number of nucleotides and nucleosides (Pearlman and Kim 1985). Finally, a test of the kappa refinement using the theoretical densities of 28 diatomic molecules proved it to be quite successful in reproducing the theoretical radial distribution of the spherical component of the atomic density (Brown and Spackman 1991).

3.2.4 The Multipole Description of the Charge Density of Aspherical Atoms

3.2.4.1 General Considerations

The κ-formalism accounts for charge transfer between atoms. However, expressions (3.7) and (3.15) demonstrate that an advanced X-ray scattering formalism must also contain nonspherical density functions. This is most successfully accomplished in models based on *atom-centered multipolar functions*. After a number of early studies proved the viability of this approach (DeMarco and Weiss 1965, Dawson 1967, Kurki-Suonio 1968), Stewart in 1969 introduced a generalized scattering formalism, based on spherical harmonic density functions centered on each of the atomic nuclei.

The atom-centered models do not account explicitly for the two-center density terms in Eq. (3.7). This is less of a limitation than might be expected, because the density in the bonds projects quite efficiently in the atomic functions, provided they are sufficiently diffuse. While the two-center density can readily be included in the calculation of a molecular scattering factor based on a theoretical density, simultaneous least-squares adjustment of one- and two-center population parameters leads to large correlations (Jones et al. 1972). It is, in principle, possible to reduce such correlations by introducing quantum-mechanical constraints, such as the requirement that the electron density corresponds to an antisymmetrized wave function (Massa and Clinton 1972, Frishberg and Massa 1981, Massa et al. 1985). No practical method for this purpose has been developed at this time.

In a non-atom-centered deformation model, due to Hellner and coworkers (Hellner 1977, Scheringer and Kotuglu 1983), the bonding density is described by "charge clouds" located between bonded atoms and in lone-pair regions.

Application to hexacyanobenzene indicates an improved fit to the 120 K experimental data (Drück and Kotuglu 1984). But interpretation of the results is not straightforward, because such a model does not deconvolute charge density and thermal motions effects, and is not well suited for comparison with theory and derivation of electrostatic properties.

A number of different atom-centered multipole models are available. We distinguish between *valence-density models*, in which the density functions represent all electrons in the valence shell, and *deformation-density* models, in which the aspherical functions describe the deviation from the IAM atomic density. In the former, the aspherical density is added to the unperturbed core density, as in the κ-formalism, while in the latter, the aspherical density is superimposed on the isolated atom density, but the expansion and contraction of the valence density is not treated explicitly.

3.2.4.2 Definition of Multipolar Density Functions

Atomic density functions are expressed in terms of the three polar coordinates r, θ, and ϕ. In the multipole formalism, the density functions are products of r-dependent *radial functions* and θ- and ϕ-dependent *angular functions*. The angular functions are the *real spherical harmonic functions* $y_{lm\pm}(\theta, \phi)$, but with a normalization suitable for density functions, further discussed below. The functions are well known as they describe the angular dependence of the hydrogenic $s, p, d, f \ldots$ orbitals.

The functions y_{lmp} are linear combinations of the *complex* spherical harmonic functions Y_{lm}. Including normalization, the latter are defined as

$$Y_{lm}(\theta, \phi) = (-1)^m \left[\frac{(2l+1)}{4\pi} \frac{(l-|m|)!}{(l+|m|)!} \right]^{1/2} P_l^{|m|} \cos(\theta) \exp im\phi \tag{3.22}$$

with $-l \leq m \leq l$. The term $(-1)^m$ is referred to as the Condon–Shortley phase factor (Condon and Shortley 1957). The functions $P_l^{|m|} \cos(\theta)$ are the associated Legendre functions, defined as (Arfken 1970)

$$P_l^m(x) = (1-x^2)^{m/2} \left(\frac{d}{dx} \right)^{l+m} \frac{1}{2^l l!} (x^2-1)^l \tag{3.23}$$

The first terms of $P_l^{|m|} \cos(\theta)$ are listed in Table 3.3.

The real spherical harmonics are given by

$$y_{l0} = Y_{l0} \tag{3.24a}$$

for $m = 0$, and, for $m > 0$, by

$$y_{lm+} = (-1)^m (Y_{lm} + Y_{l,-m})/2^{1/2} \tag{3.24b}$$

and

$$y_{lm-} = (-1)^m (Y_{lm} - Y_{l,-m})/(2^{1/2} i) \tag{3.24c}$$

Substitution of Eq. (3.22) gives

$$y_{lm+}(\theta, \phi) = \left[\frac{(2l+1)}{2\pi(1+\delta_{l0})} \frac{(l-|m|)!}{(l+|m|)!} \right]^{1/2} P_l^{|m|} \cos(\theta) \cos m\phi$$

$$\equiv N_{lm} P_l^m (\cos \theta) \cos m\phi \tag{3.25a}$$

TABLE 3.3 The Associated Legendre Functions
$(x = \cos \theta)$

$P_1^1(x) = (1 - x^2)^{1/2} = \sin \theta$
$P_2^1(x) = 3x(1 - x^2)^{1/2} = 3 \cos \theta \sin \theta$
$P_2^2(x) = 3(1 - x^2) = 3 \sin^2 \theta$
$P_3^1(x) = \frac{3}{2}(5x^2 - 1)(1 - x^2)^{1/2} = \frac{3}{2}(5 \cos^2 \theta - 1) \sin \theta$
$P_3^2(x) = 15x(1 - x^2) = 15 \cos \theta \sin^2 \theta$
$P_3^3(x) = 15(1 - x^2)^{3/2} = 15 \sin^3 \theta$
$P_4^1(x) = \frac{5}{2}(7x^3 - 3x)(1 - x^2)^{1/2} = \frac{5}{2}(7 \cos^3 \theta - 3 \cos \theta) \sin \theta$
$P_4^2(x) = \frac{15}{2}(7x^2 - 1)(1 - x^2) = \frac{15}{2}(7 \cos^2 \theta - 1) \sin^2 \theta$
$P_4^3(x) = 105x(1 - x^2)^{3/2} = 105 \cos \theta \sin^3 \theta$
$P_4^4(x) = 105(1 - x^2)^2 = 105 \sin^4 \theta$

with $0 \le m \le l$, and similarly

$$y_{lm-}(\theta, \psi) = N_{lm} P_l^m(\cos \theta) \sin m\phi \tag{3.25b}$$

3.2.4.3 Symmetry Properties of the Spherical Harmonic Functions

For $m = 0$, y_{l0+} and Y_{l0} are identical, and real. For $l = 0$, m is also zero $(0 \le m \le l)$, and the function is spherically symmetric. When $l \ne 0$ and $m = 0$, the ϕ dependence disappears and the $Y_{lm} = y_{lm}$ are cylindrically symmetric around the z axis.

The l even functions are symmetric with respect to inversion through the atomic site, while the l odd functions are antisymmetric. The functions allowed for a particular site may be symmetry restricted. An atom at a centrosymmetric site, for example, will have zero populations for the l odd dipolar and octopolar multipole functions. A full list of symmetry restrictions is given in appendix D (Kurki-Suonio 1977a). Functions with $l \le 3$ are illustrated in Figs. 3.3 and 3.4.

3.2.4.4 Real Spherical Harmonics as Density Functions

When y_{lmp} $(p = \pm)$ represent atomic orbitals, y_{lmp}^2 is a probability distribution, which should integrate to 1. The normalization condition is therefore

$$\int y_{lmp}^2 \, d\Omega = 1 \tag{3.26}$$

in which $d\Omega$ is the volume element in θ–ϕ space.

The normalization expression (3.26) is appropriate for wave functions. For a charge density function, a different normalization must be used, because the charge is given by the integral over the first power of the function. The density functions in general use are labeled d_{lmp}, and are defined by the normalization

$$\int |d_{lmp}| \, d\Omega = 2 \text{ for } l > 0 \quad \text{and} \quad \int |d_{lmp}| \, d\Omega = 1 \text{ for } l = 0 \tag{3.27}$$

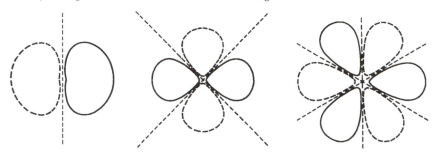

FIG. 3.3 Graphic representation of some multipolar functions: (from left to right) a dipole ($l = 1$), a quadrupole ($l = 2$), and an octupole ($l = 3$) function. Functions are negative with the dotted areas. If the x axis is horizontal and the y axis is vertical, the functions are x, $x^2 - y^2$, and $x^3 - 3xy^2$, respectively.

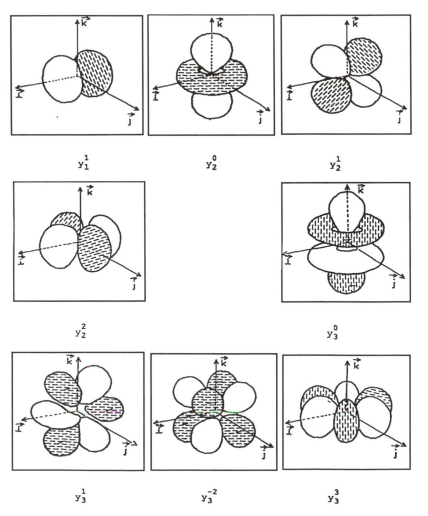

FIG. 3.4 Drawing of some dipolar, quadrupolar, and octupolar functions. *Source*: Hansen (1978).

This normalization implies that a population parameter equal to 1 corresponds to a population of one electron for the spherically symmetric function d_{00}. The nonspherical functions $(l > 0)$ represent a shift of density between regions of opposite sign. They have both positive and negative lobes which integrate to equal but opposite numbers of electrons. For these functions, the normalization expression (3.27) ensures that the population parameters represent the number of electrons shifted from the negative to the positive regions.

Analogous to Eq. (3.25), we get, for the functions d_{lmp},

$$d_{lm+} = N'_{lm}P_l^m(\cos\theta)\cos m\phi \tag{3.28a}$$

and

$$d_{lm-} = N'_{lm}P_l^m(\cos\theta)\sin m\phi \tag{3.28b}$$

where N'_{lm} is the normalization factor defined by Eq. (3.27).

The functions d_{lmp} are related to the functions c_{lmp}, in which the angular dependence is expressed in terms of the direction cosines x, y, and z in a Cartesian coordinate system, as in $c_{21+} = xz$. The Cartesian functions c_{lmp} are equal to d_{lmp} except for an lm-dependent factor L_{lm}, or

$$d_{lmp} = L_{lm}c_{lmp} \tag{3.29}$$

Expressions for c_{lmp} with $l \leq 4$ and the normalization factors L_{lm} are given in appendix D, together with the factors C_{lm}, defined by

$$C_{lm}c_{lmp} = P_l^m(\cos\theta)_{\sin m\varphi}^{\cos m\varphi} \tag{3.30}$$

Combining Eqs. (3.28) and (3.29) with Eq. (3.30) gives

$$N'_{lm} = \frac{L_{lm}}{C_{lm}} \tag{3.31}$$

The functions and their normalizations are summarized in Fig. 3.5, which can be used for the rapid derivation of equations similar to Eq. (3.31).

The spherical harmonic density functions are referred to as *multipoles*, since the functions with $l = 0, 1, 2, 3, 4$, etc., correspond to components of the charge distribution $\rho(\mathbf{r})$ which give nonzero contributions to the monopole $(l = 0)$, dipole $(l = 1)$, quadrupole $(l = 2)$, octupole $(l = 3)$, hexadecapole $(l = 4)$, etc., moments of the atomic charge distribution.

The electrostatic moments of a distribution $\rho(\mathbf{r})$ in terms of the c_{lmp} are given by

$$\mu_{lmp} = \int \rho(\mathbf{r})c_{lmp}r^l\,d\mathbf{r} \tag{3.32}$$

where μ_{lmp} is an electrostatic moment of order l. The factor r^l enters in Eq. (3.32) because the c_{lmp} are defined in terms of the direction cosines x, y, and z. The

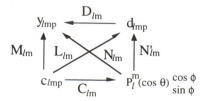

FIG. 3.5 Definition of the normalization coefficients for the spherical harmonic functions. Relations such as $y_{lmp} = D_{lm}d_{lmp}$ are implied by the direction of the arrows.

definition of the electrostatic moments and the derivation of electrostatic quantities from the electron density are treated in chapters 7–9.

For sites of cubic symmetry the point-group symmetry elements mix the spherical harmonic basis functions. As a result, linear combinations of spherical harmonic functions, referred to as *Kubic harmonics* (Von der Lage and Bethe 1947), must be used.

A more detailed discussion of the complex and real spherical harmonic functions, with explicit expressions and numerical values for the normalization factors, can be found in appendix D.

3.2.4.5 Choice of Radial Functions

To preserve the shell structure of the spherical component of the valence density, the radial function of the bonded atom may be described by the isolated atom radial dependence, modified by the κ expansion–contraction parameter.

The deformation functions, however, must also describe density accumulation in the bond regions, which in the one-center formalism is represented by the atom-centered terms. They must be more diffuse, with a different radial dependence. Since the electron density is a sum over the products of atomic orbitals, an argument can be made for using a radial dependence derived from the atomic orbital functions. The radial dependence is based on that of hydrogenic orbitals, which are valid for the one-electron atom. They have *Slater-type* radial functions, equal to exponentials multiplied by r^l times a polynomial of degree $n - l - 1$ in the radial coordinate r. As an example, the 2s and 2p hydrogenic orbitals are given by

$$R_{2s} = 2^{-1/2}(Z/a_0)^{3/2}(1 - Zr/2a_0)\,e^{-Zr/2a_0} \tag{3.33a}$$

and

$$R_{2p} = 24^{-1/2}(Z/a_0)^{5/2} r\, e^{-Zr/2a_0} \tag{3.33b}$$

where Z is the nuclear charge and a_0 is the Bohr radius, r being expressed in Å. Note that the polynomial $(1 - Zr/2a_0)$ in R_{2s} introduces a radial node, and assures orthogonality to the 1s orbital, a requirement also introduced in the orbital function Eq. (3.13). For density functions the radial node is generally omitted. Simple normalized, nodeless density functions based on hydrogenic orbitals are defined as[2]

$$R_l(r) = \kappa'^3 \frac{\zeta^{n_l + 3}}{(n_l + 2)!} (\kappa' r)^{n(l)} \exp(-\kappa' \zeta_l r) \tag{3.34}$$

in which the symbol κ' is used to describe the expansion–contraction parameters of the deformation functions. Such functions have been found to give a good fit for the nonspherical valence shell components of first- and second-row atoms, especially when an adjustable κ' parameter is included, as in Eq. (3.34). This κ' parameter is, in general, numerically different from the κ parameter applied to the spherical valence shell density, and may be selected to vary among the deformation functions on a single atom.

---

[2] The normalization follows from the integral $\int_0^\infty x^n e^{-\mu x}\,dx = n!\mu^{-n-1}$ ($n > -1, \mu > 0$), with $n = n_l + 2$ (integration over three dimensions).

TABLE 3.4 Best Single ζ Density Values
(Bohr^{-1}) for the Valence Shells of a Number
of Atoms

	2s	2p		Average
Carbon	3.216	3.136	s^2p^2	3.176
Nitrogen	3.947	3.934	s^2p^3	3.940
Oxygen	4.492	4.453	s^2p^4	4.469
Fluorine	5.128	5.110	s^2p^5	5.115

	3s	3p		Average
Silicon	3.269	2.857	s^2p^2	3.063

Source: Clementi and Raimondi (1963).

Energy-optimized, single-Slater ζ values for the electron subshells of isolated atoms have been calculated by Clementi and Raimondi (1963). For the electron density functions, such ζ values are to be multiplied by a factor of 2. Values for a number of common atoms are listed in Table 3.4, together with averages over electron shells, which are suitable as starting points in a least-squares refinement in which the exponents are subsequently adjusted by variation of κ'. A full list of the single ζ values of Clementi and Raimondi can be found in appendix F.

The coefficients n_l have to obey the condition $n_l \geq l$, imposed by Poisson's electrostatic equation, as pointed out by Stewart (1977). The radial dependence of the multipole density deformation functions may be related to the products of atomic orbitals in the quantum-mechanical electron density formalism of Eq. (3.7). The ss, sp, and pp type orbital products lead, according to the rules of multiplication of spherical harmonic functions (appendix E), to monopolar, dipolar, and quadrupolar functions, as illustrated in Fig. 3.6. The 2s and 2p hydrogenic orbitals contain, as highest power of r, an exponential multiplied by the first power of r, as in Eq. (3.33). This suggests $n_l = 2$ for all three types of product functions of first-row atoms (Hansen and Coppens 1978).

Similarly, octupoles and hexadecapoles can be thought of as arising from 2p3d and 3d3d atomic orbital products, which leads to $n_l = 3$ and 4, respectively. The

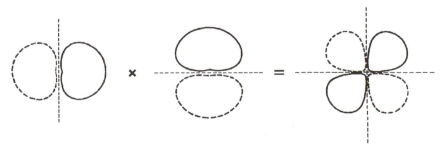

FIG. 3.6 Illustration of the product of two spherical harmonic functions: $p_x \cdot p_y = d_{xy}$.

TABLE 3.5 Position of Radial Maxima (Å) for
Different n_l ($\kappa = 1$ and ζ Values of Table 3.4)

$l =$	1	2	3	4
$n_l =$	2	2	3	4
Carbon	0.333	0.333	0.500	0.667
Nitrogen	0.276	0.276	0.414	0.551
$n_l =$	4	4	4	4
Silicon	0.691	0.691	0.691	0.691
$n_l =$	4	4	6	8
Silicon	0.691	0.691	1.037	1.382

latter reasoning is less convincing as $3d$ orbitals are not occupied for first-row atoms, and the higher multipoles solely represent the density in the bonds around an atom. Nevertheless, it provides a guideline which has proven to be useful.

The radial maxima of functions described by Eq. (3.34) are at $n_l/(\kappa\zeta)$, as found by substituting R into the equation $dR/dr = 0$. The positions of these maxima for $\kappa = 1$ for a number of atoms are given in Table 3.5.

For second-row atoms, the orbital product argument leads to $n_l = 4$ for all deformation functions. However, this model gives the same radial functions for all multipoles, and a radial maximum at 0.691 Å (Table 3.5). Since this is much less than half the 2.35 Å Si—Si bond length, such a function cannot very well describe the bonding density in an Si—Si bond. An analysis of the highly accurate Pendellösung data on silicon shows the error function $\Sigma w \Delta^2$ to be lower by a factor of better than 2 for n_l values increasing with l: $n_1 = 4, n_2 = 4, n_3 = 6, n_4 = 8$, compared with the $n_l = 4$ model. We note that a recent analysis of H_3PO_4 shows that the phosphorus density functions in this compound are better fitted with n_l values 6, 6, 7, 7 for $l = 1, 2, 3, 4$, respectively (Moss et al. 1995).

The difference in treatment between the spherical and aspherical components, described above, can be rationalized by the fact that, for covalently bonded systems, the aspherical deformation density terms also represent the accumulation of charge in bonds between atoms. But, for transition metal atoms, the asphericity of the charge distribution is mainly due to preferential occupancy of selected ligand-field stabilized d-orbital levels, as the charge accumulation in the metal–ligand bonds tends to be small. Though products of d orbitals lead to density functions with $n_l = 4$ and 6 for first- and second-row transition metal elements, respectively, a Hartree–Fock radial dependence for the metal atom deformation functions is found to give a better fit to the scattering intensities of first-row transition metal complexes (Elkaim et al. 1987).

Recommended values for n_l are summarized in Table 3.6. The example of H_3PO_4, quoted above, shows that they may not always be the optimal choice.

Other types of radial functions have been applied, including Gaussian-type functions (Stewart 1980), and harmonic oscillator wave functions (Kurki-Suonio 1977b).

TABLE 3.6 Recommended Values of n_l in the Radial Density Function Expression

	Dipole	Quadrupole	Octupole	Hexadecapole
l	1	2	3	4
First-row atoms	2	2	3	4
Second-row atoms	4	4	6	8

Source: Hansen and Coppens (1978).

3.2.4.6 The Multipole Density Formalism

Combining the angular and radial functions discussed above leads to a valence-density formalism in which the density of each of the atoms is described as (Hansen and Coppens 1978)

$$\rho_{at}(\mathbf{r}) = P_c \rho_{core}(r) + P_v \kappa^3 \rho_{valence}(\kappa r) + \sum_{l=0}^{l_{max}} \kappa'^3 R_l(\kappa' r) \sum_{m=0}^{l} P_{lm\pm} d_{lm\pm}(\theta, \phi) \qquad (3.35)$$

The two leading terms in Eq. (3.35) are identical to the κ-formalism of Eq. (3.16).

The aspherical features of the density are described by the summation added to the κ-expression. The summation includes an additional monopole, which may be omitted for first- and second-row atoms, but is necessary to describe the outer s-electron shell of transition metal atoms, which is much more diffuse than the outermost d shell.

The multipole formalism described by Stewart (1976) deviates from Eq. (3.35) in several respects. It is a deformation density formalism in which the deformation from the IAM density is described by multipole functions with Slater-type radial dependence, without the κ-type expansion and contraction of the valence shell. While Eq. (3.35) is commonly applied using local atomic coordinate systems to facilitate the introduction of chemical constraints (chapter 4), Stewart's formalism has been encoded using a single crystal-coordinate system.

The aspherical density formalism of Hirshfeld is a deformation model with angular functions which are a sum over spherical harmonics. It will be described in more detail in section 3.2.6. All three models have been applied extensively in charge density studies (for a comparison, see Lecomte 1991).

3.2.5 Aspherical Atom Scattering Factors

To obtain the atomic form factor according to the multipole formalism, we apply the Fourier transform

$$f_j(\mathbf{S}) = \int \rho_j(\mathbf{r}) \exp(2\pi i \mathbf{S} \cdot \mathbf{r}) \, d\mathbf{r} \qquad (3.36)$$

Substitution of the atomic density expression of Eq. (3.35) gives the aspherical atom scattering factor of atom j as

$$f_j(\mathbf{S}) = P_{j,c} f_{j,core}(S) + P_{j,v} f_{j,valence}(S/\kappa) + \sum_{l=0}^{l_{max}} \sum_{m=0}^{l} \sum_{p} P_{lmp} f_{lmp}(\mathbf{S}/\kappa') \qquad (3.37)$$

in which the multipole scattering factors $f_{lmp}(\mathbf{S})$ are the orientation-dependent Fourier transforms of the spherical harmonic deformation functions, and $p = \pm$.

Fourier transformation of the spherical harmonic functions is accomplished by expanding the plane wave $\exp(2\pi i \mathbf{S} \cdot \mathbf{r})$ in terms of products of the spherical harmonic functions. In terms of the complex spherical harmonics $Y_{lm}(\theta, \phi)$, and $Y_{lm}(\beta, \gamma)$, the expansion is given by (Freeman 1959, Cohen-Tannoudji et al. 1977)

$$\exp(2\pi i \mathbf{S} \cdot \mathbf{r}) = 4\pi \sum_{l=0}^{\infty} \sum_{m=-l}^{l} i^l j_l(2\pi S r) Y_{lm}(\theta, \phi) Y_{lm}^*(\beta, \gamma) \qquad (3.38)$$

where j_l is the lth order spherical Bessel function (see, e.g. Arfken 1970, p. 521), and θ and ϕ, and β and γ are the angular coordinates of $\mathbf{r}$ and $\mathbf{S}$, respectively (Fig. 3.7). The first three terms of j_l are listed in Table 3.7.

Combining terms with $m = -l$ and $m = l$, gives the expansion in terms of the real spherical harmonic functions, which we will use to evaluate the Fourier transform of the real density functions:

$$\exp(2\pi i \mathbf{S} \cdot \mathbf{r}) = \sum_{l=0}^{\infty} i^l j_l(2\pi S r)(2l + 1)$$

$$\times \sum_{m=0}^{l} (2 - \delta_{m0}) \frac{(l-m)!}{(l+m)!} P_l^m(\cos\theta) P_l^m(\cos\beta) \cos\{m(\phi - \gamma)\} \qquad (3.39)$$

We need to evaluate

$$f_{lm+}(\mathbf{S}) = \int R_l(r) d_{lm+}(\theta, \phi) \exp(2\pi i \mathbf{H} \cdot \mathbf{r}) d\mathbf{r} \qquad (3.40)$$

as well as $f_{lm-}(\mathbf{S})$.

Substitution of Eq. (3.38), and writing the explicit expression for $d_{lm+}(\theta, \phi)$, gives

$$f_{lm+}(\mathbf{S}) = N'_{lm+} \int\int\int R_l(r) P_l^m(\cos\theta) \cos m\phi \sum_{l=0}^{\infty} i^l j_l(2\pi S r)(2l + 1)$$

$$\times \sum_{m'=0}^{l} (2 - \delta_{m0}) \frac{(l-m')!}{(l+m')!} P_l^{m'}(\cos\theta) P_l^{m'}(\cos\beta) \cos\{m'(\phi - \gamma)\} r^2 \sin\theta \, d\theta \, d\phi \, dr \qquad (3.41)$$

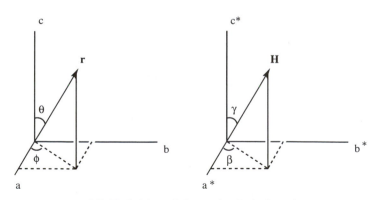

FIG. 3.7 Definition of the angles θ, ϕ, β, and γ.

TABLE 3.7 The Low-Order
Spherical Bessel Functions

$$j_0(x) = \frac{\sin x}{x}$$

$$j_1(x) = \frac{\sin x}{x^2} - \frac{\cos x}{x}$$

$$j_2(x) = \left(\frac{3}{x^3} - \frac{1}{x}\right)\sin x - \frac{3}{x^2}\cos x$$

Since the associated Legendre polynomials (and the spherical harmonics) form an orthogonal set, only terms with $l = l'$ and $m = m'$ do not vanish in the integral of Eq. (3.41). Furthermore, for the θ integration,

$$\int_0^\pi P_l^m(\cos\theta)^2 \sin\theta\, d\theta = \int_{-1}^1 P_l^m(z)^2\, dz = \frac{2}{(2l+1)}\frac{(l+|m|)!}{(l-|m|)!}$$

as implied by the normalization factor of Eq. (3.22), and for ϕ:

$$\int_0^{2\pi} \cos m\phi \cos m(\phi - \gamma)\, d\phi = \pi \cos m\gamma \text{ when } m \neq 0, \text{ and } 2\pi \text{ when } m = 0$$

Substituting these results in Eq. (3.40) gives

$$f_{lm+}(\mathbf{S}) = 4\pi i^l \langle j_l \rangle N'_{lm+} P_l^m(\cos\beta) \cos m\gamma = 4\pi i^l \langle j_l \rangle d_{lm+}(\beta, \gamma) \qquad (3.42)$$

or, in general,

$$f_{lmp}(\mathbf{S}) = 4\pi i^l \langle j_l \rangle d_{lmp}(\beta, \gamma) \qquad (3.43)$$

where $\langle j_l \rangle$, the *Fourier–Bessel transform*, is the radial integral defined as

$$\langle j_l \rangle = \int j_l(2\pi Sr) R_l(r) r^2\, dr \qquad (3.44)$$

The quantity

$$\langle j_0 \rangle = \int_0^\infty 4\pi r^2 \rho_j(r) \frac{\sin 2\pi Sr}{2\pi Sr}\, dr \equiv \int_0^\infty 4\pi r^2 \rho_j(r) j_0\, dr \qquad (3.45)$$

which we encountered in the expression (1.28) for the scattering factor of the spherical atom and in expression (3.38), is the zero-order function in this series.

Expressions (3.42) and (3.43) show that the Fourier transform of a direct-space spherical harmonic function is a reciprocal-space spherical harmonic function with the same l, m. This is summarized in the statement that the spherical harmonic functions are *Fourier-transform invariant*. It means, for example, that a dipolar density described by the function d_{10}, oriented along the $\mathbf{c}$ axis of a unit cell, will not contribute to the scattering of the $(hk0)$ reflections, for which $\mathbf{H}$ is in the $\mathbf{a}^*\mathbf{b}^*$ plane, which is a nodal plane of the function $d_{10}(\beta, \gamma)$.

The functions $\langle j_l \rangle$ for Hartree–Fock valence shells of the atoms are tabulated in scattering factor tables (*International Tables for X-ray Crystallography* 1974,

International Tables for Crystallography 1992). The function $\langle j_l \rangle$ for Slater-type radial functions can be expressed in terms of a hypergeometric series (Stewart 1980), or in closed form (Avery and Watson 1977, Su and Coppens 1990). The latter are listed in appendix G. As an example, for a first-row atom quadrupolar function ($l = 2$) with $n_l = 2$, the integral over the nonnormalized Slater function is

$$G_{4,2}(K, \zeta) = \int r^4 \exp(-\zeta r) j_2(Sr) \, dr = \frac{48K^2\zeta}{(K^2 + \zeta^2)^4} \tag{3.46}$$

with $K = 4\pi \sin \theta / \lambda$. The value of $\langle j_2 \rangle$ is subsequently obtained by multiplication with the normalization factor defined in Eq. (3.34). Since the Hartree–Fock atomic wave functions are available in terms of an expansion of Slater-type functions (Clementi and Roetti 1974), Hartree–Fock scattering factors can easily be evaluated using the closed-form expressions of appendix G.

In summary, the structure factor expression corresponding to Eq. (3.37) is given by

$$
\begin{aligned}
F(\mathbf{H}) = \sum_j \bigg[\bigg\{ & P_{j,c} f_{j,\text{core}}(H) + P_{j,v} f_{j,\text{valence}}(H/\kappa) \\
& + 4\pi \sum_{l=0}^{l_{\max}} \sum_{m=0}^{l} P_{lm\pm} i^l \langle j_l(S/\kappa') \rangle d_{lm\pm}(\beta, \gamma) \bigg\} \exp(2\pi i \mathbf{H} \cdot \mathbf{r}_j) T_j(\mathbf{H}) \bigg]
\end{aligned}
\tag{3.47}
$$

in which $T_j(\mathbf{H})$ is the temperature factor, and the terms in the deformation scattering factors are defined by Eq. (3.43) and (3.44).

3.2.6 The Aspherical Density Functions of the Hirshfeld Formalism

Hirshfeld (1971) was among the first to introduce atom-centered deformation density functions into the least squares procedure. Hirshfeld's formalism is a deformation model, in which the leading term is the unperturbed IAM density, and the deformation functions are of the form $\cos^n \theta_{jk}$, where θ_{jk} is the angle between the radius vector $\mathbf{r}_j$ and axis k of a set of $(n + 1)(n + 2)/2$ polar axes on each atom j, as defined in Table 3.8 (Hirshfeld 1977). The atomic deformation on atom j is described as

$$\Delta(\rho_j) = \sum_{n=1}^{4} \sum_k h_{nk} \equiv \sum_{n=1}^{4} \sum_k N_n P_{jnk} r_j^n \exp(-\alpha r_j) \cos^n \theta_{jk} \tag{3.48}$$

in which r_j is the distance from the atom center j, α is the exponential coefficient in the Slater-type radial function, and the P_{jnk} coefficients are the population parameters. The normalization factors N_n are defined somewhat differently from the normalization for the functions d_{lmp} [Eq. (3.27)] by

$$\int |h_{nk}| \, d\Omega = 1 \tag{3.49}$$

which gives

$$N_n = \frac{(n + 1)\alpha^{n+3}}{4\pi(n + 2)!} \tag{3.50}$$

TABLE 3.8 Direction Cosines in the Hirshfeld Formalism and their Relation to the Spherical Harmonics Density Functions

n	Directions	Corresponding Spherical Harmonics and their Number
0		d_{00} (1)
1	100, 010, 00$\bar{1}$	d_{1m} (3)
2	110, 1$\bar{1}$0, 101, 10$\bar{1}$, 011, 01$\bar{1}$	d_{2m} (5) + d_{00} (1)
3	110, 1$\bar{1}$0, 101, 10$\bar{1}$, 011, 01$\bar{1}$, 111, 1$\bar{1}\bar{1}$, $\bar{1}$1$\bar{1}$, $\bar{1}\bar{1}$1	d_{3m} (7) + d_{1m} (3)
4	100, 010, 001, α11, 1α1, 11α, 1$\bar{\alpha}$1, 1$\bar{\alpha}$1, 11$\bar{\alpha}$, $\alpha\bar{1}$1, 1$\alpha\bar{1}$, $\bar{1}$1α, α1$\bar{1}$, $\bar{1}\alpha$1, 1$\bar{1}\alpha^a$	d_{4m} (9) + d_{2m} (5) + d_{00} (1)

^{a} $\alpha = \sqrt{2} - 1$.

For a given value of n, the functions h_{nk} are identical to a sum of spherical harmonics with $l = n,\ n - 2, n - 4, \ldots, (0, 1)$ for $n > 1$. The relationships are summarized in Table 3.8. For $n = 0, 1$, the Hirshfeld functions are identical to the spherical harmonics with $l = 0, 1$, but, starting with the $n = 2$ functions, lower-order spherical harmonics are included for each n value. Unlike the spherical harmonics, the h_{nl} functions are therefore not mutually orthogonal. As the radial functions in Eq. (3.48) contain the factor r^n, quite diffuse s, p, and d functions are included in the $n = 2, 3$, and 4 sets. For $n \leq 4$ there are 35 deformation functions on each atom, compared with 25 valence-shell density functions with $l \leq 4$ in the multipole expansion of Eq. (3.35).

The Hirshfeld functions give an excellent fit to the density, as illustrated for tetrafluoroterephthalonitrile in chapter 5 (see Fig. 5.12). But, because they are less localized than the spherical harmonic functions, net atomic charges are less well defined. A comparison of the two formalisms has been made in the refinement of pyridinium dicyanomethylide (Baert et al. 1982). While both models fit the data equally well, the Hirshfeld model leads to a much larger value of the molecular dipole moment obtained by summation over the atomic functions using the equations described in chapter 7. The multipole results appear in better agreement with other experimental and theoretical values, which suggests that the latter are preferable when electrostatic properties are to be evaluated directly from the least-squares results. When the evaluation is based on the density predicted by the model, both formalisms should perform well.

4

Least-Squares Methods and Their Use in Charge Density Analysis

4.1 Least-Squares Equations

4.1.1 Background

The number of reflection intensities measured in a crystallographic experiment is large, and commonly exceeds the number of parameters to be determined. It was first realized by Hughes (1941) that such an overdetermination is ideally suited for the application of the least-squares methods of Gauss (see, e.g., Whittaker and Robinson 1967), in which an error function S, defined as the sum of the squares of discrepancies between observation and calculation, is minimized by adjustment of the parameters of the observational equations. As least-squares methods are computationally convenient, they have largely replaced Fourier techniques in crystal structure refinement.

In addition to the positional and thermal parameters of the atoms, least-squares procedures are used to determine the scale of the data, and parameters such as mosaic spread or particle size, which influence the intensities through multiple-beam effects (Becker and Coppens 1974a, b, 1975). It is not an exaggeration to say that modern crystallography is, to a large extent, made possible by the use of least-squares methods. Similarly, least-squares techniques play a central role in the charge density analysis with the scattering formalisms described in the previous chapter.

4.1.2 General Formalisms for Linear Least-Squares

The following description follows closely the treatment given by Hamilton (1964). Suppose we have n experimental observations

$$f_1, f_2, \ldots, f_n \qquad (4.1)$$

each of which is known to depend *linearly* on a set of m parameters

$$x_1, \ldots, x_m \tag{4.2}$$

with $m \leq n$.

If each observation f_i is subject to a random error ε_i, the observational equations may be written as

$$f_1 = a_{11}x_1 + a_{12}x_2 + \cdots + a_{1m}x_m + \varepsilon_1$$

$$f_2 = a_{21}x_1 + a_{22}x_2 + \cdots + a_{2m}x_m + \varepsilon_2$$

$$\cdots \cdots \cdots \cdots \cdots \cdots \cdots \cdots \cdots \cdots$$

$$f_n = a_{n1}x_1 + a_{n2}x_2 + \cdots + a_{nm}x_m + \varepsilon_n \tag{4.3a}$$

Or, in matrix notation

$$\mathbf{F}_{n,1} = \mathbf{A}_{n,m}\mathbf{X}_{m,1} + \mathbf{E}_{n,1} \tag{4.3b}$$

The matrix $\mathbf{A}$ is called the *design matrix*. Its elements are the derivatives $\partial f_i/\partial x_j = a_{ij}$.

Given the n observations, our aim is to obtain the best estimates $\hat{\mathbf{X}}$ of the m unknown parameters to be determined. Gauss proposed the minimization of the sum of the squares of the discrepancies, defining the *error function S*, which, after assignment of a weight w_i to each of the observations, is given by

$$S = \sum_{i=1}^{n} w_i(f_i - \hat{f}_i)^2 = \sum_{i=1}^{n} w_i\Delta_i^2 \tag{4.4}$$

where the $\hat{f}_i$ are the values for the observations based on the estimates $\hat{\mathbf{X}}$.

Suppose the observations are correlated with the correlation coefficient γ_{ij} describing the correlation between the ith and jth observations. The variances and covariances of the observations will then be given by the variance–covariance matrix $\mathbf{M}_f$ with elements $\sigma_i\sigma_j\gamma_{ij}$, where σ_i is the standard deviation of the ith observation. The error function S for a set of correlated observations is defined as

$$S = \mathbf{V}^T\mathbf{M}_f^{-1}\mathbf{V} \tag{4.5}$$

where $\mathbf{V}$ is a column matrix, with elements $f_i - \hat{f}_i$, or

$$\mathbf{V} \equiv \mathbf{F} - \hat{\mathbf{F}} \equiv \mathbf{F} - \mathbf{A}\hat{\mathbf{X}} \tag{4.6}$$

Thus,

$$S = \mathbf{V}^T\mathbf{M}_f^{-1}\mathbf{V} = (\mathbf{F} - \mathbf{A}\hat{\mathbf{X}})^T\mathbf{M}_f^{-1}(\mathbf{F} - \mathbf{A}\hat{\mathbf{X}})$$

$$= \mathbf{F}^T\mathbf{M}_f^{-1}\mathbf{F} + \hat{\mathbf{X}}^T\mathbf{A}^T\mathbf{M}_f^{-1}\mathbf{A}\hat{\mathbf{X}} - \mathbf{F}^T\mathbf{M}_f^{-1}\mathbf{A}\hat{\mathbf{X}} - \hat{\mathbf{X}}^T\mathbf{A}^T\mathbf{M}_f^{-1}\mathbf{F}$$

$$= \mathbf{F}^T\mathbf{M}_f^{-1}\mathbf{F} + \hat{\mathbf{X}}^T\mathbf{A}^T\mathbf{M}_f^{-1}\mathbf{A}\hat{\mathbf{X}} - 2\mathbf{F}^T\mathbf{M}_f^{-1}\mathbf{A}\hat{\mathbf{X}} \tag{4.7}$$

The best estimate of the unknowns is obtained when $\partial S/\partial x_i = 0$ for each of the m unknowns. In matrix notation, introducing the differential operator δ, these conditions are written as

$$\delta S = \delta(\mathbf{V}^T\mathbf{M}_f^{-1}\mathbf{V}) = 0 \tag{4.8a}$$

or, substituting Eq. (4.7),

$$\delta(\mathbf{V}^T\mathbf{M}_f^{-1}\mathbf{V}) = 2(\delta\hat{\mathbf{X}})^T(\mathbf{A}^T\mathbf{M}_f^{-1}\mathbf{A}\hat{\mathbf{X}} - \mathbf{A}^T\mathbf{M}_f^{-1}\mathbf{F}) = 0 \tag{4.8b}$$

The *m normal equations* are defined by the solutions

$$(A^T M_f^{-1} A)\hat{X} = A^T M_f^{-1} F \tag{4.9}$$

or, substituting $B \equiv A^T M_f^{-1} A$,

$$B\hat{X} = A^T M_f^{-1} F \tag{4.10}$$

where **B** is known as the matrix of the normal equations. In the case that the observations are not correlated, M_f^{-1} is the diagonal weight matrix, and the elements of **B** are equal to

$$\sum_{i=1}^{n} w_i \frac{\partial f_i}{\partial x_j} \frac{\partial f_i}{\partial x_k}$$

According to Eq. (4.10), the best estimates of the unknowns $\hat{X}$ are given by

$$\hat{X} = B^{-1} A^T M_f^{-1} F \tag{4.11}$$

The vector **F** and the corresponding variance–covariance matrix M_f are the only parts of this equation that depend on the measurements; the other quantities are derived from the observational equations.

4.1.3 Explicit Expressions for Structure Factor Least-Squares

In the least-squares treatment of diffraction intensity data, the experimental quantities f_i are usually defined as either the square of the structure factor, F^2, or the structure factor, F. The discrepancy between observation and calculation is then

$$\Delta = |F(\text{obs}) - k|F(\text{calc})|| \tag{4.12}$$

or

$$\Delta = |F^2(\text{obs}) - k^2|F(\text{calc})|^2| \tag{4.13}$$

where the scale factor k is needed to bring experiment and observation to a common scale.

Unlike the treatment described in the previous section, the structure factors are not linear functions of the unknowns x_j, as was the case for the observations described by expression (4.3). The equations of the preceding section can only be retained by the approximation that all the second and higher derivatives are zero. The price paid for this assumption is that the equations are no longer exact, so that a single calculation no longer leads to the minimum of the error function. If the deviations from linearity are not pronounced, the minimum may still be reached, but a number of iterations may be necessary to achieve convergence.

To linearize the equations, the slope of the error function S, S_j' may be developed as a Taylor series in the current values of the unknowns x_j. For the jth unknown, we obtain

$$S_j' = S_j'(X_0) + \sum_k \frac{\partial S_j'(X_0)}{\partial x_k} \delta x_k + \frac{1}{2} \sum_k \sum_p \frac{\partial^2 S_j'(X_0)}{\partial x_k \partial x_p} \delta x_k \, \delta x_p + \dots \tag{4.14}$$

with $k, p = 1, \ldots, m$ for m unknowns. The subscript in $\mathbf{X}_0$ indicates that current values of the unknowns are used. The first term is the slope of the error function along the parameter axis x_j at $\mathbf{X}_0$, and is obtained from Eq. (4.4) as

$$S_j'(\mathbf{X}_0) = \frac{\partial S}{\partial x_j} = 2 \sum_i^n w_i \Delta_i \frac{\partial \Delta_i}{\partial x_j} \qquad (4.15a)$$

Linearization is accomplished by truncating the series of Eq. (4.14) after the first derivative, or

$$S_j' = S_j'(\mathbf{X}_0) + \sum_k \frac{\partial S_j'(\mathbf{X}_0)}{\partial x_k} \delta x_k \qquad (4.16)$$

We will treat the case that the observations are chosen to be the structure factors F. The expression relevant in this case, Eq. (4.12), gives

$$\frac{\partial \Delta_i}{\partial x_j} = -\frac{\partial |kF_{i,c}|}{\partial x_j} \frac{(F_{\text{obs}} - |kF_{\text{calc}}|)}{|F_{\text{obs}} - |kF_{\text{calc}}||}$$

On substitution into Eq. (4.15a), one obtains

$$S_j'(\mathbf{X}_0) = -2 \sum_{i=1}^n w_i (F_{i,o} - |kF_{i,c}|) \frac{\partial |kF_{i,c}|}{\partial x_j} \qquad (4.15b)$$

which gives, for Eq. (4.16),

$$S_j' = -2 \sum_{i=1}^n w_i (F_{i,o} - |kF_{i,c}|) \frac{\partial |kF_{i,c}|}{\partial x_j}$$
$$+ 2 \sum_k \left\{ \sum_{i=1}^n w_i \left(\frac{\partial |kF_{i,c}|}{\partial x_j} \right) \left(\frac{\partial |kF_{i,c}|}{\partial x_k} \right) - \sum_{i=1}^n w_i \Delta_i \frac{\partial^2 |kF_{i,c}|}{\partial x_k \partial x_j} \delta x_k \right\} \qquad (4.17)$$

The last summation in Eq. (4.17) which contains second derivatives is again omitted. This is consistent with the original approximation of setting all second derivatives equal to zero, but implies that even the first-order term in the Taylor expansion of Eq. (4.15a) is not fully taken into account. The resulting m normal equations $S_j' = 0$ ($j = 1, \ldots, m$) are

$$\sum_k \left\{ \left[\sum_{i=1}^n w_i \left(\frac{\partial |kF_{i,c}|}{\partial x_j} \right) \left(\frac{\partial |kF_{i,c}|}{\partial x_k} \right) \right] \delta x_k \right\} = \sum_{i=1}^n w_i \{ F_{i,o} - |kF_{i,c}| \} \frac{\partial |kF_{i,c}|}{\partial x_j} \qquad (4.18)$$

This result is equivalent to the linear least-squares normal equations, Eq. (4.10), with a diagonal variance–covariance matrix $\mathbf{M}_f$,

$$\mathbf{B}\hat{\mathbf{X}} = \mathbf{A}^T \mathbf{M}_f^{-1} \mathbf{F} \qquad (4.10)$$

with the elements of $\mathbf{A}(nm)$ given by $A_{ij} = \partial |kF_{i,c}|/\partial x_j$, those of $\mathbf{F}(n1)$ by $F_i = w_i |F_{i,o} - |kF_{i,c}||$, and those of the symmetric matrix $\mathbf{B}(nn)$ by

$$B_{jk} = \sum_{i=1}^n w_i \left(\frac{\partial |kF_{i,c}|}{\partial x_j} \right) \left(\frac{\partial |kF_{i,c}|}{\partial x_k} \right)$$

The calculation of the matrices $\mathbf{A}$ and $\mathbf{B}$, the inversion of $\mathbf{B}$, and the matrix multiplications, are the major steps in a least-squares iteration.

As is evident from Eq. (4.18), the elements of the vector $\mathbf{X}$ in Eq. (4.10) are defined as $X_i = \delta x_i$. The replacement of x_i by δx_i shows that the solution of Eq. (4.11) now leads to the *shifts* in the unknowns, rather than directly to the best values of the unknowns.

Since the truncation of Eq. (4.14) is an approximation, the shifts may be under- or overestimated. When the shifts are added to the original parameter values, that is, $\mathbf{X}_2 = \mathbf{X}_1 + \delta\mathbf{X}$, the elements of the design matrix $\mathbf{A}$, which depend on $\mathbf{X}$, change. Many iterations may be needed before convergence can be reached, if it is indeed reachable. Strategies for coping with "difficult" refinements have been discussed by Watkin (1994).

The agreement between observations and calculations is the basis for judging the success of the refinement as described in the following section.

4.1.4 Variances and Covariances of the Least-Squares Parameter Estimates

The matrix $\mathbf{M}_x$, describing the variances and covariances in the best values of the unknowns $\hat{\mathbf{X}}$, is written as (Hamilton 1964)

$$\mathbf{M}_x = \varepsilon\{(\hat{\mathbf{X}} - \mathbf{X}^0)(\hat{\mathbf{X}} - \mathbf{X}^0)^T\} \tag{4.19}$$

in which the superscript 0 labels the true values, and the symbol ε indicates an estimate. Substitution of Eq. (4.19) into Eq. (4.11), $\hat{\mathbf{X}} = \mathbf{B}^{-1}\mathbf{A}^T\mathbf{M}_f^{-1}\mathbf{F}$, gives

$$\mathbf{M}_x = \varepsilon\{\mathbf{B}^{-1}\mathbf{A}^T\mathbf{M}_f^{-1}(\mathbf{F} - \mathbf{F}^0)(\mathbf{F} - \mathbf{F}^0)^T\mathbf{M}_f^{-1}\mathbf{A}\mathbf{B}^{-1}\}$$
$$= \mathbf{B}^{-1}\mathbf{A}^T\mathbf{M}_f^{-1}\varepsilon\{(\mathbf{F} - \mathbf{F}^0)(\mathbf{F} - \mathbf{F}^0)^T\}\mathbf{M}_f^{-1}\mathbf{A}\mathbf{B}^{-1} \tag{4.20}$$

The estimate $\varepsilon\{(\mathbf{F} - \mathbf{F}^0)(\mathbf{F} - \mathbf{F}^0)^T\}$ is the variance–covariance matrix of the observations $\mathbf{M}_f$ defined earlier. Thus,

$$\mathbf{M}_x = \mathbf{B}^{-1}\mathbf{A}^T\mathbf{M}_f^{-1}\mathbf{M}_f\mathbf{M}_f^{-1}\mathbf{A}\mathbf{B}^{-1} = \mathbf{B}^{-1}\mathbf{A}^T\mathbf{M}_f^{-1}\mathbf{A}\mathbf{B}^{-1} \tag{4.21a}$$

As $\mathbf{B}$ is defined as $\mathbf{B} \equiv \mathbf{A}^T\mathbf{M}_f^{-1}\mathbf{A}$, we obtain

$$\mathbf{M}_x = \mathbf{B}^{-1} \tag{4.21b}$$

In other words, the variance–covariance matrix of the unknowns is the inverse of the matrix $\mathbf{B}$. As described in the preceding section, the elements of $\mathbf{B}$ are the sums over the products of the derivatives of the observational equations with respect to the unknowns. Expression (4.21b) shows what we might have anticipated intuitively. As the elements of $\mathbf{B}^{-1}$ will tend to be inversely proportional to those of $\mathbf{B}$, an element of $\mathbf{B}^{-1}$ mainly related to small derivatives will be large. Consequently, when the observations are not sensitive to an unknown parameter, the errors in the unknown parameter are large, and vice versa.

It is quite common that only the relative values of the variances and covariances in the observations can be estimated. We may then write

$$\mathbf{M}_f = \sigma^2\mathbf{N}_f \tag{4.22}$$

where σ^2 is a scale factor required to reduce $\mathbf{M}_f$ to the matrix $\mathbf{N}_f$ of the absolute-scale variances and covariances. The scale factor σ^2 is described as the *variance of an observation of unit weight*, that is, when an element of $\mathbf{N}$ equals 1, σ^2 is the variance of the corresponding variable.

We obtain

$$\mathbf{M}_f^{-1} = \frac{1}{\sigma^2} \mathbf{N}_f^{-1} \equiv \frac{1}{\sigma^2} \mathbf{P}_f \tag{4.23}$$

and

$$\mathbf{B} = \mathbf{A}^T \mathbf{M}_f^{-1} \mathbf{A} = \frac{\mathbf{A}^T \mathbf{P}_f A}{\sigma^2} \tag{4.24}$$

It follows that the variances–covariances of the unknowns are given by

$$\mathbf{M}_x \equiv \mathbf{B}^{-1} = \sigma^2 (\mathbf{A}^T \mathbf{P}_f A)^{-1} \tag{4.25}$$

where $\mathbf{P}_f$ is commonly taken as a diagonal matrix with the weights of the observations as elements. We note that with area detectors, groups of reflections are measured on a single frame, so correlations between observations may no longer be negligible.

The best estimate $\hat{\sigma}^2$ of σ^2 is related to the magnitude of the discrepancies $\mathbf{V}$. The value of $\hat{\sigma}^2$ is an average of the weighted squares of the discrepancies, taking into account that the fit will progressively improve as the number of unknowns m approaches the number of observations n. When $n = m$ the solution of the observational equations is exact, but the variances and covariances are indeterminate. The best estimate $\hat{\sigma}^2$ is obtained from

$$\hat{\sigma}^2 = \frac{\mathbf{V}^T \mathbf{P}_f V}{n - m} \tag{4.26}$$

or, in the case of a diagonal matrix $\mathbf{P}_f$,

$$\hat{\sigma}^2 = \frac{\sum_{i=1}^{n} w_i \Delta_i^2}{n - m} \tag{4.27}$$

When realistic weights are assigned to the observations, and the model is adequate, then $w_i \approx \Delta_i^{-2}$. If these conditions are fulfilled, *the goodness of fit $\hat{\sigma}^2$* will be close to unity.

Substitution of Eq. (4.27) into (4.25) gives the variance–covariance matrix of the unknowns as

$$\mathbf{M}_x = \frac{\sum_{i=1}^{n} w_i \Delta_i^2}{n - m} (\mathbf{A}^T \mathbf{P}_f A)^{-1} \tag{4.28}$$

The element M_{ij} of $\mathbf{M}_x$ is often written as $\sigma_i \sigma_j \gamma_{ij}$. Thus, the *correlation coefficient γ_{ij}* may be obtained from

$$\gamma_{ij} = \frac{M_{ij}}{\sqrt{M_{ii} M_{jj}}} \tag{4.29}$$

Correlation can be illustrated in a two-dimensional section of m-dimensional parameter space, as shown in Fig. 4.1. The two parameters, x_1 and x_2, are positively correlated; if x_1 is too large, the error in x_2 is likely to be in the same direction.

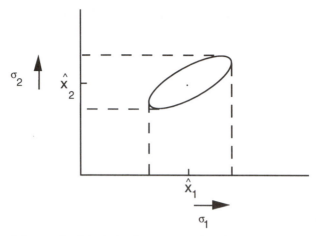

FIG. **4.1** Ellipse of the standard deviation for two correlated parameters, with best estimates $\hat{x}_1$ and $\hat{x}_2$. The standard deviations σ_1 and σ_2 are the horizontal and vertical dimensions of the ellipsoid at the point $\hat{x}_1, \hat{x}_2$. The projections of the ellipsoid on the parameter axes are referred to as the conditional standard deviations.

4.1.4.1 Propagation of Errors

Suppose we want to calculate p derived quantities u_i, dependent on the m unknowns x_j, determined by a least-squares procedure. If **D** is the $p \times m$ matrix of derivatives, defined by

$$D_{ij} = \frac{\partial u_i}{\partial x_j} \tag{4.30}$$

the variances and covariances of the elements of **U** are obtained as

$$\mathbf{M}_u = \mathbf{DM}_x \mathbf{D}^T \tag{4.31}$$

A simple example is a bond length defined as $x_1 - x_2$. The 1×2 matrix **D** has two elements equal to 1 and -1, and the errors in the difference of the coordinates are given by

$$\sigma^2(x_1 - x_2) = \sigma^2(x_1) + \sigma^2(x_2) - 2\sigma(x_1)\sigma(x_2)\gamma_{12}$$

In the special case that the two atoms are related by a center of symmetry, the correlation coefficient will be -1, and we obtain

$$\sigma^2(x_1 - x_2) = \{\sigma(x_1) + \sigma(x_2)\}^2$$

Thus, the error in the difference of the unknowns is larger than it would be without correlation, unlike the example given in Fig. 4.1 for which the error in the sum of the unknowns is increased.

Expression (4.31) is widely applied to calculate the error in properties derived from the least-squares variables. We will use it in chapter 7 for the calculation of the standard deviations of the electrostatic moments derived from the parameters of the multipole formalism.

4.1.5 Uncorrelated Linear Combinations of Variables

When the parameters are strongly correlated, it is still possible to define a set of mutually uncorrelated combinations of the parameters. This can be shown as follows. If $\mathbf{T}$ represents the matrix of the eigenvectors of the variance–covariance matrix $\mathbf{M}_x$, then $\mathbf{M}_x$ is diagonalized by the transformation

$$\Lambda = \mathbf{T}\mathbf{M}_x\mathbf{T}^T \tag{4.32}$$

We may define a set of linear combinations of the unknowns as

$$\mathbf{X}' = \mathbf{T}\mathbf{X} \tag{4.33}$$

Because of Eq. (4.31), Λ is the variance–covariance matrix of the set of unknowns X', which we will refer to as the *eigenparameter*. The eigenparameters $\mathbf{X}'$ are, by the definition of the variance–covariance matrix, not correlated.

Some of the linear combinations will be well defined and others poorly defined. The latter may be eliminated in a filtering procedure, referred to in the literature under the names *characteristic value filtering*, *eigenvalue filtering*, and *principal component analysis*. If the parameter set is not homogeneous, but includes different types, relative scaling is important. Watkin (1994) recommends that the unit be scaled such that similar shifts in all parameters lead to similar changes in the error function S.

Diamond (1966) has applied a filtering procedure in the refinement of protein structures, in which poorly determined linear combinations are not varied. In charge density analysis, the principal component analysis has been tested in a refinement of theoretical structure factors on diborane, B_2H_6, with a formalism including both one-center and two-center overlap terms (Jones et al. 1972). Not unexpectedly, it was found that the sum of the populations of the $2s$ and spherically averaged $2p$ shells on the boron atoms constitutes a well-determined eigenparameter, while the difference is very poorly determined. Correlation between one- and two-center terms was also evident in the analysis.

The value of Eq. (4.32) is that it shows exactly which features of the structure are well determined and which are poorly determined in the fitting procedure. For the two-dimensional example of Fig. 4.1, the eigenparameters correspond to the principal axes of the variance–covariance ellipsoid in the figure. In general, they define the principal axes of the hyper-ellipsoid in m-dimensional parameter space which represents the variance–covariance matrix.

4.2 The Least-Squares Parameters in Charge Density Analysis

4.2.1 The Parameters in a Charge Density Refinement

The nature of the charge density parameters to be added to those of the structure refinement follows from the charge density formalisms discussed in chapter 3. For the atom-centered multipole formalism as defined in Eq. (3.35), they are the valence shell populations, $P_{i,\text{val}}$, and the populations $P_{i,lmp}$ of the multipolar density functions on each of the atoms, and the κ expansion–contraction parameters for

TABLE 4.1 Summary of Least-Squares Variables

Conventional Variables	Charge Density Variables
Scale factor k	*Valence-shell parameters*
Positional parameters x_i, y_i, z_i	Population parameters $P_{i,\text{val}}$
Thermal parameters	Expansion–contraction parameters κ_i
$\quad$ harmonic β_i, β_{ijk}	
$\quad$ higher-order quasimoments $c_i^{jkl\cdots}$	*Deformation parameters*
$\quad$ or	Population parameters $P_{i,lmp}$
$\quad$ higher-order cumulants $\kappa_i^{jkl\cdots}$	Expansion–contraction parameters κ_i'
Extinction parameters	
$\quad$ isotropic or anisotropic	
Occupancy parameters	

both the valence shell and the deformation functions of the atoms. These are summarized in Table 4.1.

4.2.2 Parameter Restrictions Imposed by Site and Local Symmetry and Chemical Equivalence

The number of multipole parameters is reduced by the requirements of symmetry. As discussed in chapter 3, the only allowed multipolar functions are those having the symmetry of the site, which are invariant under the local symmetry operations. For example, only $l =$ even multipoles can have nonzero populations on a centrosymmetric site, while for sites with axial symmetry the dipoles must be oriented along the symmetry axis. For a highly symmetric site having 6 *mm* symmetry, the lowest allowed $l \neq 0$ is d_{66+}; all lower multipoles being forbidden by the symmetry. The *index-picking rules* listed in appendix D give the information required for selection of the allowed parameters.

It is often found that multipoles which violate local symmetry are not significantly populated. If this is the case, the number of variables can be reduced significantly by application of local-symmetry restrictions. Examples are the mirror symmetry of aromatic rings, and the symmetry at the metal site of many transition metal complexes.

Many molecules contain *chemically* equivalent atoms, which, though in a different crystal environment, have, to a good approximation, the same electron distribution. Such atoms may be linked, provided equivalent local coordinate systems are used in defining the multipoles. In particular, for the weakly scattering hydrogen atoms, abundant in most organic molecules, this procedure can lead to more precisely determined population parameters.

We demonstrate the use of local coordinate systems with the molecule of tetrasulfur tetranitride, S_4N_4, (Fig. 4.2) as an example. It occupies a general position in its crystal's space group, with one molecule in the asymmetric unit. Thus, there are eight crystallographically independent atoms. If multipoles up to and including the hexadecapoles are included, the number of population parameters

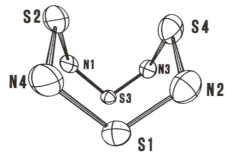

FIG. 4.2 Molecular structure of tetrasulfur tetranitride with 50% thermal probability ellipsoids. *Source*: DeLucia (1977).

is $8(1 + 3 + 5 + 7 + 9) = 200$, plus 16 κ and κ' parameters, for a total of 216 charge density variables.

The sulfur atoms are located in molecular but not crystallographic symmetry planes, and the nitrogen atoms are located on molecular two-fold axes, passing through N1 and N2, and through N3 and N4. The equivalent local coordinate systems are shown in Fig. 4.3. The multipoles allowed in the specified coordinate systems are listed in Table 4.2. The introduction of local symmetry and chemical equivalence of the four S and four N atoms reduced the total number of spherical-harmonic density functions to 15 for S, and 13 for N. The restrictions result in 28 population parameters, plus two each κ and κ' values. The dramatic reduction from 216 to 32 charge density parameters greatly improves the stability of the charge density refinement of this molecule.

4.2.3 The Scale Factor

Since reflection intensities are commonly measured on a relative rather than an absolute scale, a scale factor is required to bring observations and calculations on

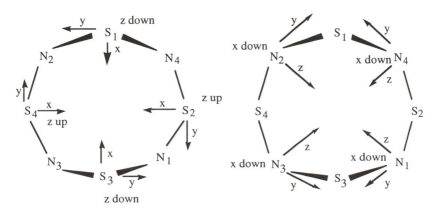

FIG. 4.3 Local-symmetry-adapted coordinate systems for S (left) and N (right) in the S_4N_4 molecule. *Source*: Stevens (1980).

TABLE 4.2 Local-Symmetry-Allowed Multipole Population Parameters in S_4N_4

(a) Nitrogen (two-fold axis along z axis)

$l =$	0, 1	2	3	4
	P_{00}	P_{20}	P_{30}	P_{40}
	P_{10}	P_{22+}	P_{32+}	P_{42+}
		P_{22-}	P_{32-}	P_{42-}
				P_{44+}
				P_{44-}

(b) Sulfur (mirror plane $\perp$ to y axis)

$l =$	0, 1	2	3	4
	P_{00}	P_{20}	P_{30}	P_{40}
	P_{11+}	P_{21+}	P_{31+}	P_{41+}
	P_{10}	P_{22-}	P_{32+}	P_{42+}
			P_{33+}	P_{43+}
				P_{44+}

a common scale. As introduced in Eq. (4.12), the scale factor k is defined by

$$F_{obs}(\mathbf{H}) = k|F_{calc}(\mathbf{H})| \qquad (4.34)$$

Since the scale factor is considered an unknown in the least-squares procedure, its estimate is dependent on the adequacy of the scattering model. Other parameters that correlate with k may be similarly affected. In particular, the temperature factors are positively correlated with k, the correlation being more pronounced the smaller the $\sin \theta/\lambda$ range of the data set, as for a small range the scale factor k and the temperature factor $\exp(-B \sin \theta^2/\lambda^2)$ affect the structure factors in identical ways.

As discussed in the following chapter, difference electron density maps, representing $\Delta\rho = \rho_{obs} - \rho_{calc}$, are based on the Fourier transform of the complex difference structure factors ΔF, defined as

$$\Delta F = F_{obs}(\mathbf{H})/k - F_{calc}(\mathbf{H}) \qquad (4.35)$$

The difference density is strongly dependent on k, especially where the electron density ρ is large, as it is near the nuclear positions. When k is overestimated, the difference density will be underestimated, and pronounced negative holes will occur in the nuclear regions.

The scale factor can be measured experimentally by a number of techniques, using either single crystal or powder samples (Stevens and Coppens 1975). Measurement for a number of crystals, including orthorhombic sulfur (S_8) and α-deutero-glycylglycine, and comparison with least-squares values, indicate that scale factors from spherical-atom refinements are subject to a positive bias of typically several percent. The magnitude of the bias depends on the composition of the material and on the high-order cut-off of the diffraction data. It is much reduced by limiting the refinement to the high-order data; even a cut-off of $\sin \theta / \lambda > 0.65 \text{ Å}^{-1}$ appears adequate in the cases studied. Improved scattering models, as discussed in chapter 3, are an effective means of reducing scale factor bias, though accurate experimental k values remain preferable.

That the positive bias in the scale factors correlates with an increase in thermal parameters is evident from comparison of X-ray and neutron results (Coppens 1968). The apparent increase in thermal parameters of some of the atoms may be interpreted as the response of the spherical-atom model to the existence of overlap density. Because of the positive correlation between the temperature parameters and k, this increase is accompanied by a positive bias in k.

4.3 Physical Constraints of the Electron Density

4.3.1 The Electroneutrality Constraint

Since a crystal is neutral, the total electron population must equal the sum of the nuclear charges of the constituent atoms, or

$$\sum_i (P_{i,\,\text{valence}} + P_{i,\,00}) = \sum_i Z_i \tag{4.36}$$

where Z_i is the nuclear charge of the ith atom.

Equivalently, starting from a neutral crystal,

$$\sum_i \sum_i q_{ij}\, \delta P_{ij} = 0 \tag{4.37a}$$

where δP_{ij} are the calculated shifts for population parameter j on atom i, and q_{ij} is the charge integrated over the corresponding density function. With the normalizations discussed in chapter 3, q_{ij} will be either one *for $l = 0$* or zero *for $l \neq 0$*. Thus, Eq. (4.37a) can be rewritten as

$$\sum_i (\delta P_{i,\,\text{valence}} + \delta P_{i,\,00}) = 0 \tag{4.37b}$$

Several ways to introduce the constraint into the refinement are discussed in the following sections.

4.3.1.1 The Hamilton Method

In a convenient method, due to Hamilton (1964), the Lagrangian multipliers representing the constraint are algebraically eliminated from the least-squares expressions. The linear constraints are defined as

$$\mathbf{Q}_{bm} \mathbf{X}_{m1} = \mathbf{Z}_{b1} \tag{4.38}$$

where values of $\mathbf{X}$ are the parameter shifts δx_i, $\mathbf{Q}_{bm}$ is a coefficient matrix, and the column vector $\mathbf{Z}_{b1}$ contains the values to which the linear combinations are constrained. Hamilton has shown that the best constrained estimates $\bar{\mathbf{X}}$ are obtained from the shifts without the constraints, using the equation

$$\bar{\mathbf{X}}^T = \hat{\mathbf{X}}^T + (\mathbf{Z}^T - \hat{\mathbf{X}}^T\mathbf{Q}^T)(\mathbf{QB}^{-1}\mathbf{Q}^T)^{-1}\mathbf{QB}^{-1} \tag{4.39}$$

In the case of the electroneutrality constraint, $\mathbf{Z}$ in Eq. (4.38) equals 0, so that

$$\bar{\mathbf{X}}^T = \hat{\mathbf{X}}^T - \hat{\mathbf{X}}^T\mathbf{Q}^T(\mathbf{QB}^{-1}\mathbf{Q}^T)^{-1}\mathbf{QB}^{-1} \tag{4.40}$$

where $\mathbf{Q}$ is a row vector with a 1 for every element representing the population of a normalized monopole, and 0 otherwise.

The variance–covariance matrix with the constraint is modified in a similar manner, according to the expression

$$\mathbf{M}_x = \mathbf{B}^{-1} - \mathbf{B}^{-1}\mathbf{Q}^T(\mathbf{QB}^{-1}\mathbf{Q}^T)^{-1}\mathbf{QB}^{-1} \tag{4.41}$$

Equations (4.40) and (4.41) are easily implemented in an existing least-squares program and give both the constrained and the unconstrained results in a single refinement cycle. However, the method fails if the unconstrained refinement corresponds to a singular matrix, as would be the case, for example, if all population parameters, including those of the core functions, were to be refined in addition to the scale factor k.

4.3.1.2 The Use of Independent Variables

The introduction of dependent variables is an often-used approach in handling constraints. It is, for example, quite common in crystals that one site is occupied by two different atoms, or one atom is distributed over two sites. In this case, a single population parameter P'_1 may be introduced, such that $P_1 = P'_1$, and $P_2 = 1 - P'_1$.

More generally, the shifts in the new set of independent variables $\mathbf{Y}$ are related to the shifts in the dependent variables $\mathbf{X}$ by (Raymond 1972)

$$\delta\mathbf{X} = \mathbf{J}\delta\mathbf{Y} \qquad \text{with } J_{ij} = \partial x_i/\partial y_j \tag{4.42}$$

The matrix $\mathbf{J}$ represents the constraints. If there are b constraints, $\mathbf{J}$ will be an $m \times (m - b)$ matrix, where m is the number of variables before the introduction of the constraint. The derivatives of the observations with respect to the independent parameters $\mathbf{Y}$ are obtained from

$$\frac{\partial\mathbf{F}}{\partial\mathbf{Y}} = \mathbf{J}^T\frac{\partial\mathbf{F}}{\partial\mathbf{X}} \tag{4.43}$$

For example, in a refinement of three population parameters, satisfying the constraint $\sum P_j = n_e$, P_1 and P_2 may be chosen as the set $\mathbf{Y}$. Equation (4.42) then becomes

$$\begin{pmatrix} \delta P_1 \\ \delta P_2 \\ \delta P_3 \end{pmatrix} = \begin{pmatrix} 1 & 0 \\ 0 & 1 \\ -1 & -1 \end{pmatrix} \begin{pmatrix} \delta P_1 \\ \delta P_2 \end{pmatrix} \tag{4.44}$$

The independent-variable method leads to a reduction in the size of the matrix $\mathbf{B}$ from $m \times m$ to $(m - b) \times (m - b)$.

4.3.1.3 The $F(000)$ Constraint

Electroneutrality may also be implemented by imposing the requirement that $F(000)$ equal the number of electrons in the unit cell. The equation $F(000) = n_e$ can be treated as an observation, with a weight sufficient to keep the crystal practically neutral, but sufficiently small such as not to dominate the least-squares treatment. This *slack constraint* (Pawley 1972) has been applied in electron density analysis by Hirshfeld (1977).

4.3.1.4 The Use of a Core Scale Factor

When the core populations are not varied, a somewhat different approach can be used, in which the scale factor k multiplies only the functions with fixed electron population, rather than all the electron functions (Stewart 1976). The observational equations are of the form

$$F_{calc}(\mathbf{H}) = kF_{core} + F_{valence} = \sum_i \left\{ kF_{i,\,core} + \sum_i P'_{ij} f_{ij,\,valence} \right\} \exp\left(2\pi i \mathbf{H} \cdot \mathbf{r}_i\right) T_i$$

(4.45)

The valence-shell populations are scaled after completion of the refinement by use of $P_{ij} = P'_{ij}/k$. If the procedure is successful, the sum of P_{ij} should be close to the total number of valence electrons. The difference between these two quantities is used as a test of the adequacy of the valence density functions.

4.3.2 The Hellmann–Feynman Constraint

The electrostatic Hellmann–Feynman theorem states that for an exact electron wave function, and also of the Hartree–Fock wave function, the total quantum-mechanical force on an atomic nucleus is the same as that exerted classically by the electron density and the other nuclei in the system (Feynman 1939, Levine 1983). The theorem thus implies that the forces on the nuclei are fully determined once the charge distribution is known. As the forces on the nuclei must vanish for a nuclear configuration which is in equilibrium, a constraint may be introduced in the X-ray refinement procedure to ensure that the Hellmann–Feynman force balance is obeyed (Schwarzenbach and Lewis 1982).

In a bond A—B between first-row atoms, the internuclear repulsion is electrostatically balanced by the buildup of charge density in the bond, that is due to either σ- or π-valence electrons (Hirshfeld and Rzotkiewicz 1974). But this charge migration is accompanied by a polarization of the σ-valence electrons within the 2s radial nodal surface, and, through exchange interaction between the σ electrons and the core, by a very sharp polarization of the core electron density in the regions very close to the nuclei. Such deformations correspond to density functions with large values of the orbital exponent coefficient ζ. Functions of this type are normally not included in the X-ray model, because limited resolution and thermal smearing hamper determination of their parameters. Since they are located very close to the nuclei, the occupancy of the polarization functions correlates strongly with the atomic coordinates. The Hellmann–Feynman force-balance constraint indirectly provides the missing information. Enforcing the constraint will adjust the parameters such as to maintain force balance.

The Hellmann–Feynman constraint has been applied successfully to the exocyclic fluorine, carbon, and nitrogen atoms in tetrafluoroterephthalonitrile (1,4-dicyano-2,3,5,6-tetrafluorobenzene). Charge balance is achieved without deterioration of the least-squares agreement factors, though the resulting changes in the density maps are very small (Hirshfeld 1984) (see chapter 5).

We note that the constraint applies to the static density. Its application therefore requires adequate deconvolution of thermal motion and electron density effects.

4.4 Joint Refinement of X-ray and Neutron Data

4.4.1 The Use of Complementary Information

The use of complementary data from different techniques can be a powerful tool to reduce parameter correlation in least-squares methods and to obtain the best estimates compatible with all available physical information.

The joint use of X-ray and neutron diffraction data is particularly expedient. Firstly, the interaction between the magnetic moments of neutrons and electrons is the basis for polarized-neutron diffraction, from which the unpaired spin density in a system can be derived. The diffraction of spin-polarized neutrons is an important technique, beyond the scope of this volume. Secondly, the interaction between neutrons and the atomic nuclei, which is the basis for structure determination by neutron diffraction, leads directly to information on the positions and mean-square vibrations of the nuclei.

Neutron diffraction is especially important for the location of hydrogen atoms, as the pronounced effect of bonding on the hydrogen-atom charge density leads to a systematic bias in the X-ray positions, as discussed in chapter 3. If the charge density in a hydrogen-containing molecule is to be studied, independent information on positions and thermal vibrations of the H atoms is invaluable.

The combination of data from different techniques raises several issues, as the experimental conditions may not have been identical, and the data no longer form a homogeneous set of observations. Several of these issues are discussed below.

4.4.2 Differences in Temperature Parameters

Early studies, which did not include many high-order reflections, revealed systematic differences between spherical-atom X-ray- and neutron-temperature factors (Coppens 1968). Though the spherical-atom approximation of the X-ray treatment is an important contributor to such discrepancies, differences in data-collection temperature (for studies at nonambient temperatures) and systematic errors due to other effects cannot be ignored. For instance, thermal diffuse scattering (TDS) is different for neutrons and X-rays. As the effect of TDS on the Bragg intensities can be mimicked by adjustment of the thermal parameters, systematic differences may occur. Furthermore, since neutron samples must be larger, because the beams are weaker, extinction and multiple diffraction tend to be more pronounced. On the other hand, absorption of neutrons within the crystal is generally lower, notwithstanding the larger specimen sizes.

Such effects will contribute to the discrepancies between X-ray (X) and neutron (N) temperature parameters, which have been found to exist even room-temperature studies, for which temperature ambiguities are minimal (Craven and McMullan 1979). The effect of TDS can be especially pronounced in room-temperature studies of often soft molecular crystals, and, if not recognized, can lead to an artificial enhancement of features in difference maps based on a combination of the two techniques (Scheringer et al. 1978) (see chapter 5).

To account for temperature factor differences, a temperature scale factor k_T multiplying the neutron temperature parameters may be introduced, as defined by the expression (Coppens et al. 1981)

$$F_{\text{neutron}}(\mathbf{H}) = \sum_{i}^{\substack{\text{all} \\ \text{atoms}}} \left\{ b_i \exp(2\pi i \mathbf{H} \cdot \mathbf{r}_i) \exp\left(-2\pi^2 k_T \sum_j \sum_k U_{ijk,N} h_j h_k a_j^* a_k^* \right) \right\}$$

(4.46)

This formulation is appropriate in the "high-temperature limit" at which temperature factors are proportional to absolute temperature (chapter 2). For most molecular crystals, this limit is reached even at liquid-nitrogen temperature.

Other alternatives exist, such as,

$$F_{\text{neutron}}(\mathbf{H}) = \sum_{i}^{\substack{\text{all} \\ \text{atoms}}} \left\{ b_i \exp(2\pi i \mathbf{H} \cdot \mathbf{r}_i) \exp\left[-2\pi^2 \sum_j \sum_k (U_{ijk,N} + \Delta U_{jk}) \cdot h_j h_k a_j^* a_k^* \right] \right\}$$

(4.47)

where ΔU_{jk}, common to all atoms, may be chosen either to follow the symmetry transformation of U_{ij} for symmetry-equivalent atoms, or to be similarly oriented for all atoms in the unit cell. The latter choice would correct for systematic errors due to incorrect allowance for absorption or anisotropic extinction in either of the two data sets.

Blessing (1995) has tested a number of alternative models for describing the discrepancy between X-ray and neutron thermal vibration parameters, using X-ray and neutron U_{ij} values from the IUCr oxalic acid project (section 12.1.3, Coppens 1984). The expressions $U_{ijk,X} = U_{ijk,N} + \Delta U_{ij}$ [i.e., Eq. (4.47)] and $U_{ijk,X} = k_T U_{ijk,N} + \Delta U_{ij}$ [a combination of Eqs. (4.46) and (4.47)] were found to be the most effective in providing reliable corrections. Blessing points out that the corrections calculated for the nonhydrogen atoms of a crystal can be used for adjusting neutron U_{ij}'s of the hydrogen atoms, to provide a set of fixed hydrogen thermal parameters for use in an X-ray analysis of the charge density.

4.4.3 Relative Weighting of the X-ray and Neutron Data

Because the error function $S = \sum_{i=1}^{n} w_i \Delta_i^2$ is a weighted sum of the squared discrepancies between observed and calculated values, the relative weighting of the two data sets affects their relative importance in the minimalization procedure. Since the weights should be based on the experimental uncertainties, they may be derived from the agreement between symmetry-related reflections in each of the data sets, and, counting statistical considerations, by using expressions described in the literature (McCandlish et al. 1975).

4.4.4 Estimate of the Goodness of Fit

The goodness of fit achieved in a refinement is defined by Eq. (4.27). Its evaluation for each of the subsets of data requires partitioning the unknowns between the two sets. Some, such as the scale factors, will be dependent on one subset only; other, such as heavy-atom positional parameters, will be determined by both sets of data. Information on the relative dependency is contained in the matrix $\mathbf{B}$, the elements of which are the sums of the contributions from the X-ray and neutron observations. We may write

$$\mathbf{B} = \mathbf{B}_X + \mathbf{B}_N \tag{4.48}$$

For each unknown u_j the fractional dependence φ can be based on the relative contributions of the two subsets to the diagonal elements of $\mathbf{B}$, according to (Coppens et al. 1981)

$$\varphi_{X,j} = \frac{\sum\limits_{X} w \left(\frac{\partial k E_c}{\partial u_j}\right)^2}{\sum\limits_{X} w \left(\frac{\partial k E_c}{\partial u_j}\right)^2 + \sum\limits_{N} w \left(\frac{\partial k E_c}{\partial u_j}\right)^2}$$

$$\varphi_{N,j} = \frac{\sum\limits_{N} w \left(\frac{\partial k E_c}{\partial u_j}\right)^2}{\sum\limits_{X} w \left(\frac{\partial k E_c}{\partial u_j}\right)^2 + \sum\limits_{N} w \left(\frac{\partial k E_c}{\partial u_j}\right)^2} \tag{4.49}$$

where E is either F or F^2 depending on the function minimized, and $\varphi_{X,j} + \varphi_{N,j} = 1$. In case a diagonal approximation to B is used in the least-squares refinement, φ_X and φ_N, as defined by Eq. (4.40), are inversely proportional to the squared standard deviations of the separate X-ray and neutron analyses. Some parameters will be entirely dependent on one of the two sets, in which case $\varphi = 1$; and for the other set, $\varphi = 0$.

For the case that $E = F(\mathbf{H})$, the relative weight of the two data sets in the analysis can be approximated by

$$W_X = n_X \langle (w h_j \hat{f}_i)^2 \rangle / \{n_X \langle (w h_j \bar{f}_i)^2 \rangle + n_N \langle w(h_j \bar{b}_i)^2 \rangle\} \tag{4.50a}$$

for the positional parameters u_j, and

$$W_X = n_X \langle (w h_j h_k \hat{f}_i)^2 \rangle / \{n_X \langle (w h_j h_k \bar{f}_i)^2 \rangle + n_N \langle w(h_j h_k \bar{b}_i)^2 \rangle\} \tag{4.50b}$$

for the vibrational parameters U_{jk}. The expressions for W_N are analogous. In Eq. (4.50), n_X and n_N represent the number of X-ray and neutron observations, respectively, and f and b are the thermally attenuated X-ray and neutron scattering

TABLE 4.3 Relative Dependence ($\varphi_X \cdot 100$) of Parameters in the Joint $X-N$ Refinement on the X-ray Data On α-Oxalic Acid Dihydrate (100 K). The Neutron Dependence φ_N is Given by $1 - \varphi_X$.

	x	y	z	U_{11}	U_{22}	U_{33}	U_{12}	U_{13}	U_{23}
C(1)	83	80	80	81	75	79	83	81	83
O(1)	91	90	89	89	86	88	90	89	89
O(2)	91	90	90	90	86	87	90	89	89
O(3)	91	88	90	90	87	88	90	90	89
H(1)	2	2	2	1	1	1	1	1	1
H(2)	4	4	3	2	3	2	2	3	2
H(3)	4	4	4	2	2	2	2	2	2

factors, respectively. Expressions (4.50a) and (4.50b) illustrate the effect of the number of reflections and their relative weighting, and the decreasing relative importance of the X-ray data in the high-angle region, where the neutron scattering factors dominate.

Finally, we obtain, for the goodness of fit of the X-ray data, in analogy to Eq. (4.27),

$$
\hat{\sigma}_X^2 = \frac{\displaystyle\sum_{i=1}^{n_X} w_{i,X}\,\Delta_{i,X}^2}{n_X - \displaystyle\sum_{j=1}^{m_X} \varphi_{X,j}}
\tag{4.51}
$$

and an analogous expression for the neutron data set.

4.4.4.1 An example

The joint refinement of low-temperature (≈ 100 K) X-ray and neutron data on oxalic acid dihydrate (Coppens et al. 1981) is an example of the combined use of different experimental techniques. The temperature scale factor according to Eq. (4.46) was found to be 0.892, indicating a lower temperature of the neutron experiment, which was performed on a sample in a cryostat, rather than in a cold gas stream. The main improvements compared with the X-ray-only refinement are in the hydrogen positional and thermal parameters, which are not properly reproduced in the X-ray-only refinement, even when the multipole model is used. The φ_X values for the positional and thermal parameters are listed in Table 4.3. As expected, the hydrogen parameters are dominated by the neutron observations.

5

Fourier Methods and Maximum Entropy Enhancement

Image formation in diffraction is no different from image formation in other branches of optics, and it obeys the same mathematical equations. However, the nonexistence of lenses for X-ray beams makes it necessary to use computational methods to achieve the Fourier transform of the diffraction pattern into the image. The phase information required for this process is, in general, not available from the diffraction experiment, even though progress has been made in deriving phases from multiple-beam effects. This is the phase problem, the paramount issue in crystal structure analysis, which also affects charge density analysis of noncentrosymmetric structures. For centrosymmetric space groups, the independent-atom model is a sufficiently close approximation to allow calculation of the signs for all but a few very weak reflections.

Images of the charge density are indispensable for qualitative understanding of chemical bonding, and play a central role in charge density analysis. In this chapter, we will discuss methods for imaging the experimental charge density, and define the functions used in the imaging procedure.

5.1 General Expressions

5.1.1 The Total Density

According to Eq. (1.22), the structure factor $F(\mathbf{H})$ is the Fourier transform of the electron density $\rho(\mathbf{r})$ in the crystallographic unit cell. The electron density $\rho(\mathbf{r})$ is then obtained by the inverse Fourier transformation, or

$$\rho(\mathbf{r}) = \int F(\mathbf{H}) \exp(-2\pi i \mathbf{H} \cdot \mathbf{r}) \, d\mathbf{H} \qquad (5.1)$$

in which $F(\mathbf{H})$ are the (complex) structure factors corrected for the anomalous scattering discussed in chapter 1.

Since $F(\mathbf{H})$ is defined at the discrete set of reciprocal lattice points $\mathbf{H}$, the integral in Eq. (5.1) can be replaced by a summation:

$$\rho(\mathbf{r}) = V^* \sum_{\mathbf{H}} F(\mathbf{H}) \exp(-2\pi i \mathbf{H} \cdot \mathbf{r}) \tag{5.2}$$

where V^*, the volume associated with a reciprocal lattice point, equals $1/V$, V being the unit-cell volume. Thus,

$$\rho(\mathbf{r}) = \frac{1}{V} \sum_{\mathbf{H}} F(\mathbf{H}) \exp(-2\pi i \mathbf{H} \cdot \mathbf{r}) \tag{5.3}$$

The electron density is a real function. The right-hand side of Eq. (5.3) must therefore be real also. This can be shown as follows. Writing

$$F(\mathbf{H}) = |F(\mathbf{H})| \exp i\varphi(\mathbf{H}) \equiv A(\mathbf{H}) + iB(\mathbf{H}) \tag{5.4}$$

where $\varphi(\mathbf{H})$ is the phase of the structure factor (see Fig. 5.9). Combining the contributions to the summation of $F(\mathbf{H})$ and $F(-\mathbf{H})$, using $A(\mathbf{H}) = A(-\mathbf{H})$ and $B(\mathbf{H}) = -B(-\mathbf{H})$, gives, after cancellation of terms,

$$\rho(\mathbf{r}) = \frac{2}{V} \sum_{1/2} \{A(\mathbf{H}) \cos(2\pi \mathbf{H} \cdot \mathbf{r}) + B(\mathbf{H}) \sin(2\pi \mathbf{H} \cdot \mathbf{r})\} \tag{5.5}$$

or, with

$$A(\mathbf{H}) = |F(\mathbf{H})| \cos \varphi \qquad \text{and} \qquad B(\mathbf{H}) = |F(\mathbf{H})| \sin \varphi \tag{5.6}$$

$$\rho(\mathbf{r}) = \frac{2}{V} \sum_{1/2} [|F(\mathbf{H})| \cos\{2\pi \mathbf{H} \cdot \mathbf{r} - \varphi(\mathbf{H})\}] \tag{5.7}$$

In other words, each structure factor contributes a plane wave to the total density with wavevector $\mathbf{H}$ and phase φ. As noted in the introductory paragraphs to this chapter, formation of the image, which is the density, requires knowledge of the phases of the structure factors. Once an approximation to the scattering density is known, $\varphi(\mathbf{H})$ may be calculated on the basis of this approximation, and an admittedly imperfect image of the structure can be obtained. At the same time, anomalous scattering can be corrected for, which can be done by subtracting the calculated contributions $\Delta A_{\text{calc}}^{\text{anomalous}}$ and $\Delta B_{\text{calc}}^{\text{anomalous}}$ from A and B, respectively, using the anomalous scattering factors f' and f'' discussed in section 1.3.

The period of the plane wave with amplitude $F(\mathbf{H})$, in the direction of the wavevector $\mathbf{H}$, equals $1/H$. The period is therefore shorter for higher-order reflections, which thus add resolution to the image. As more higher-order reflections are included in the summation, the resolution of the image improves, as illustrated in Fig. 5.1. The improvement is analogous to the increase in resolution in an optical image obtained with shorter-wavelength radiation.

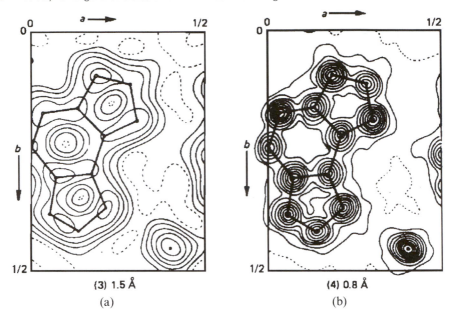

FIG. 5.1 Increase in resolution as a function of the $(\sin\theta/\lambda)_{max}$ of the data in the Fourier summation. Section containing a condensed ring system: (a) $\sin\theta/\lambda < 0.333$ Å^{-1}; (b) $\sin\theta/\lambda < 1$ Å^{-1}. *Source*: Glusker and Trueblood (1985)

5.1.2 The Residual Density

The difference $\Delta\rho(\mathbf{r})$ between the total electron density $\rho(\mathbf{r})$ and a reference density $\rho_{ref}(\mathbf{r})$, is a measure for the adequacy of the reference density in representing the system. *Difference densities $\Delta\rho(\mathbf{r})$ are obtained by Fourier summation in which the coefficients $\Delta\mathbf{F}$ are equal to the difference between the observed and calculated structure factors.* If k is the scale factor, as defined in chapter 4, the difference structure factor $\Delta\mathbf{F}$ is given by

$$\Delta\mathbf{F} = \mathbf{F}_{obs}(\mathbf{H})/k - \mathbf{F}_{calc}(\mathbf{H}) \tag{5.8}$$

while the difference density is obtained by the Fourier transformation

$$\Delta\rho(\mathbf{r}) = \rho_{obs}(\mathbf{r}) - \rho_{calc}(\mathbf{r}) = \frac{1}{V}\sum_{\mathbf{H}}\Delta\mathbf{F}\exp\left(-2\pi\mathbf{H}\cdot\mathbf{r}\right) \tag{5.9a}$$

We have introduced the boldface notation to underline that $\Delta\mathbf{F}$ is a vector in the complex plane (see Fig. 5.9), because both F_{obs} and F_{calc} are, in general, complex quantities, as is evident from Eq. (5.6). The phase angles φ of the two vectors are not necessarily the same, as is further discussed in section 5.2.5. In a different notation we may write, like Eq. (5.5),

$$\Delta\rho(\mathbf{r}) = \frac{2}{V}\left\{\sum_{1/2}(A_{obs} - A_{calc})\cos 2\pi\mathbf{H}\cdot\mathbf{r} + \sum_{1/2}(B_{obs} - B_{calc})\sin 2\pi\mathbf{H}\cdot\mathbf{r}\right\} \tag{5.9b}$$

When the model used for F_{calc} is that obtained by least-squares refinement of the observed structure factors, and the phases of F_{calc} are assigned to the observations, the map obtained with Eq. (5.9) is referred to as a *residual density map*. The residual density is a much-used tool in structure analysis. Its features are a measure for the shortcomings of the least-squares minimization, and the functions which constitute the least-squares model for the scattering density.

5.1.3 Least-Squares Minimization and the Residual Density

The relation between the least-squares minimization and the residual density follows from the Fourier convolution theorem (Arfken 1970). It states that the Fourier transform of a convolution is the product of the Fourier transforms of the individual functions: $\hat{F}(f * g) = \hat{F}(f)\hat{F}(g)$. If $G(y)$ is the Fourier transform of $g(x)$:

$$G(y) = \frac{1}{\sqrt{2\pi}} \int g(x) \exp{(ixy)} \, dx \qquad (5.10)$$

and $F(y)$ is the Fourier transform of $f(x)$, then, according to the convolution theorem,

$$(f * g) = \widehat{F^{-1}(\hat{F}(f)\hat{F}(g))} \quad \text{or} \quad \int f(t)g^*(t - x) \, dt = \int F(y)G^*(y) \exp{(-ixy)} \, dy \qquad (5.11)$$

For $f(t) = g(t)$, and the special case of $x = 0$, this reduces to *Parseval's theorem*:

$$\int_{-\infty}^{\infty} |f(x)|^2 \, dx = \int_{-\infty}^{\infty} |F(y)|^2 \, dy \qquad (5.12)$$

which states that the space integrals over the square of a function $f(x)$ and over the square of its Fourier transform $F(y)$ are equal.

Since $\Delta\rho$ is the Fourier transform of ΔF, Eq. (5.12) implies that minimization of $\int (\rho_{obs} - \rho_{calc})^2 \, d\mathbf{r}$ and of $\int (F_{obs} - F_{calc})^2 \, d\mathbf{S}$ are equivalent. Thus, the structure factor least-squares method also minimizes the features in the residual density. Since the least-squares method minimizes the sum of the squares of the discrepancies in reciprocal space, it also minimizes the features in the difference density. The flatness of residual maps, which in the past was erroneously interpreted as the insensitivity of X-ray scattering to bonding effects, is an intrinsic result of the least-squares technique. If an inadequate model is used, the resulting parameters will be biased such as to produce a flat $\Delta\rho(\mathbf{r})$.

It is of importance that expression (5.12) holds even when $f(x)$ is known only in part of space, as is the case in a crystallography experiment at finite resolution determined by $\mathbf{H}_{max}$. Using the Fourier convolution theorem, we may write (Dunitz and Seiler 1973)

$$\frac{2}{V} \sum_{1/2} (\Delta F)^2 \cos{2\pi\mathbf{H} \cdot \mathbf{u}} = \int \Delta\rho(\mathbf{r})\Delta\rho(\mathbf{r} - \mathbf{u}) \, d\mathbf{r} \equiv D(\mathbf{u}) \qquad (5.13)$$

On the left is the Fourier transform of the even function $(\Delta F)^2$; on the right is the autoconvolution of the Fourier transform of ΔF. The term $D(\mathbf{u})$ is the Patterson function of the difference density of Eq. (5.9).

At $u = 0$, Eq. (5.13) becomes

$$D(0) = \frac{2}{V} \sum_{1/2} (\Delta F)^2 = \int \Delta\rho(\mathbf{r})^2 \, d\mathbf{r} \qquad (5.14)$$

This result is equivalent to Eq. (5.12), except that the left-hand side is no longer an integral over all space, but a summation up to the limit of resolution.

The conclusion on the equivalence of direct-space and reciprocal-space minimization is not completely flawless, because weights are assigned to the observations in the least-squares refinement, so a weighted difference density is minimized.

Dunitz and Seiler (1973) have used the equivalence to modify least-squares weighting, such as to emphasize the fit near the density peak positions, in order to obtain parameters less biased by bonding effects. The resulting weights emphasize high-order reflections, similar to the higher-order refinement method, but with a smoothly varying cut-off rather than a sharp $\sin \theta/\lambda$ limit.

5.2 Deformation Densities

5.2.1 Definition of the Deformation Density

As discussed in the previous section, a residual density calculated after least-squares refinement will have minimal features. This is confirmed by experience (Dawson 1964, O'Connell et al. 1966, Ruysink and Vos 1974). Least-biased structural parameters are needed if the adequacy of a charge density model is to be investigated. Such parameters can be obtained by neutron diffraction, from high-order X-ray data, or by using the modified scattering models discussed in chapter 3.

The *deformation density* is defined as the difference between the total density and the density calculated with a reference model based on unbiased positional and thermal parameters. The deformation density is obtained by Fourier transform, like the residual density [Eq. (5.9)], but with F_{calc} from the reference state with which the experimental density is to be compared.

When observed structure factors are used, the thermally averaged deformation density, often labeled the *dynamic deformation density*, is obtained. An attractive alternative is to replace the observed structure factors in Eq. (5.8) by those calculated with the multipole model. The resulting *dynamic model deformation map* is model dependent, but any noise not fitted by the multipole functions will be eliminated. It is also possible to plot the model density directly using the model functions and the experimental charge density parameters. In that case, thermal motion can be eliminated (subject to the approximations of the thermal motion formalism!), and an image of the *static model deformation density* is obtained, as discussed further in section 5.2.4.

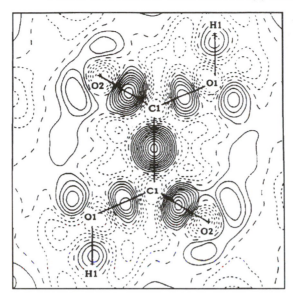

FIG. 5.2 Standard deformation density in the molecular plane of the oxalic acid molecule at 15 K, calculated with structural parameters based on the high-order data. Contours are at 0.05 Å^{-3}. Zero and negative lines are dashed ($\sin \theta/\lambda \leq 0.71$ Å^{-1}). *Source*: Zobel et al. (1992).

5.2.2 The Choice of a Model for the Reference State

5.2.2.1 The Standard Deformation Density

A common reference density, first used by Roux and Daudel (1955), is the superposition of spherical ground-state atoms, centered at the nuclear positions. It is referred to as the *promolecule density*, or simply the *promolecule*, as it represents the ensemble of randomly oriented, independent atoms prior to interatomic bonding. It is a hypothetical entity that violates the Pauli exclusion principle. Nevertheless, the promolecule is electrostatically binding; if only the electrostatic interactions would exist, the promolecule would be stable (Hirshfeld and Rzotkiewicz 1974). The difference density calculated with the promolecule reference state is commonly called the deformation density, or the *standard deformation density*. It is the difference between the total density and the density corresponding to the sum of the spherical ground-state atoms located at the positions $\mathbf{R}_i$:

$$\Delta\rho(\mathbf{r}) = \rho(\mathbf{r}) - \rho_{\mathrm{pro}}(\mathbf{r}) = \rho(\mathbf{r}) - \sum_{\substack{\text{all} \\ \text{atoms}}} \rho(\mathbf{R}_i) \tag{5.15}$$

An example of a standard deformation density, for oxalic acid dihydrate, obtained at limited resolution using parameters from a high-order refinement, is shown in Fig. 5.2. Oxalic acid dihydrate has been the subject of an extensive study aimed at calibrating the techniques used in different laboratories. The map shows density in the bonds and in the lone-pair regions near the oxygen atoms of the molecule.

While the presence of density accumulation in the bonds in an atom-deformation map is indicative of covalent bonding, the opposite statement cannot be made. The absence of a bond peak in the deformation density does not imply the absence of covalent contributions to bonding, because, for elements with more than half-filled shells, the neutral spherical atoms which are subtracted have more than one electron per orbital. This is most easily illustrated through an example. When a spherical oxygen atom with the configuration $(1s)^2(2s)^2(2p)^4$ is subtracted from a molecular density, 1.333 electrons are removed per valence orbital. The extra one-third electron subtracted out in the bond region more than compensates for the accumulation of density due to bond formation, and therefore causes a depletion of density in the bond region relative to the spherical-atom reference state. This explains the lack of density in the O—O bond in hydrogen peroxide as observed by Savariault and Lehmann (1980), and the appearance of the bonds in the deformation density of 1,2,7,8-tetraaza-4,5,10,11-tetraoxatricyclo[6.4.1.1^{2,7}] tetradecane (Dunitz and Seiler 1983). Sections through the C—C, N—N, and O—O bonds in the latter molecule (Fig. 5.3) shows the decrease in bond peak heights towards the right of the periodic table, in agreement with the oxygen-atom example given above.

The X-ray deformation maps of tetraazaoxatricyclotetradecane are in good agreement with ab-initio results (Kunze and Hall 1987). Analysis of the theoretical results confirms that the deformation features can be understood by viewing bond formation as a two-step process: atom preparation followed by bond formation. In the standard deformation density, the effects of atomic orientation, promotion to an atomic bonding state, hybridization, charge transfer, and covalent bond formation are superimposed. In other words, its features represent both the effect of the orientation of atoms with a nonspherical ground state, such as C (^3P), O (^3P), or F (^2P), and the effect of bond formation. Some of these are opposite, leading to complications in the interpretation of the maps. For this reason, alternative reference states can be useful. It is often opportune to use different deformation densities in conjunction, rather than rely on a single reference state.

5.2.2.2 Oriented Atomic Reference States

The use of alternate reference states makes it possible to separate the steps which lead from the nonoriented spherical atom to the atom in a molecule, and gives a better understanding of the process of bond formation. In early studies on diatomic molecules, Bader and coworkers used an atomic reference state for first- and second-row atoms, such as N, O, F, S, and Cl, with a single vacancy in the $p\sigma$ bonding orbital, and the remaining p electrons averaged over the $p\pi$ orbitals (Bader et al. 1967, Bader and Bandrauk 1968, Cade et al. 1969). For fluorine, this corresponds to a $(1s^2 2s^2 2p_x^2 2p_y^2 2p_z^1)$ reference state, with the z axis along the bond direction. Hall and coworkers further developed the partitioning and introduced a reference state corresponding to a hybrid atom, defined by the (truncated) localized molecular orbital (LMO) from an SCF calculation (Kunze and Hall 1986). The bonding hybrid in this LMO is described by the configuration $sp^{9.3}$,

FIG. 5.3 Selected sections through the deformation density of 4,5,10,11-tetraazaoxatricyclo-[6.4.1.1^{2,7}] tetradecane. Contours are at 0.075 eÅ^{-3}. Full line for positive, dashed line for negative and dotted line for zero density. Top: C—N—C; middle: C—N—N; bottom: O—O—C. *Source*: Dunitz and Seiler (1983).

which represents the hybrid atomic orbital $10.3^{-1/2} (\phi_s + \sqrt{9.3}\phi_p)$. The orthogonal, fully occupied σ hybrid then has some p character and points away from the bonded neighboring atom. Since the bonding hybrid is singly occupied in the reference state, which is subtracted, the deformation density has a stronger buildup of density in the bond region. The difference between the three functions is

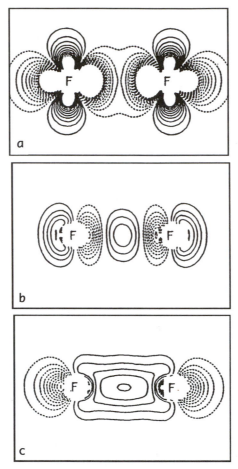

FIG. 5.4 Deformation densities for the F_2 molecule. (a) Total deformation density: molecule minus spherical atoms. (b) Oriented atom deformation density: molecule minus oriented atoms. (c) Hybrid deformation density: molecule minus hybridized atoms. *Source*: Kunze and Hall (1986).

illustrated for F_2 in Fig. 5.4. It shows a depleted bonding region in the total deformation density, some density in the oriented ground-state deformation density, and a large extended bond region in the hybrid-atom deformation density.

A further subdivision of the steps involved in bond formation has been made by Low and Hall (1990). Though their theoretical analysis cannot easily be applied to experimental densities, it provides insight into the effects that contribute to the features in an experimental deformation density. A distinction is made between *hybridization effects*, including atomic orientation, polarization, electron promotion, and orbital hybridization; *delocalization effects*, including charge transfer between atoms; and the effect of *constructive interference*, which is the energy-stabilizing combination of two atomic bonding orbitals, as in $1s_A + 1s_B$ of Eq. (3.11). Subtraction of hybridized atoms from the electron

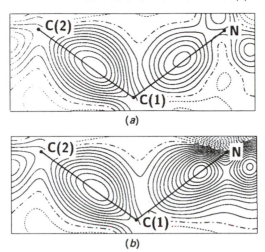

FIG. 5.5 Model deformation density in a plane containing a C—N and C—C bond in the ligand of *meso*-[Co(hexaazacyclooctadecane)]Cl$_3$ at 106 K. (a) Independent-atom model reference state. (b) Reference state of oriented-atom model with one electron in the N-sp^3 lobe pointing toward the carbon atom. ($\sin \theta/\lambda)_{max} = 1.3$ Å^{-1}. Contour intervals are at 0.05 eÅ^{-3}. *Source*: Morooka et al. (1992).

densities of first-row hydrides produces maps in which the values of the highest contour correlate with the energies of the X–H bonds. Such a relation between the deformation density features and the strength of a bond cannot be achieved with standard deformation densities.

In the *chemical deformation densities* introduced by Schwarz and collaborators, the density and the orientation of the atoms is quantitatively defined by variation of the atomic orientation and orbital population such as to minimize the space integral over the squared deformation density (Schwarz et al. 1989, Mensching et al. 1989).

For HF, the F atom in the oriented reference state of the chemical deformation density has 1.414 e (rather than $5/3 = 1.67$ e in the spherical atom, or 1 e in the oriented atom) in the $p\sigma$ orbital, and 1.793 e (rather than 1.67 e in the spherical atom, or 2 e in the oriented atom) in each of the $p\pi$ orbitals. As in Fig. 5.4(b) and (c), the trough along the bond axis of the total deformation density disappears, and the overlap density becomes evident.

An experimental example is shown in Fig. 5.5. The peak height in the C—N bond increases from 0.32 eÅ^{-3} to 0.68 eÅ^{-3} by the introduction of a prepared reference state on the nitrogen atom.

5.2.2.3 The Fragment Deformation Density

If we are particularly interested in the nature of the bonding between the fragments of a molecular complex, a *fragment deformation* density may be calculated by subtracting fragment densities from the total distribution. In the case of a transition metal complex, the fragment may be a metal atom plus ligand, or just the density

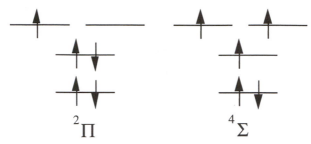

FIG. 5.6 The ground state (left) and the low-lying excited state (right) of the C—H fragment.

of the isolated ligand molecule. The former case is demonstrated by the subtraction of three $Co(CO)_3$ fragments from the density of the triangular metal complex $Co_3CCl(CO)_9$ (nonacarbonyl-μ_3-chloromethylidene-triangulocobalt) (Hall 1982). In this complex, the cobalt atoms are "terminated" by CO ligands, and the Co_3 triangle is "capped" by the CCl ligand, oriented perpendicular to the triangle plane. The bonding of the metal to the carbonyl groups removes density from the regions of the metal–metal bonds. As a result, the metal–metal overlap density is masked in the standard deformation density, but appears in the fragment deformation density. Other theoretical fragment deformation densities reported include those of several hydrogen-bonded dimers (Yamabe and Morokuma 1975, Hermansson 1985) and organometallic compounds (Heijser et al. 1980, Hall 1986).

An experimental example is the subtraction of thermally smeared theoretical CO molecular densities from the experimental density of chromium hexacarbonyl $Cr(CO)_6$. The map shows a decrease in σ-density and an increase in π-density, in agreement with the generally accepted σ-donation, π-back-donation description of metal–ligand bonding (Rees and Mitschler 1976). In a second example, the fragment deformation density was used differently. In the complex $Co_3CH(CO)_9$ molecule, the CH radial has a $^2\Pi$ ground state, and a low-lying $^4\Sigma$ excited state (Coppens 1985). The two states differ by promotion of an electron from a carbon σ to a carbon p_π orbital (Fig. 5.6). The standard deformation map reproduced in Fig. 5.7(a), as expected, shows density in the C—H bond. The fragment deformation densities obtained by subtraction of the two alternate ligand densities have peaks in the π and σ regions of the ligand C atom, respectively [Fig. 5.7(b) and (c)]. The difference is a result of the promotion of an electron from the σ to the p_π orbital in the $^4\Sigma$ reference state, subtracted to obtain Fig. 5.7(c). Since different features remain upon subtraction of the two different states, the density in the bonded C—H ligand must be intermediate between the densities of the $^2\Pi$ and $^4\Sigma$ states of the isolated C—H radical.

The appearance of a deformation density depends crucially on the definition of the reference state used in its calculation. This has occasionally been interpreted as an ambiguity and an argument against the use of the deformation density as an analytical tool. More precisely, a deformation density is meaningful only in terms of its reference state, which must be taken into account in the interpretation. The different deformation functions are complementary, and when used properly, they provide detailed understanding of the steps in the bond formation process.

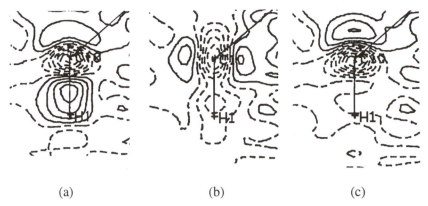

(a)   (b)   (c)

FIG. 5.7 Deformation density sections through the C—H ligand in $Co_3CH(CO)_9$. Contours at 0.1 e$Å^{-1}$. (a) Standard deformation density. (b) Fragment deformation density subtracting $^2\Pi$ density. (c) Fragment deformation density subtracting $^4\Sigma$ density. *Source*: Coppens (1985)

5.2.3 The Choice of Structural Parameters in the Calculation of the Deformation Density

5.2.3.1 Deformation Densities

Let us suppose that the atomic positional parameters $\mathbf{r}_i$ and the thermal smearing factors T_i have been obtained by neutron diffraction. We can then calculate the X-ray scattering of the procrystal density with spherical-atom X-ray scattering factors, using

$$F_{calc,N} = \sum_{atoms} f_{i,X} \exp{(2\pi i \mathbf{H} \cdot \mathbf{r}_{i,N})} T_{i,N} \qquad (5.16)$$

where the symbol N indicates that neutron parameters are used. The $X-N$ deformation density is then, in analogy to Eq. (5.9a), given by

$$\rho_{deformation}^{X-N}(\mathbf{r}) = \frac{1}{V} \sum_{\mathbf{H}} (\mathbf{F}_{obs,X} - \mathbf{F}_{calc,N}) \exp{(-2\pi i \mathbf{H} \cdot \mathbf{r})} \qquad (5.17)$$

A first map of this kind is shown in Fig. 5.8. The $X-N$ deformation density is thermally averaged, and has limited resolution as the summation in Eq. (5.17) is truncated at the limit of the experimental observations. Since both F_{obs} and F_{calc} are complex for an acentric structure, the structure factor phases are continuously variable, and must be considered. Expression (5.17) can be rewritten as

$$\rho_{deformation}^{X-N}(\mathbf{r}) = \frac{1}{V} \sum_{\mathbf{H}} \{|F_{obs,X}| \cos{[2\pi \mathbf{H} \cdot \mathbf{r} - \varphi_X(\mathbf{H})]}$$

$$- |F_{calc,N}| \cos{[2\pi \mathbf{H} \cdot \mathbf{r} - \varphi_N(\mathbf{H})]}\} \qquad (5.18)$$

In centrosymmetric crystals, with very few exceptions, $\varphi_X(\mathbf{H}) = \varphi_N(\mathbf{H})$ ($=0$

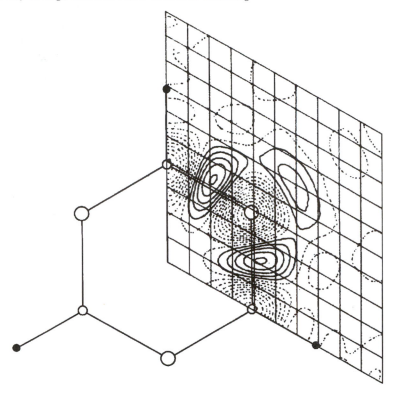

○ NITROGEN
○ CARBON
● HYDROGEN

FIG. 5.8 Room-temperature $X-N$ deformation density in the plane of the molecule of s-triazine, showing features in the bond and lone-pair regions. Contours at 0.05 eÅ^{-1}. *Source*: Coppens (1967).

or π) so that Eq. (5.18) reduces to

$$\rho^{X-N}_{\text{deformation}}(\mathbf{r}) = \frac{1}{V} \sum_{\mathbf{H}} \left(|F_{\text{obs},X}| \frac{|F_{\text{calc},N}|}{F_{\text{calc},N}} - F_{\text{calc},N} \right) \cos 2\pi \mathbf{H} \cdot \mathbf{r} \qquad (5.19)$$

which is readily evaluated. But in acentric crystals, $\varphi_N(\mathbf{H}) \neq \varphi_X(\mathbf{H})$. Thus, the phase of $(\mathbf{F}_{\text{obs},X} - \mathbf{F}_{\text{calc},N})$ in Eq. (5.17) will, in general, not be equal to $\varphi_N(\mathbf{H})$, as illustrated in Fig. 5.9.

In support of this conclusion, Hanson et al. (1973) found, for a large data set on sucrose (with $\sin \theta / \lambda < 0.81$ Å^{-1}), a mean difference between the phases φ_X from a spherical-atom refinement and φ_N from the spherical-atom calculation with neutron parameters of $1.8°$. A better approximation to the "true" phases is

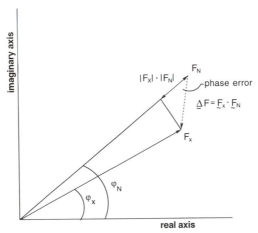

FIG. 5.9 Phase angles in an acentric $X-N$ analysis: φ_N is the phase angle as calculated with spherical-atom form factors and neutron positional and thermal parameters; φ_X is the unknown phase of the X-ray structure factors which must be estimated for the calculation of the vector ΔF. Use of $|F_X| - |F_N|$ introduces a large phase error. *Source*: Coppens (1974).

obtained with the aspherical atom formalisms, which were not fully developed at the time of the sucrose analysis. The phase contribution to the deformation density is analyzed in more detail in section 5.2.6.

The $X-N$ technique is sensitive to systematic errors in either data set. As discussed in chapter 4, thermal parameters from X-ray and neutron diffraction frequently differ by more than can be accounted for by inadequacies in the X-ray scattering model. In particular, in room-temperature studies of molecular crystals, differences in thermal diffuse scattering can lead to artificial discrepancies between the X-ray and neutron temperature parameters. Since the neutron parameters tend to be systematically lower, lack of correction for the effect leads to sharper atoms being subtracted, and therefore to larger holes at the atoms, but increases in peak height elsewhere in the $X-N$ deformation maps (Scheringer et al. 1978).

The $X-N$ deformation densities are important for the study of the charge density distribution in and around hydrogen atoms. Without the extra effort required for a neutron experiment, assumptions on the hydrogen atom location and vibrations must be made which introduce a considerable uncertainty in the results.

5.2.3.2 $X-X$ Deformation Densities

As discussed in chapter 3, the valence electrons scatter mainly in the low-order region. Consequently, refinement of high-order data, first proposed by Jeffrey and Cruickshank (1953), yields parameters less biased by bonding effects. The $X-X$ deformation density is calculated with the high-order X-ray parameters, and is defined as

$$\rho^{X-X}_{\text{deformation}}(\mathbf{r}) = \frac{1}{V} \sum_{\mathbf{H}} (F_{\text{obs},X} - F_{\text{calc},X\text{ high order}}) \exp(-2\pi i \mathbf{H} \cdot \mathbf{r}) \qquad (5.20)$$

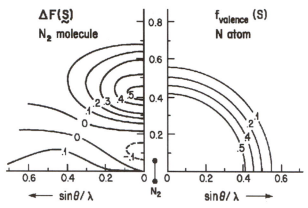

FIG. 5.10 Transform of the theoretical *deformation* density of the nitrogen molecule (oriented with its axis along the vertical direction) (Feil 1977), compared with the *valence* density scattering of the spherical nitrogen atom.

What is a proper lower cut-off for a high-order refinement? Assuming the frozen-core approximation to be valid, the answer is dependent on the persistence of valence scattering with increasing values of H. Examination of spherical-atom scattering factors would suggest that beyond $\sin \theta / \lambda = 0.6 \text{ Å}^{-1}$, valence scattering is insignificant. But such a conclusion ignores the effect of bonding, which tends to concentrate the density in certain regions of space and thus leads to scattering at higher angles, as illustrated in Fig. 5.10 for the nitrogen molecule. Further information is obtained by examination of the height of the density features in deformation maps upon inclusion of additional high-order reflections in the Fourier summation. Figure 5.11 shows the variation with $(\sin \theta / \lambda)_{max}$ of the average peak heights in the bond- and lone-pair regions of an $X-N$ deformation map of the *p*-nitropyridine-*N*-oxide molecule. Very-high-order reflections are available in this 30 K study, in which the effect of thermal motion is minimized, though not fully eliminated. The peak heights in the bond regions level off at about 1.05 Å^{-1}, but for the lone-pair peaks contributions persist to much higher angles. The extrapolated values marked on the y axis of Fig. 5.11 are obtained with the assumption of isotropic Gaussian shape of the bond- and lone-pair peaks.

The result implies that $X-X$ maps will systematically underestimate lone-pair peak heights, especially when the lower cut-off in the high-order X-ray refinement is less than 0.75 Å^{-1}. The separation of structural and electronic effects in $X-X$ maps is therefore incomplete. On the positive side, the exclusive use of X-ray data has the advantage over the $X-N$ technique that systematic errors tend to cancel when only one data set is used. A third alternative, the use of structural parameters from a charge density refinement of the X-ray data, avoids the arbitrary cut-off of the $X-X$ method, though it remains dependent on the adequacy of the scattering model. For acentric structures, introduction of the aspherical atom least-squares technique has the great advantage of providing an improved estimate of the phases of the structure factor amplitudes.

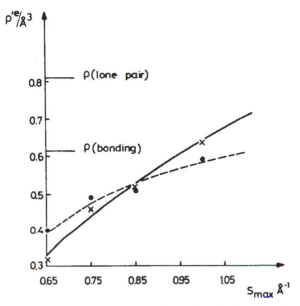

FIG. 5.11 Average peak heights in the $X-N$ deformation density of p-nitropyridine-N-oxide in the bonding and lone-pair regions as a function of the cut-off value S_{max} in the Fourier summation ($S = \sin \theta/\lambda$). Circles are average values for the maxima in the C—C and C—N bonds; crosses are average values in the lone-pair regions. The extrapolated limits for infinite resolution $\rho(0)$ are indicated, as well as the line describing the best least-squares fit. (*Source:* Coppens and Lehmann 1976, Lehmann and Coppens 1977).

5.2.4 Combining Fourier and Least-Squares Methods: The Model Deformation Density

The implications of a charge density least-squares refinement can be visualized by calculation of the deformation density corresponding to the least-squares fitted model.

The *dynamic model deformation density* is defined as

$$\Delta\rho_{model}(\mathbf{r}) = \rho_{model}(\mathbf{r}) - \rho_{reference}(\mathbf{r}) \tag{5.21}$$

in which, in analogy to Eq. (5.3), the total model density is calculated with

$$\rho_{model}(\mathbf{r}) = \frac{1}{V}\sum_{\mathbf{H}} F_{calc,\,model}(\mathbf{H}) \exp(-2\pi i \mathbf{H} \cdot \mathbf{r}) \tag{5.22}$$

and the analogous expression with $F_{calc,\,reference}$ is used for $\rho_{reference}$. When the summation is over all reciprocal lattice points within the experimental resolution limit, and the calculated structure factors include the effect of thermal motion, $\Delta\rho_{model}(\mathbf{r})$ as defined by Eq. (5.21) will be both thermally averaged and resolution limited.

Alternatively, the model functions may be plotted directly, in which case a static density is obtained, provided thermal effects and chemical bonding have been successfully deconvoluted. The image now is at infinite resolution; however, detail beyond the resolution of the experiment will be that of the model functions of the least-squares fitting.

The *static model deformation density* corresponding to the multipole refinement results is given by

$$\Delta\rho_{model}(\mathbf{r}) = \sum_{i}^{\text{all atoms}} \left\{ P_{i,c}\rho_{core}(r) + P_{i,v}\kappa^3\rho_{valence}(\kappa_i r) \right.$$

$$\left. + \sum_{l=0}^{l_{max}} \kappa_i'^3 R_{i,l}(\kappa_i' r) \sum_{m=0}^{l} P_{i,lm\pm} d_{lm\pm}(\mathbf{r}/r) \right\} - \rho_{reference}(\mathbf{r}) \quad (5.23)$$

Figure 5.12 shows both the dynamic and the static model deformation densities in the plane of the oxalic acid molecule, based on the data set also used for Fig. 5.2. The increase in peak height, due to higher resolution, and reduction in background noise relative to the earlier maps is evident. The model acts as a noise filter because the noise is generally not fitted by the model functions during the minimalization procedure.

Static deformation density maps can be compared directly with theoretical deformation densities. For tetrafluoroterephthalonitrile (1,4-dicyano-2,3,5,6-tetra-fluorobenzene) (Fig. 5.13), a comparison has been made between the results of a density-functional calculation (see chapter 9 for a discussion of the density-functional method), and a model density based on 98 K data with a resolution of $(\sin\theta/\lambda)_{max} = 1.15$ Å^{-1} (Hirshfeld 1992). The only significant discrepancy is in the region of the lone pairs of the fluorine and nitrogen atoms, where the model functions are clearly inadequate to represent the very sharp features of the density distribution.

Hirshfeld (1984) found the electrostatic charge balance at the F nuclei, based on the experimental deformation density, to be several times more repulsive (i.e., anti-bonding) than that of the promolecule. Very sharp dipolar functions at the exocyclic C, N, and F atoms, oriented along the local bonds, were introduced in a new refinement in which the coefficients of the sharp functions were constrained to satisfy the electrostatic Hellmann–Feynman theorem (chapter 4). The electro-static imbalance was corrected with negligible changes in the other parameters of the structure. The model deformation maps were virtually unaffected, except for the innermost contour around the nuclear sites.

Some minor discrepancies between theory and experiment on tetra-fluoroterephthalonitrile remain to be resolved. The peak densities in the bonds are slightly but systematically lower in the theoretical than in the ex-perimental maps. Analysis of the second moments of the pseudoatoms from the Hirshfeld space partitioning (chapter 6) indicate a greater contraction into the molecular plane in the theoretical than in the experimental study. Whether such discrepancies are artifacts of the refinement model, the result of inter-molecular interactions, or have another origin, is a question of considerable interest.

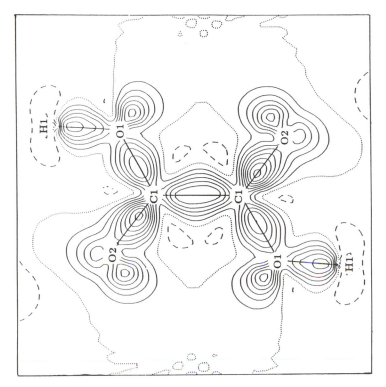

(b)

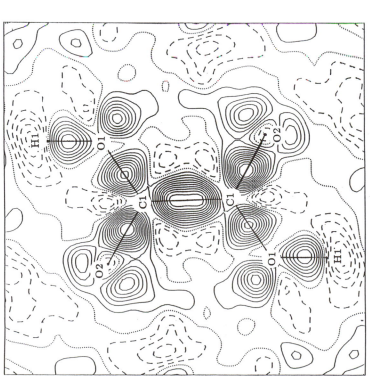

(a)

FIG. 5.12 Standard model deformation densities in the plane of the oxalic acid molecule at 15 K. (a) Dynamic model density ($\sin \theta / \lambda \leq 1.08 \text{ Å}^{-1}$). Contours at 0.05 eÅ^{-3}; zero level dotted lines; negative contours dashed lines. (b) Static model density. Contours at 0.10 eÅ^{-3}. *Source:* Zobel et al. (1992)

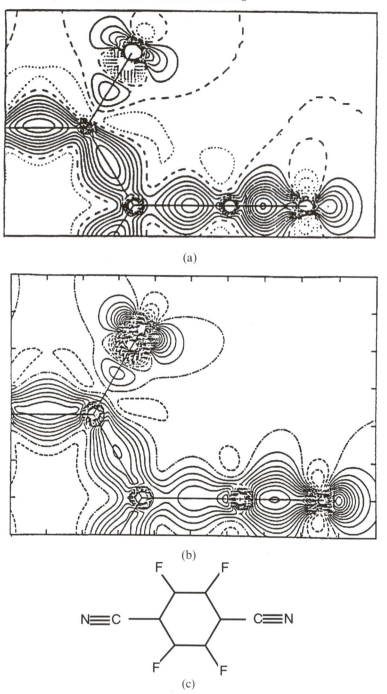

(a)

(b)

(c)

FIG. 5.13 Standard deformation density of tetrafluoroterephthalonitrile in the mean molecular plane. Contour interval is 0.1 eÅ^{-3}, terminated at 1.5 eÅ^{-3}. (a) Static model density from multipole refinement. (b) Model from density functional calculation. (c) Molecular diagram. *Source*: Hirshfeld (1984).

5.2.5 Phase Contributions to the Deformation Density

The true phases of the structure factors will, in general, be different from the phases calculated with the independent-atom model. In centrosymmetric structures, with phases restricted to 0 or π, only very few weak reflections are affected. In acentric structures, only the reflections of centrosymmetric projections, such as the $hk0$, $h0l$, and $0kl$ reflections in the space group $P2_12_12_1$, are invariant.

The effect of phase differences on dynamic deformation density may be estimated as follows (Souhassou et al. 1991). The amplitudes of the deformation density Fourier series may be written as (Fig. 5.9)

$$\Delta|F| = |F_{obs}| - |F_{ref}| \tag{5.24}$$

The deformation density Fourier series can thus be written as

$$\Delta\rho_{deformation}(\mathbf{r}) = \frac{1}{V}\sum_{\mathbf{H}}(|F_{obs}|\exp i\varphi_{true}(\mathbf{H}) - |F_{ref}|\exp i\varphi_{ref}(\mathbf{H}))\exp(-2\pi i\mathbf{H}\cdot\mathbf{r}) \tag{5.25}$$

where φ_{true} $(=\tan^{-1}B/A)$ is the correct phase angle. Substitution of $|F_{obs}| = \Delta|F| + |F_{ref}|$ from Eq. (5.24) gives

$$\Delta\rho_{def}(\mathbf{r}) = \frac{1}{V}\sum[(\Delta|F| + |F_{ref}|)\exp i\varphi_{true} - |F_{ref}|\exp i\varphi_{ref}]\exp(-2\pi i\mathbf{H}\cdot\mathbf{r})$$

$$= \frac{1}{V}\sum\Delta|F|\exp i\varphi_{true}\exp(-2\pi i\mathbf{H}\cdot\mathbf{r})$$

$$+ \frac{1}{V}\sum|F_{ref}|[\exp i\varphi_{true} - \exp i\varphi_{ref}]\exp(-2\pi i\mathbf{H}\cdot\mathbf{r}) \tag{5.26}$$

In other words, the deformation density can be separated into structure-factor magnitude and structure-factor phase contributions:

$$\Delta\rho = \Delta\rho(\Delta|F|) + \Delta\rho(\Delta\varphi) \tag{5.27}$$

With $\Delta\varphi = \varphi_{true} - \varphi_{ref}$, the phase-factor difference in the phase contribution can be rewritten as

$$\exp i\varphi_{true} - \exp i\varphi_{ref} = \exp i(\varphi_{ref} + \Delta\varphi) - \exp i\varphi_{ref}$$

$$= \exp i\varphi_{ref}[\exp(i\Delta\varphi - 1)]$$

$$= \exp i\varphi_{ref}\exp i\Delta\varphi/2[\exp i\Delta\varphi/2 - \exp(-i\Delta\varphi/2)]$$

$$= \exp[i(\varphi_{ref} + \Delta\varphi/2)][2i\sin(\Delta\varphi/2)]$$

where $i = \sqrt{-1}$ and $2i\sin(\Delta\varphi/2)$ constitutes a phase factor, which may be written as $\exp(i\pi/2)$ to give

$$\exp i\varphi_{true} - \exp i\varphi_{ref} = \exp[i(\varphi_{ref} + \Delta\varphi/2)]\cdot[2\exp(i\pi/2)\sin(\Delta\varphi/2)]$$

$$= 2\sin(\Delta\varphi/2)\exp i(\varphi_{ref} + \Delta\varphi/2 + \pi/2)$$

$$= 2\sin(\Delta\varphi/2)\exp[i(\varphi_{ref} + \varphi_{true} + \pi)/2] \tag{5.28}$$

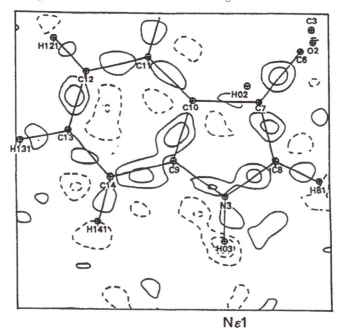

FIG. 5.14 Phase contribution to the deformation density in the indole ring of *N*-acetyl-L-tryptophan methylamide. Contours at 0.05 eÅ^{-3}; negative contours dashed lines; zero contour omitted. Peak heights in the full deformation density are 0.4–0.5 Å. *Source:* Souhassou et al. (1991).

Thus, the phase contribution equals

$$\Delta\rho(\Delta\varphi) = \frac{1}{V}\sum 2|F_{\text{ref}}| \sin(\Delta\varphi/2) \exp\left[i(\varphi_{\text{ref}} + \varphi_{\text{true}} + \pi)/2\right] \exp(-2\pi i\mathbf{H}\cdot\mathbf{r})$$

(5.29)

or, for small $\Delta\varphi$,

$$\Delta\rho(\Delta\varphi) \approx \frac{1}{V}\sum |F_{\text{ref}}| \Delta\varphi \exp\left[i(\varphi_{\text{ref}} + \varphi_{\text{true}} + \pi)/2\right] \exp(-2\pi i\mathbf{H}\cdot\mathbf{r}) \quad (5.30)$$

Because of the addition of $\pi/2$ in the exponent of Eq. (5.30), the waves of the phase contribution are shifted by $\pi/2$ relative to those of the amplitude contribution, while their amplitudes are proportional to the phase difference.

The effect of the neglect of Eq. (5.30) can be quite large. For the peptide *N*-acetyl-α,β-dehydrophenylalanine methylamide (space group *Cc*), for example, the underestimate of the density is reported to be as large as 0.19 eÅ^{-1}. In *N*-acetyl-L-tryptophan methylamide, which crystallizes in the space group $P2_12_12_1$ with centrosymmetric projections, the features are somewhat smaller (Fig. 5.14), but still important.

5.2.6 The Variances and Covariances of the Experimental Density

5.2.6.1 Errors in the Observed Density

To assess the significance of the features of the charge density, the propagation of observational errors must be examined. Given the standard deviations in the observations, and assuming a diagonal variance–covariance matrix of the observations, we may write, for the covariance between the densities at points A and B,

$$\text{cov}(\rho_{\text{obs},A}\rho_{\text{obs},B}) \cong \sum \frac{\partial \rho_{\text{obs},A}}{\partial |F_{\text{obs}}(\mathbf{H})|} \frac{\partial \rho_{\text{obs},B}}{\partial |F_{\text{obs}}(\mathbf{H})|} \sigma^2[|F_{\text{obs}}(\mathbf{H})|]$$

$$+ \sum \frac{\partial \rho_{\text{obs},A}}{\partial \varphi_{\text{obs}}(\mathbf{H})} \frac{\partial \rho_{\text{obs},B}}{\partial \varphi_{\text{obs}}(\mathbf{H})} \sigma^2[\varphi_{\text{obs}}(\mathbf{H})] \qquad (5.31)$$

where F_{obs} is the structure factor on an absolute scale, and errors in the scale factor, considered below, have been neglected.

Though Eq. (5.31) can be evaluated directly, provided the phase errors can be estimated, it can be reduced to a simpler, computationally more convenient equation for the average covariance in a density map (Rees 1976). For the space group $P\bar{1}$, Eq. (5.31) becomes

$$\text{cov}(\rho_{\text{obs},A}\rho_{\text{obs},B}) \cong \frac{4}{V^2} \sum_{1/2} \sigma^2[F_{\text{obs}}(\mathbf{H})] \cos (2\pi \mathbf{r}_A \cdot \mathbf{H}) \cos (2\pi \mathbf{r}_B \cdot \mathbf{H})$$

$$\cong \frac{2}{V^2} \sum_{1/2} \sigma^2(F_{\text{obs}})[\cos 2\pi(\mathbf{r}_A + \mathbf{r}_B)\cdot\mathbf{H} + \cos 2\pi(\mathbf{r}_A - \mathbf{r}_B)\cdot\mathbf{H}]$$

$$(5.32)$$

The summation is over a hemisphere in reciprocal space.

The first term of Eq. (5.32) rapidly averages to zero as H_{max} increases, except for points close to one of the centers of symmetry, where $|\mathbf{r}_A + \mathbf{r}_B|$ has close to integer value. If $\sigma^2(F_{\text{obs}})$ does not vary systematically with H, the second term in the square brackets may be replaced by its average over all directions and values of $\mathbf{H}$ for a given distance $|\mathbf{r}_A - \mathbf{r}_B|$:

$$\langle \cos 2\pi(\mathbf{r}_A - \mathbf{r}_B)\cdot\mathbf{H} \rangle = 3(\sin u - u \cos u)/u^3 \equiv C(u) \qquad (5.33)$$

with $u = 2\pi|\mathbf{r}_A - \mathbf{r}_B|H_{\text{max}}$. Substitution into Eq. (5.32) gives

$$\text{cov}(\rho_{\text{obs},A}\rho_{\text{obs},B}) \cong \frac{2}{V^2} C(u) \sum_{1/2} \sigma^2(F_{\text{obs}}) \qquad (5.34)$$

The function $C(u)$, plotted in Fig. 5.15, is a measure for the correlation between points A and B. As u is proportional to H_{max} for a given value of $|\mathbf{r}_A - \mathbf{r}_B|$, the correlation between adjacent points decreases with H_{max}, a manifestation of the increased resolution on adding high-order data. The correlation between adjacent points is generally positive, but it becomes slightly negative at $u \approx 5.8$.

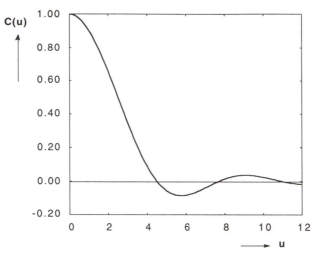

FIG. 5.15 Coefficient $C(u)$ in the expression for the correlation coefficient between the observed electron density at adjacent points A and B [Eq. (5.33)]. The variable $u = 2\pi|\mathbf{r}_A - \mathbf{r}_B|H_{max}$. *Source*: Rees (1976).

The average variance follows from Eq. (5.34) by setting $C(u) = 1$:

$$\sigma^2(\rho_{obs}) \cong \frac{2}{V^2} \sum_{1/2} \sigma^2(F_{obs}) \qquad (5.35)$$

which is a well-known relation first reported by Cruickshank (1949), valid at a sufficient distance from any symmetry element.

For other centric space groups, the most convenient way to derive the covariance between $\rho(\mathbf{r}_A)$ and $\rho(\mathbf{r}_B)$ is to assume that the densities are calculated as for $P\bar{1}$, and then averaged over the n symmetry-equivalent positions. This leads, for the averaged density $\bar{\rho}_{obs}$, to

$$\text{cov}(\bar{\rho}_{obs, A}, \bar{\rho}_{obs, B}) = \frac{1}{n}\sigma^2(\rho_{obs, G}) \sum_{i=1}^{n} C(2\pi|\mathbf{r}_{A1} - \mathbf{r}_{Bi}|H_{max}) \qquad (5.36)$$

for the covariance, and for the variance

$$\sigma^2(\bar{\rho}_{obs}) = \frac{1}{n}\sigma^2(\rho_{obs, G})\left[1 + \sum_{i=2}^{n} C(2\pi|\mathbf{r}_1 - \mathbf{r}_i|H_{max})\right] \qquad (5.37)$$

in which $\rho_{obs, G}$ is the standard deviation at a general position before averaging, and the sum is over all symmetry-equivalent positions. Equation (5.37) shows that when $\mathbf{r}_1 - \mathbf{r}_i$ is small, that is, in the vicinity of a symmetry element, even the variance depends on the correlation coefficient C.

5.2.6.2 Errors in the Deformation Density

The variances and covariances of the deformation density result both from the errors in the observations, and from the uncertainties in the refined parameters, including the scale factor. We will assume, for the sake of simplicity, that the errors in the observations and the parameters are not correlated, as would be the case for an $X-N$ deformation density. We will assume further that the density functions of the model are not subject to error, and that the true phases are known from a least-squares refinement with an appropriate scattering model.

We start with

$$\sigma^2(\Delta\rho) = \sigma^2(\rho_{\text{obs}} - \rho_{\text{calc}}) \tag{5.38}$$

The scaled total electron density $\rho_{\text{obs}} = \rho'_{\text{obs}}/k$, where ρ'_{obs} is the density on the experimental scale. Using the expression for propagation of errors given in Eq. (4.31):

$$\sigma^2(\rho_{\text{obs}}) = \sigma^2(\rho'_{\text{obs}}/k) = \frac{1}{k^2}\sigma^2(\rho'_{\text{obs}}) + (\rho'_{\text{obs}})^2\frac{\sigma^2(k)}{k^4} = \sigma_0^2(\rho_{\text{obs}}) + (\rho_{\text{obs}})^2\frac{\sigma^2(k)}{k^2}$$

$$\tag{5.39}$$

where the subscript zero indicates that errors in the scale factor have been factored out.

Taking into account the correlation between ρ_{obs} and ρ_{calc} due to the interdependence of the structural parameters and the scale factor, the error in the calculated density is given by

$$\sigma^2(\rho_{\text{calc}}) = \sigma_0^2(\rho_{\text{calc}}) + \sum_i \frac{\partial\rho_{\text{calc}}}{\partial u_i}\frac{\partial\rho_{\text{obs}}}{\partial k}\sigma(u_i)\sigma(k)\gamma(u_i, k) \tag{5.40}$$

where u_i is a positional or thermal parameter, and $\gamma(u_i, k)$ are the correlation coefficients between the scale factor and other parameters (Stevens and Coppens 1976, Rees 1978).

The variance of the deformation density is obtained by combining Eqs. (5.39) and (5.40):

$$\sigma^2(\Delta\rho) = \sigma_0^2(\rho_{\text{obs}}) + \sigma_0^2(\rho_{\text{calc}}) + (\rho_{\text{obs}})^2\frac{\sigma^2(k)}{k^2} + \sum_i \frac{\partial\rho_{\text{calc}}}{\partial u_i}\rho_{\text{obs}}\sigma(u_i)\frac{\sigma(k)}{k}\gamma(u_i, k)$$

$$\tag{5.41}$$

The standard deviation in the observed density, $\sigma_0(\rho_{\text{obs}})$, is derived as described in the preceding section. The term $\sigma[\rho_{\text{calc}}(\mathbf{r})]$ follows from the errors in the parameters and their correlations. Using Eq. (4.31), we obtain

$$\sigma_0^2(\rho_{\text{calc}}) = \sum_i\left(\frac{\delta\rho_{\text{calc}}}{\delta u_i}\right)^2\sigma^2(u_i) + 2\sum_i\sum_{j>i}\left(\frac{\delta\rho_{\text{calc}}}{\delta u_i}\right)\left(\frac{\delta\rho_{\text{calc}}}{\delta u_j}\right)\sigma(u_i)\sigma(u_j)\gamma(u_i, u_j)$$

$$\tag{5.42}$$

The derivatives in Eqs. (5.41) and (5.42) can be readily evaluated from the structure factor expression. The contribution to $\sigma[\Delta\rho(\mathbf{r})]$ due to the error in the scale factor depends on the magnitude of $\rho_{\text{obs}}(\mathbf{r})$. It is large wherever ρ_{obs} is large, which is the case in the vicinity of the nuclei of heavier atoms. As a result, the deformation density in these regions is quite unreliable.

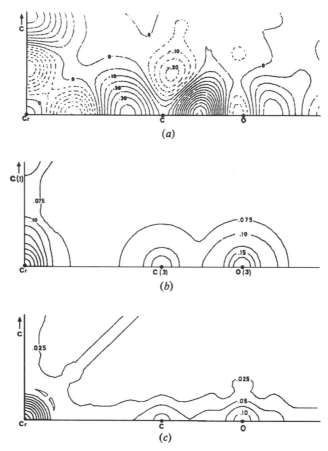

FIG. 5.16 Deformation electron density $\Delta\rho = \rho_{obs}/k - \rho_c$ in $Cr(CO)_6$ at a resolution $\lambda/(2\sin\theta)_{max} = 0.66$ Å. (a) Deformation density averaged over symmetry-equivalent planes in the molecule. Contour intervals at 0.05 $e\text{Å}^{-3}$; negative contours dotted. (b) Standard deviation (not including the error in the scale factor) before averaging. Cr and C(1) are on the mirror plane. Contour intervals at 0.025 $e\text{Å}^{-3}$. (c) Standard deviation (not including the error in the scale factor) after averaging over symmetry-equivalent planes. Contours as in part (b). *Source*: Rees (1976).

The covariance between the deformation densities at points A and B is given by the analogous expression

$$\text{cov}(\Delta\rho_A, \Delta\rho_B) = \text{cov}(\rho_{obs,A}\rho_{obs,B})/k^2 + \text{cov}(\rho_{calc,A}\rho_{calc,B}) + \rho_{obs,A}\rho_{obs,B}[\sigma(k)/k]^2$$

$$(5.43)$$

in which the correlation between the scale factor and the other parameters has been neglected.

An example of a deformation density and the associated error function is given in Fig. 5.16. The complex $Cr(CO)_6$ has octahedral symmetry, but only one diagonal mirror plane is retained in the crystal. The deformation density averaged over chemically equivalent, but crystallographically independent, sections through the Cr—C—O atoms is shown in Fig. 5.16(a). The standard deviation before averaging [Fig. 5.16(b)], peaks at ≈ 0.18 eÅ^{-3}, except at the Cr nucleus, where it is larger. After averaging, the errors are reduced [Fig. 5.16(c)], especially in regions away from the nuclei, where the errors in ρ_{obs}, rather than those in the least-squares parameters, dominate. Away from the nuclei, the standard deviation is less than 0.025 eÅ^{-3}, a value reached in many high-quality studies.

5.3 Maximum Entropy Enhancement of Electron Densities

5.3.1 The Mathematics of Entropy Maximization

The maximum entropy method (MEM) is an information–theory-based technique that was first developed in the field of radioastronomy to enhance the information obtained from noisy data (Gull and Daniell 1978). The theory is based on the same equations that are the foundation of statistical thermodynamics. Both the statistical entropy and the *information entropy* deal with the most probable distribution. In the case of statistical thermodynamics, this is the distribution of the particles over position and momentum space ("phase space"), while in the case of information theory, the distribution of numerical quantities over the ensemble of pixels is considered.

The probability of a distribution of N identical particles over m boxes, each populated by n_i particles, is given by

$$P = \frac{N!}{n_1! n_2! n_3! \ldots n_m!} \tag{5.44}$$

As in statistical thermodynamics, the entropy is defined as $\ln P$. Since the numerator is constant, the entropy is, apart from a constant, equal to

$$S = -\sum_i n_i \ln n_i \tag{5.45}$$

where Stirlings' formula ($\ln N! \approx N \ln N - N$) has been used.

In case there is a prior probability q_i for box i to contain n_i particles, expression (5.45) becomes

$$P = \frac{N!}{n_1! n_2! n_3! \ldots n_m!} \times q_1^{n_1} q_2^{n_2} \ldots q_m^{n_m} \tag{4.46}$$

which gives, for the entropy expression,

$$S = -\sum_i n_i \ln n_i + \sum_i n_i \ln q_i = -\sum_{i=1}^{m} n_i \ln \frac{n_i}{q_i} \tag{5.47}$$

The maximum entropy method was first introduced into crystallography by Collins (1982), who, based on Eq. (5.47), expressed the information entropy of the electron density distribution as a sum over M grid points in the unit cell, using the entropy formula (Jaynes 1968)

$$S[\rho(\mathbf{r})] = -\sum_{j=1}^{M} p(\mathbf{r}) \ln \frac{p(\mathbf{r})}{m(\mathbf{r})} \tag{5.48}$$

where both $p(\mathbf{r})$ and $m(\mathbf{r})$ are fractional quantities defined as

$$p(\mathbf{r}_j) = p_j = \frac{\rho(\mathbf{r}_j)}{\sum_{j=1}^{M} \rho(\mathbf{r}_j)} \quad \text{and} \quad m(\mathbf{r}_j) = m_j = \frac{\rho_0(\mathbf{r}_j)}{\sum_{j=1}^{M} \rho_0(\mathbf{r}_j)} \tag{5.49}$$

The subscript zero refers to the prior density. Note that $p(\mathbf{r})$ is proportional to the probability of finding an electron at $\mathbf{r}$, and $m(\mathbf{r})$ is proportional to the prior probability of finding an electron at $\mathbf{r}$.

Expression (5.48) is applied in an iterative procedure. The entropy $S[\rho(\mathbf{r})]$ is maximized subject to the constraint

$$C[\rho(\mathbf{r})] = \chi^2 = \sum_{k=1}^{N} \left| \frac{F_k^{obs}(\mathbf{H}) - F_k^{calc}(\mathbf{H})}{\sigma(F_k)} \right|^2 = N \tag{5.50}$$

where N is the number of unique observations, and F^{calc} is obtained by summation over the M grid points:

$$F_k^{calc}(\mathbf{H}) = \frac{V_{unit cell}}{M} \sum_{j=1}^{M} \rho(\mathbf{r}_j) \exp \{2\pi i \mathbf{H}_k \cdot \mathbf{r}_j\} \tag{5.51}$$

with suitable scaling to F^{obs}. If the constraint of Eq. (5.50) were not introduced, maximizing the entropy would invariably produce a uniform distribution, which corresponds to the maximum entropy.

The constraint is enforced by introducing a Lagrangian multiplier λ in the minimization function given by

$$L(\lambda) = S(\rho(\mathbf{r})) - \lambda\chi^2 \tag{5.52}$$

When convergence is reached, the gradient of the minimization function equals zero:

$$\nabla_\rho(L) = \nabla_\rho(S) - \lambda\nabla_\rho(\chi^2) = 0 \tag{5.53a}$$

or, equivalently,

$$\nabla_\rho(S) = \lambda\nabla_\rho(\chi^2) \tag{5.53b}$$

For each grid point j, this corresponds to

$$\frac{\partial S}{\partial \rho_j} = \lambda \frac{\partial \chi^2}{\partial \rho_j} \tag{5.54}$$

In the case of a uniform prior, $\rho_0(\mathbf{r}_j) = \rho_{0j} = \rho_0$ for all grid points j, and differentiation of S, using Eq. (5.48), gives

$$\frac{\partial S}{\partial \rho_j} = -\frac{1}{\sum_{j=1}^{M} \rho_j} \ln(\rho_j/A) \tag{5.55}$$

where A is a constant. Substitution into Eq. (5.54) leads to

$$\rho_j = A \exp\left\{-\lambda(\sum \rho_j)\frac{\partial C}{\partial \rho_j}\right\} \tag{5.56}$$

A can be selected as

$$A \approx \exp\left\{\sum p_j \ln \rho_j\right\} \tag{5.57}$$

The quantity A is a *weighted logarithmic average* of the converged entropic density $\rho(\mathbf{r})$ over the unit cell. The selection of an optimal value of A is discussed in the literature (Papoular et al. 1992). The value of A corresponds to the expected density far away from any atom, and reconstructed density values smaller than A are considered unreliable.

Since

$$\frac{\partial F_{\text{calc}}(\mathbf{H})}{\partial \rho_j} = \exp 2\pi i \mathbf{H} \cdot \mathbf{r}_j$$

the density at $\mathbf{r}_j$ in the $(n + 1)$th iteration is obtained with Eq. (5.56) as

$$\rho(\mathbf{r}_j, n + 1) = \exp\left[\ln \sum_j p_j \ln \rho_j(n) + \lambda F(0) \sum_{\mathbf{H}} \frac{2}{\sigma(\mathbf{H})^2}\right.$$

$$\left. \times |F_{\text{obs}}(\mathbf{H}) - F_{\text{calc}}(\mathbf{H})| \exp 2\pi i \mathbf{H} \cdot \mathbf{r}_j\right] \tag{5.58}$$

with

$$F(0) = \frac{V_{\text{unit cell}}}{M} \sum_{j=1}^{M} \rho(\mathbf{r}_j)$$

The algorithm of entropy maximization is nonlinear, and must therefore be applied iteratively. It is possible to solve for both $\lambda(n + 1)$ and $\rho_j(n + 1)$, starting from $\lambda(n)$ and $\rho_j(n)$ at each iteration n. The starting values are $\lambda(0) \approx 0$, and $\rho_j(0)$ equal to the prior density. Achieving convergence involves a two-step process, in which first the $\chi^2 = N$ constraint is satisfied, and subsequently the entropy S is maximized. In a variation of the method, λ is kept fixed at a small value adequate to ensure convergence (Sakata and Sato 1990).

After completion of the MEM enhancement, it becomes possible to evaluate the reflections missing from the summation. In a Fourier summation, the amplitudes of the unobserved reflections are assumed to be equal to zero, while the MEM technique provides the most probable values.

When extinction is present in the data set, it must be corrected for before the MEM procedure is started. The structure factors must similarly be corrected for anomalous scattering, if present. Both corrections require a model for their evaluation. The independent-atom model is usually adequate for this purpose.

5.3.2 Application of MEM to Single-Crystal Data on Silicon

A particularly interesting application of the MEM is the analysis by Sakata and Sato (1990) of the very accurate Pendellösung data of Saka and Kato (1968). With

TABLE 5.1 MEM Prediction of some of the Weak Reflections of Silicon

Reflection	MEM Value	Experimental Value	Reference
222	1.5270	1.456 (8)	Alkire et al. (1982)
442	−0.0349	−0.0370 (23)	Tischler and Batterman (1984)
622	−0.0112	+0.0088 (11)	Tischler and Batterman (1984)
644	−0.0126		
662	0.0121		
842	−0.0130		

the known phases of the silicon structure, convergence is achieved rapidly. The data set does not contain the weak spherical-atom-forbidden reflections with $h + k + l = 4n + 2$, but it is clear from inspection of the MEM map that the bonding density, to which the weak reflections make the major contribution, is well accounted for. Calculation of the missing reflections shows that the values for the (222) and (442) reflections are very close to the observed ones. This is not the case for the very weak (662) reflection, which, though of proper magnitude, has the incorrect sign (Table 5.1).

When it is assumed that the phases of the structure factors are unknown, the analysis proceeds well, after fixing the origin of the cubic unit cell by choosing the sign of the strong (111) reflection. This corresponds to a direct structure determination without any prior knowledge of the structure, and supports the value of the maximum entropy method in the early stages of structure determination.

The silicon data are of unusual accuracy, and the result may not be typical of those obtained with more noisy data. Even for Si, it is found that the distribution of the discrepancies between F^{obs} and F^{calc} after the MEM procedure often deviates greatly from the ideal Gaussian distribution, with large discrepancies being observed for a few strong low-order reflections (Jauch and Palmer 1993, Jauch 1994). This points to the weakness that optimization of the entropy, subject to the constraint $\chi^2 = N$, constrains the variance of the distribution, but not its shape. Assignment of a weighting factor equal to $|H|^{-4}$ to each of the terms in the summation of Eq. (5.47) has been reported to give improved distribution of the residual errors (De Vries et al. 1994).

5.3.3 The Two-Channel Maximum Entropy Method

There are ample indications that the maximum entropy method is of limited use when the densities to be reconstructed have a large dynamic range and the aim is to recover the fine detail of the distribution. In the analysis of the γ-ray data on MnF_2 and NiF_2, the atomic peaks are hardly affected by the choice of starting values of the density function, but the low-density bonding regions tend to be contaminated by artifacts (Jauch and Palmer 1993, Jauch 1994). For a simulated promolecule data set for the molecular crystal α-glycine, containing 1205 structure factors, the general tendency of the maximum entropy method to sharpen strong features, but flatten weak features, leads to a significant flattening of the maxima at the hydrogen-atom positions, and the appearance of sharp spikes at the positions of the heavier atoms (Papoular et al. 1996).

The dynamic range can be reduced greatly by applying the maximum entropy method to the deformation density. As the entropy functional $S[\rho(\mathbf{r})]$ defined by Eq. (5.48) requires a positive density everywhere in the unit cell, a *two-channel method* is used in which $\Delta\rho(\mathbf{r})$ is defined as the difference between two positive functions, $\rho^+(\mathbf{r})$ and $\rho^-(\mathbf{r})$, representing the densities of excess and lack of electrons, respectively. The two-channel method was first applied to magnetization densities (Papoular and Gillon 1990a, b) and to neutron diffraction results involving atoms with scattering lengths of opposite signs (Sakata et al. 1993). In both cases, positive as well as negative scattering density occurs.

At a given $\mathbf{r}_j$ in the unit cell, either $\rho^+(\mathbf{r})$ or $\rho^-(\mathbf{r})$ is significant, excess and lack of scattering density being mutually exclusive. As described by Papoular et al. (1966), the related probabilities $p^+(\mathbf{r}_j)$ and $p^-(\mathbf{r}_j)$ are defined as

$$p^+(\mathbf{r}_j) = \frac{\rho^+(\mathbf{r}_j)}{\sum \{\rho^+(\mathbf{r}_j) + \rho^-(\mathbf{r}_j)\}} \quad \text{and} \quad p^-(\mathbf{r}_j) = \frac{\rho^-(\mathbf{r}_j)}{\sum \{\rho^+(\mathbf{r}_j) + \rho^-(\mathbf{r}_j)\}} \quad (5.59)$$

In analogy to Eq. (5.49), the prior models $m^+(\mathbf{r})$ and $m^-(\mathbf{r})$ are introduced, and the two-channel entropy is defined by

$$S[\Delta\rho] = -\sum_{j=1}^{M} \left\{ p_j^+ \ln \frac{p_j^+}{m_j^+} + p_j^- \ln \frac{p_j^-}{m_j^-} \right\} \quad (5.60)$$

The entropic densities then follow from equations equivalent to Eqs. (5.56) and (5.57):

$$\rho_j^+ = A \exp\left\{ -\lambda \left(\sum \{\rho_j^+ + \rho_j^-\} \right) \frac{\partial C[\rho]}{\partial \rho_j^+} \right\} \quad j = 1, M \quad (5.61a)$$

and

$$\rho_j^- = A \exp\left\{ -\lambda \left(\sum \{\rho_j^+ + \rho_j^-\} \right) \frac{\partial C[\rho]}{\partial \rho_j^-} \right\} \quad j = 1, M \quad (5.61a)$$

with

$$A = \exp\left(\sum_{j=1}^{M} [p_j^+ \ln \rho_j^+ + p_j^- \ln \rho_j^-] \right) \quad (5.62)$$

Since

$$\frac{\partial C[\rho]}{\partial \rho_j^+} = -\frac{\partial C[\rho]}{\partial \rho_j^-}$$

the positive and negative scattering densities are related by

$$\rho_j^+ \rho_j^- = A^2 \quad (5.63)$$

It follows that in a given pixel $\{\mathbf{r}_j\}$, either ρ_j^+ or ρ_j^- can have a value larger than A, which is taken as a measure of the significance level. The relation of Eq. (5.63) reduces the number of unknowns to one per pixel as in the one-channel maximum entropy method.

In the application of the two-channel method to α-glycine, use of a uniform prior density sharpens and enhances the bond peaks relative to the observed deformation density, but suppresses the lone-pair peaks to much lower levels. The use of the multipole refinement deformation density as a nonuniform prior gives better results and some increase in detail.

The electrostatic properties of the molecule may be used as a criterion for judging the MEM enhancement. Using the uniform prior density, the MEM molecular dipole moment derived by the discrete boundary partitioning of space (chapter 6) is only 1.3 D, compared with 9.1 D based on the experimental density, 13.8 D from the multipole population parameters, and a solution value of 11.6 D. With a nonuniform prior, a more acceptable, but still low, MEM value of 7.8 D is obtained. While this physical criterion shows the nonuniform prior to be preferable, the validity of the MEM enhancement in charge density studies remains to be assessed.

6

Space Partitioning and Topological Analysis of the Total Charge Density

In partitioning space in the analysis of a continuous charge distribution, the requirement of locality, formulated by Kurki-Suonio (Kurki-Suonio 1968, 1971; Kurki-Suonio and Salmo 1971), should be preserved. It states that *density at a point should be assigned to a center in the proximity of that point*. In *discrete boundary* partitioning schemes, the density at each point is assigned to a specific basin, while in *fuzzy boundary* partitioning, the density at the point may be assigned to overlapping functions centered at different locations.

The least-squares formalisms described in chapter 3 implicitly define a space partitioning scheme, based on the density functions used in the refinement that are each centered on a specific nucleus. Since the density functions are continuous, they overlap, so the fragments interpenetrate rather than meet at a discrete boundary. Such fuzzy boundaries correspond to smoothly varying functions, both in real and reciprocal space, and therefore to well-behaved fragment scattering factors, and reasonable fragment electrostatic moments. The interpenetrating-fragment partitioning schemes are related to the Mulliken and Løwdin population analyses of theoretical chemistry.

The topological analysis of the total density, developed by Bader and coworkers, leads to a scheme of natural partitioning into atomic basins which each obey the virial theorem. The sum of the energies of the individual atoms defined in this way equals the total energy of the system. While the Bader partitioning was initially developed for the analysis of theoretical densities, it is equally applicable to model densities based on the experimental data. The density obtained from the Fourier transform of the structure factors is generally not suitable for this purpose, because of experimental noise, truncation effects, and thermal smearing.

The topological analysis of the density leads to a powerful classification of bonding based on the electron density. It is discussed in the final sections of this chapter.

6.1 Methods of Space Partitioning

6.1.1 The Stockholder Partitioning Concept

The stockholder partitioning concept is one of the important contributions to charge density analysis made by Hirshfeld (1977b). It defines a continuous sampling function $w_i(\mathbf{r})$, which assigns the density among the constituent atoms. The sampling function is based on the spherical-atom promolecule density—the sum of the spherically averaged ground-state atom densities. The sampling function $w_i(\mathbf{r})$ for atom i is defined by the relative contribution of atom i to the promolecule density:

$$w_i(\mathbf{r}) = \rho_i^{\text{spherical atom}}(\mathbf{r}) \bigg/ \sum_i \rho_i^{\text{spherical atom}}(\mathbf{r}) = \rho_i^{\text{spherical atom}}(\mathbf{r}) / \rho^{\text{promolecule}}(\mathbf{r}) \qquad (6.1)$$

The density assigned to atom i is given by

$$\rho_i^{\text{at}}(\mathbf{r}) = w_i(\mathbf{r})\rho^{\text{total}}(\mathbf{r}) \qquad (6.2)$$

or, equivalently,

$$\rho_i^{\text{at}}(\mathbf{r}) = w_i(\mathbf{r})\,\Delta\rho(\mathbf{r}) + \rho_i^{\text{spherical}}(\mathbf{r}) \qquad (6.3)$$

Accordingly, each atom is assigned a fraction of the charge density at a point proportional to its "investment" in the promolecule density at that point. This is the basis of the *stockholder concept*. We note that Eq. (6.3) can be reformulated as

$$\Delta\rho_i^{\text{at}}(\mathbf{r}) = w_i(\mathbf{r})\,\Delta\rho(\mathbf{r}) \qquad (6.4)$$

Figure 6.1 shows the stockholder decomposition of the theoretical deformation density of the cyanoacetylene molecule, H—C≡C—C≡N (Hirshfeld 1977b). The overlap density in the bonds is distributed between the bonded atoms. The assignment of part of the density near the hydrogen nucleus to the adjacent carbon atom manifests the difference between fuzzy and discrete boundary partitioning methods.

Once the partitioning has been accomplished, net atomic charges, atomic electrostatic moments, and other physical properties are obtained by straight-forward integration using the expectation value expression

$$\langle O \rangle = \int_{v_T} \hat{O}\rho(\mathbf{r})\,d\mathbf{r} \qquad (6.5)$$

in which $\hat{O}$ is the operator for the desired property, and V_T is the volume of integration.

Net atomic charges based on the stockholder partitioning of theoretical densities for a number of linear molecules containing N, C, and H are listed in Table 6.1. For these molecules, the charge transfer between atoms is relatively small. Much larger values are obtained for more electronegative atoms, such as oxygen and fluorine bonded to carbon atoms.

The stockholder recipe partitions the density according to each atom's contribution to the promolecule density. The partitioned fragment distributions

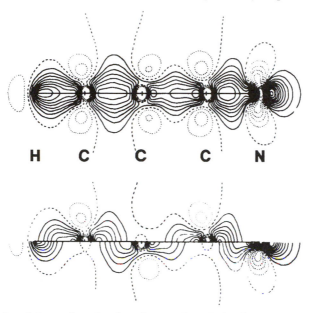

FIG. 6.1 Molecular deformation density: (upper figure) $\Delta\rho$ in cyanoacetylene, derived from SCF wave function of McLean and Yoshimine (1967), resolved into bonded-atom fragments; (lower figure) $\Delta\rho$ for H, C, and N shown below symmetry axis, two other C atoms above axis. Contour interval is 0.1 $e\text{Å}^{-3}$; zero contours are dashed lines, negative contours are dotted lines; inner contours around heavy nuclei have been omitted (Hirshfeld 1977a).

TABLE 6.1 Net Charges q (e) from the Stockholder Partitioning.[a] Charges in the Second Row are from a Discrete Boundary Partitioning by Politzer (1971) and Politzer and Reggio (1972)

HCN	H	C	N		
q	+0.133	+0.066	−0.201		
	+0.18	0.0	−0.18		
HCCCN	H	C	C	C	N
q	+0.124	−0.015	−0.031	+0.096	−0.176
	+0.18	−0.06	−0.05	+0.09	−0.16
HCCH	H	C	C	H	
q	+0.094	−0.094	−0.094	+0.094	
	+0.14	−0.14	−0.14	+0.14	
NCCN	N	C	C	N	
q	−0.126	+0.126	+0.126	−0.126	
	−0.10	+0.10	+0.10	−0.10	

[a] Small Charge imbalances reflect errors in the numerical integration. *Source*: Hirshfeld (1977b).

therefore tend toward their values in the promolecule. As a result, the stockholder charges and higher moments are often somewhat smaller than those from other partitioning methods.

6.1.2 Space Partitioning Based on the Atom-Centered Multipole Expansion

The atom-centered multipole expansion used in the density formalisms described in chapter 3 implicitly assigns each density fragment to the nucleus at which it is centered. Since the shape of the density functions is fitted to the observed density in the least-squares minimalization, the partitioning is more flexible than that based on preconceived spherical atoms.

Two disadvantages of multipole partitioning should be mentioned. The first is that any density not fitted by the model is discarded in the partitioning process. Examination of the residual density is required to ensure the completeness of the set of modeling functions. The second is that very diffuse functions of the model, if included, violate the requirement of locality discussed above, and may lead to counterintuitive results.

Nevertheless, multipole partitioning leads to very acceptable molecular electrostatic moments, as fully discussed in chapter 7.

6.1.3 Atomic Fragments Defined by Discrete Boundaries

The integration of the charge density over a region defined by discrete boundaries fits conventional ideas about area partitioning, and obeys the requirement of locality. Kurki-Suonio and coworkers examined the charge density integrated over a sphere centered on the ion, as a function of the radius of the sphere. The results for NH_4Cl (Vahvaselkä and Kurki-Suonio 1975), based on powder data, are shown in Fig. 6.2. The minimum in the radial density is defined as the *radius of best separation*. The 18-electron sphere for Cl^- terminates somewhat beyond this radius of best separation, indicating incomplete charge transfer from the cation to the anion. The electronic charges at best separation are found to be 17.55 and 9.55 e. Similar analyses of a series of alkali halides and metal oxides like MnO, CoO, and NiO lead to quite reasonable charges. But charge neutrality is not maintained because voids between the atomic spheres remain unassigned, and ionic spheres may overlap, as is the case for NH_4Cl. For NH_4Cl, the radii of best separation are $R_{Cl} = 1.75$ Å and $R_{NH_4} = 1.76$ Å, the sum of which exceeds the interionic distance by 4.6%. As a result, the sum of the electron counts, which is 27.1 e, is slightly larger than the number of electrons in the complex.

The violation of electroneutrality can be avoided by using a *space-filling* model. Such a model must fulfill the condition

$$\sum_i V_i = V_{\text{asymmetric unit}} \qquad (6.6)$$

that is, all of space must be accounted for.

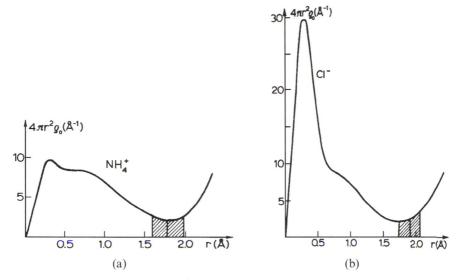

(a)   (b)

FIG. 6.2 Radial charge density $4\pi r^2 \rho(r)$ as a function of the radius of the sphere, based on powder data on NH_4Cl: (a) NH_4^+; (b) Cl^-. The shaded regions represent the boundaries which lead to populations of 10.0 ± 5 and 18.0 ± 5 electrons for NH_4^+ and Cl^-, respectively. *Source*: Vahvaselkä and Kurki-Suonio (1975).

In the theory of metals and alloys, the *Wigner–Seitz cell* is defined by planes perpendicular to the interatomic vectors. Analogously, the boundary between two molecules or molecular fragments can be defined by using the relative sizes R_A and R_B of atom A in molecule I and the adjacent atom B in molecule II.

Let $\mathbf{r}_{AB}$ be a unit vector pointing from atom A to atom B (Fig. 6.3). To achieve the partitioning of the space between the two molecules, the vectors from atoms A and B to the point i are projected on the interatomic vector, and the ratio of the two projections is compared with the ratio of the Van der Waals radii of the two atoms. The selection criterion is

$$\frac{(\mathbf{r}_i - \mathbf{r}_A)\cdot\mathbf{r}_{AB}}{R_A} \underset{>}{\overset{?}{\lessgtr}} \frac{(\mathbf{r}_i - \mathbf{r}_B)\cdot\mathbf{r}_{BA}}{R_B} \tag{6.7}$$

If the ratio on the left is the smallest, the point i belongs to atom A, and thus to molecule I, and vice versa. The *discrete boundary Van-der-Waals-ratio partition-*

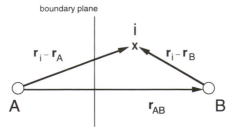

FIG. 6.3 Definition of vectors used in discrete boundary space partitioning.

ing results in boundary planes perpendicular to the interatomic vectors $\mathbf{r}_{AB}$. Their exact location depends only on the ratios R_A/R_B.

The result of the integration will be less sensitive to the exact position of the boundary when the density in the boundary area is small, which means that integration over the deformation density is preferable. Because covalently bonded atoms overlap significantly, the method is suitable for obtaining ionic or molecular charges but not for separation of atoms in a molecule. It should not be used for the latter purpose.

The molecular boundary for the formamide crystal, defined according to Eq. (6.7), is shown in Fig. 6.4. The molecular volume is defined by the sum of the volumes of small parallelepipeds around each of the points assigned to the molecule.

The integration over discrete units of space can be performed directly from the structure factors. Starting from Eq. (6.5) and substituting for $\rho(\mathbf{r})$ the Fourier summation over the structure factors, we obtain, for the thermally averaged density,

$$\langle O \rangle = \frac{1}{V} \int_{V_T} \hat{O}(\mathbf{r}) \sum_H F(\mathbf{H}) \exp(2\pi i \mathbf{H} \cdot \mathbf{r}) \, d\mathbf{r} \tag{6.8}$$

or

$$\langle O \rangle = \frac{1}{V} \int_{V_T} \hat{O}(\mathbf{r}) \left\{ \langle \rho \rangle_{\text{promolecule}} + \sum_H \Delta F(\mathbf{H}) \exp(2\pi i \mathbf{H} \cdot \mathbf{r}) \right\} d\mathbf{r} \tag{6.9}$$

in which the angle brackets indicate that the density is thermally averaged.

For the charge and dipole moment operators

$$\int_{V_T} \hat{O} \langle \rho \rangle_{\text{promolecule}} \, d\mathbf{r} = 0 \tag{6.10}$$

The higher moments are generally not zero for the promolecule, but can be readily derived as discussed in chapter 7.

Series truncation effects due to the experimental resolution limit are reduced when the core- or spherical-atom densities are subtracted from the Fourier summation, as in Eq. (6.9).

Suppose the volume of integration is centered at $\mathbf{r}_i$. Expression (6.8) can then be rewritten as

$$\langle O \rangle = \frac{1}{V} \sum_H \left[F(\mathbf{H}) \exp(2\pi i \mathbf{H} \cdot \mathbf{r}_i) \int_{V_T} \hat{O}(\mathbf{r}) \exp(2\pi i \mathbf{H} \cdot (\mathbf{r} - \mathbf{r}_i)) \, d\mathbf{r} \right] \tag{6.11}$$

In other words, the desired property is obtained by multiplying each structure factor in the conventional summation by the Fourier transform of the operator, integrated over the volume element of interest. For the net charge, the operator $\hat{O} = 1$, and the integral is referred to as the *shape transform* of the volume of integration.

When the molecular properties are to be evaluated, the volume of interest is irregularly shaped, except for the simplest molecules. To facilitate integration, the volume may be subdivided into integrable subunits of volume v_i, with $\sum v_i = V_T$. Since the Fourier transformation is additive, the sought-after result may be

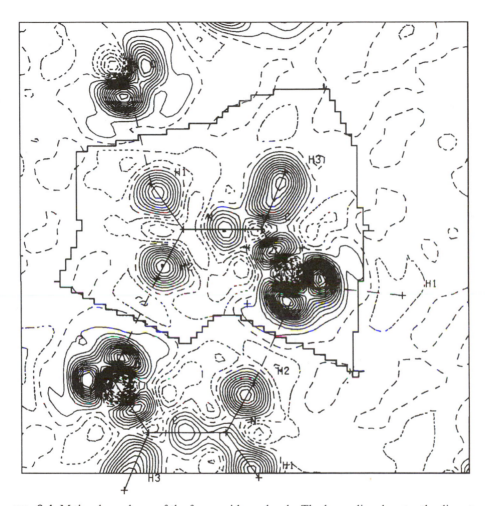

FIG. 6.4 Molecular volume of the formamide molecule. The heavy line denotes the discrete molecular boundary obtained with Eq. (6.7) and van der Waals radii: O, 1.4; N, 1.5; C, 1.7; and H, 1.2 Å. The density is a theoretical difference density in the plane of the molecule according to a wave function given by Snyder and Basch (1972). Contours are at 0.05 eÅ^{-3} intervals. Negative contours are denoted by short dashed lines and the zero contour by the long dashed line. *Source*: Moss and Coppens (1980).

obtained by summation over the subunits:

$$\langle O \rangle = \frac{1}{V} \left[F(\mathbf{H}) \sum_i \left\{ \left(\int_{v_i} \hat{O}(\mathbf{r}) \exp\left(2\pi i \mathbf{H}(r - r_i)\, d\mathbf{r}\right) \right) \exp\left(2\pi i \mathbf{H} \cdot \mathbf{r}_i\right) \right\} \right] \quad (6.12)$$

or, substituting

$$s_i(\mathbf{H}) = \int_{v_i} \hat{O}(\mathbf{r}) \exp\left(2\pi i \mathbf{H}(\mathbf{r} - \mathbf{r}_i)\right) d\mathbf{r} \tag{6.13}$$

$$\langle O \rangle = \frac{1}{V} \sum_{\mathbf{H}} F(\mathbf{H}) \sum_i s_i(\mathbf{H}) \exp\left(2\pi i \mathbf{H} \cdot \mathbf{r}_i\right) \equiv \frac{1}{V} \sum_{\mathbf{H}} F(\mathbf{H}) S_T(\mathbf{H}) \tag{6.14}$$

where

$$S_T(\mathbf{H}) \equiv \sum_i s_i(\mathbf{H}) \exp\left(2\pi i \mathbf{H} \cdot \mathbf{r}_i\right) \tag{6.15}$$

For the higher moments, s_i depends on the location of the subunit i. But in the case of the net charge, all s_i are equal, provided the subunits have identical shape. When the subunits are parallelepipeds with edges of length $2\delta_x$, $2\delta_y$, and $2\delta_z$ parallel to the crystallographic axes, as in Fig. 6.4, the shape transform of the subunit, $s_0(\mathbf{H})$, is of a particularly simple form (Weiss 1966, Coppens and Hamilton 1968):

$$s_0(\mathbf{H}) = v \frac{\sin 2\pi h \left(\frac{\delta_x}{a}\right)}{2\pi h \left(\frac{\delta_x}{a}\right)} \frac{\sin 2\pi k \left(\frac{\delta_y}{b}\right)}{2\pi k \left(\frac{\delta_y}{b}\right)} \frac{\sin 2\pi l \left(\frac{\delta_z}{c}\right)}{2\pi l \left(\frac{\delta_z}{c}\right)}$$

$$= v j_0\left(\frac{2\pi h \delta_x}{a}\right) j_0\left(\frac{2\pi k \delta_y}{b}\right) j_0\left(\frac{2\pi l \delta_z}{c}\right) \tag{6.16}$$

where v is the volume of the subunit. As the Bessel function $j_0(x)$ decreases rapidly with x (Fig. 6.5), high-order reflections contribute relatively little to the integrated charge. This is because their contributions to the density vary rapidly in space, and thus will integrate to values close to zero when integration is over a large volume. Therefore, series-truncation effects are less important than in the calculation of the total density.

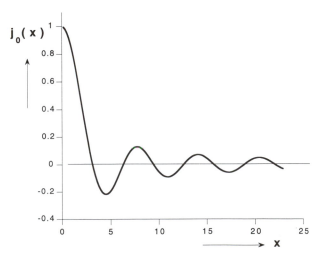

FIG. 6.5 The spherical Bessel function $j_0(x)$.

The error in a derived property $\langle O \rangle$ follows directly from the experimental standard deviations in the structure factors:

$$\sigma^2(\langle O \rangle) = \frac{1}{V^2} \sum_H S(\mathbf{H})^2 \sigma^2[F(\mathbf{H})] \qquad (6.17)$$

assuming that the errors in the structure factors are not correlated (i.e., the variance–covariance matrix $\mathbf{M}_f$ of chapter 4 is a diagonal matrix). The direct space integration for the dipole and higher moments is discussed in chapter 7.

6.1.3.1 An Example: Charge Transfer in TTF-TCNQ

An example of discrete boundary charge integration is the study on the organic charge transfer salt tetrathiofulvalene-tetracyanoquinodimethanide (TTF-TCNQ) (Coppens 1975, Coppens et al. 1987), which is a well-known low-dimensional conducting organic solid. The crystal contains homogeneous stacks containing either TTF or TCNQ molecules (Fig. 6.6). As the two molecules have strongly different electronegativities, charge transfer occurs from the donor TTF to the acceptor TCNQ. Less than a full electron is transferred per molecule, leading to mixed molecular valence in each of the stacks. For a time-averaged experiment, all molecules are equivalent, as the electrons hop rapidly between the molecules. As a result of the incompleteness of the charge transfer, the one-dimensional conduction bands along the direction of the homogeneous molecular stacks are not fully filled, and the solid has a high conductivity in the stacking axis (the monoclinic b axis) direction.

Since the molecular volumes of neither the TTF nor the TCNQ molecule are easily described by a regular volume of integration, the parallelepiped subunit method was used and integration of the valence density was performed using Eqs.

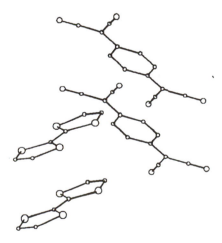

FIG. 6.6 Homogeneous stacking of TTF and TCNQ molecules along the short b axis in the monoclinic crystals of TTF-TCNQ. *Source*: Coppens et al. (1987).

TABLE 6.2 Results of Charge Integration of the TTF-TCNQ Data

Van der Waals Radii Used (Å)				Uncorrected		Valence Electron Population: With 2% Scale Correction	
S	N	C	H	P_{TCNQ}	P_{TTF}	P_{TCNQ}	P_{TTF}
1.85	1.50	1.75	1.1	72.45	51.56	72.57	51.44
1.85	1.55	1.65	1.2	72.50	51.51	72.62	51.39

Source: Coppens (1975).

(6.12)–(6.15). The results show only a modest variation when the van der Waals radii are changed within reasonable bounds (Table 6.2). As the data were not refined with the aspherical atom formalism, the scale of the observed structure factors may be biased, an effect estimated on the basis of other studies (Stevens and Coppens 1975) to correspond to a maximal lowering of the scale by 2%. Values corrected for this effect are listed in the last two columns of Table 6.2. Since neutral TTF and TCNQ have, respectively, 72 and 52 *valence* electrons, the results imply a charge transfer close to 0.60 e.

After publication of the X-ray study, the charge transfer was obtained from the reciprocal-space position of the satellite reflections, which occur in the diffraction pattern at temperatures below the Peierls-type metal–insulator transition at 53 K (Pouget et al. 1976). Assuming that the gap in the band structure occurs at twice the Fermi wavevector, that is, at $2k_F$, the position of the satellite reflections corresponds to a charge transfer of 0.59 e, in excellent agreement with the direct integration. The agreement confirms the assumption that the gap in the band structure occurs at $2k_F$.

6.2 Space Partitioning Based on the Topology of the Total Electron Density

6.2.1 The Definition of Critical Points

Bader (1990) has emphasized the necessity to base the definition of atoms on the physical structure exhibited by the electronic charge distribution, which is, in Bader's words, "a physical manifestation of the forces acting within the system." The dominant feature in the topology of the charge density is the occurrence of local maxima at the positions of the nuclei as a consequence of the attractive interaction of the electronic density and the nuclei. The maxima of the electron density are critical points, at which the first derivatives of the density are equal to zero, and the curvatures of the density in all directions are negative, that is, the slope of the density decreases along any path passing through the position of a maximum.

The gradient vector of the density in the Cartesian coordinate system **i**, **j**, and **k** is defined as

$$\nabla \rho(\mathbf{r}) = \mathbf{i}\frac{\partial \rho(\mathbf{r})}{\partial x} + \mathbf{j}\frac{\partial \rho(\mathbf{r})}{\partial y} + \mathbf{k}\frac{\partial \rho(\mathbf{r})}{\partial z} \tag{6.18}$$

At a critical point, $\nabla \rho(\mathbf{r})$ equals zero because each of the three contributions to Eq. (6.18) are zero. The classification of critical points is based on the *second* derivatives, which as noted above are all negative for a density maximum, but have different signs for saddle points and minima of the distribution.

The *Hessian* matrix $\mathbf{H}(\mathbf{r})$ is defined as the symmetric matrix of the nine second derivatives $\partial^2 \rho / \partial x_i \, \partial x_j$. The eigenvectors of $\mathbf{H}(\mathbf{r})$, obtained by diagonalization of the matrix, are the principal axes of the curvature at $\mathbf{r}$. The *rank w* of the curvature at a critical point is equal to the number of nonzero eigenvalues: the *signature* σ is the algebraic sum of the signs of the eigenvalues. The critical point is classified as (w, σ). There are four possible types of critical points in a three-dimensional scalar distribution:

$(3, -3)$ Peaks: all curvatures are negative and ρ is a local maximum at $\mathbf{r}_c$.

$(3, -1)$ Passes or saddle points: two curvatures are negative, and, at $\mathbf{r}_c$, ρ is a maximum in the plane defined by the axes corresponding to the negative curvatures; ρ is a minimum at $\mathbf{r}_c$ along the third axis which is perpendicular to this plane. The $(3, -1)$ critical points are found between every pair of nuclei considered linked by a chemical bond.

$(3, +1)$ Pales: two curvatures are positive, and, at $\mathbf{r}_c$, ρ is a minimum in the plane defined by the axes corresponding to the positive curvatures; ρ is a maximum at $\mathbf{r}_c$ along the third axis which is perpendicular to this plane. The $(3, +1)$ critical points are found at the center of a ring of bonded atoms.

$(3, +3)$ Pits: all curvatures are positive and ρ is a local minimum at $\mathbf{r}_c$.

In an isolated molecule, or cluster of atoms, the Poincaré–Hopf relationship

$$N(\text{peaks}) - N(\text{passes}) + N(\text{pales}) - N(\text{pits}) = 1 \tag{6.19}$$

holds. This can be checked quickly for simple molecules: if there is no ring, the number of bonds (number of passes) is one less than the number of atoms (number of peaks), and there are no pales or pits. For each ring, the number of bonds is increased by one, but a pale is created at the same time.

In an assembly of molecules such as a molecular crystal, there are minima, that is, pits, in the voids between the molecules, and the Poincaré–Hopf relation is replaced by the Morse equation (Johnson 1992)

$$N(\text{peaks}) - N(\text{passes}) + N(\text{pales}) - N(\text{pits}) = 0 \tag{6.20}$$

The function $\nabla \rho(\mathbf{r})$ defines a field of vectors directed at each point along the gradient of the charge density. The gradient vectors originate at critical points with positive curvature, and terminate at points with negative curvature. Thus, the gradient vectors terminate at the maxima of the distribution (these, in general, coincide with the atomic nuclei), which are therefore called the *attractors* of the distribution.

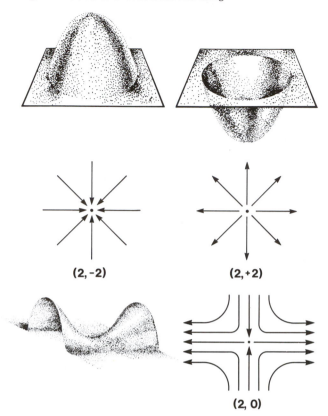

FIG. 6.7 Illustration of $(2, -2)$, $(2, +2)$, and $(2, 0)$ critical points in a two-dimensional distribution, representing a maximum, a minimum, and a saddle point, respectively. Gradient vectors originating and terminating in the critical points are shown. The bond path in the lower figure corresponds to the horizontal line containing the two gradient vectors emanating from the $(2, 0)$ critical point. *Source*: Bader (1990), Bader and Laidig (1991).

The saddle point between two atoms is a $(3, -1)$ critical point. The saddle point is the origin of the gradient vectors along the direction in which the density is a minimum. The gradient vectors in this direction link the $(3, -1)$ critical point with the atoms, and constitute the *bond path*, connecting the atoms. In the plane perpendicular to the bond path at the $(3, -1)$ critical point, the gradient vectors terminate as illustrated for the two-dimensional case in Fig. 6.7.

6.2.2 The Surface of Zero Flux

The points at which the nuclei are located are attractors of the gradient vectors of the electron distribution. The region containing all gradient paths terminating at the attractor defines a basin which is associated with the nucleus. Any gradient path originating in the basin terminates at the attractor. The space of the distribution is thus partitioned into regions which each contain one attractor. They

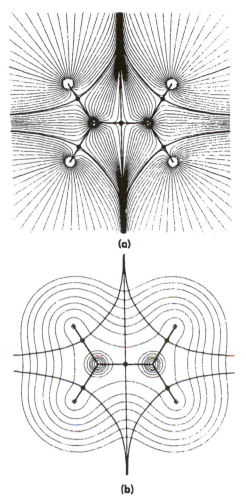

FIG. 6.8 (a) Gradient vector field of the charge density in the plane of the ethylene molecule. Each line represents a trajectory of $\nabla\rho(r)$. Trajectories which terminate at the positions of the nuclei and trajectories which terminate and originate at the $(3, -1)$ critical point in the charge distribution are shown. The full circles indicate the positions of the $(3, -1)$ critical points. Heavy lines represent the gradient paths originating at the $(3, -1)$ critical points and defining the bond paths. (b) A superposition of the trajectories associated with the $(3, -1)$ critical points on a contour map of the charge density. *Source*: Bader et al. (1981).

are defined as the *atomic basins*. This is illustrated in Fig. 6.8 for the ethylene molecule. The boundaries of the basins are never crossed by a gradient vector. Thus, if at each point the normal of the surface is given by $\mathbf{n}(\mathbf{r})$, the total boundary surface of the basin Ω is defined by

$$\nabla\rho(\mathbf{r})\cdot\mathbf{n}(\mathbf{r}) = 0 \qquad (6.21)$$

Since the surface is not crossed by any gradient lines, it is referred to as the *surface of zero flux*. As further discussed below, the virial theorem is satisfied for each of the regions of space satisfying the zero-flux boundary condition.

TABLE 6.3 Atomic Charges (lel) in Urea Based on Theoretical MP2 (Møller–Plesset 2) and HFS (Hartree–Fock–Slater) Densities, According to Different Partitioning Methods

Method/Basis Set		Mulliken	Løwdin	Hirshfeld	Bader
MP2/6-31G**	C	0.810	0.195	0.175	2.188
	O	−0.567	−0.349	−0.343	−1.270
	N	−0.735	−0.320	−0.166	−1.356
	H1	0.320	0.209	0.130	0.468
	H2	0.294	0.189	0.120	0.429
HFS/TZD	C	0.771		0.183	1.590
	O	−0.602		−0.318	−1.061
	N	−0.106		−0.200	−1.090
	H1	0.021		0.139	0.429
	H2	0.001		0.128	0.397

Source: Velders 1992.

Though rare, there are cases in which the total density shows minor maxima at non-nuclear positions. As all $(3, -3)$ critical points are attractors of the gradient field, basins occur which do not contain an atomic nucleus. These non-nuclear basins (which have been found in Si—Si bonds[1] in Li metal, and some other cases, distinguish the zero-flux partitioning from other space partitioning methods.

Velders has compared the integrated atomic charges obtained by Bader partitioning of a number of theoretical densities to those obtained by the stockholder definition and the theoretical Mulliken and Løwdin partitioning schemes (Velders 1992). In agreement with other studies (Bachrach and Streitweiser 1989), it is found that the Bader charges tend to be much larger than those from other space partitioning methods, as illustrated in Table 6.3 for the urea molecule. Since the center of gravity of the electrons in the nuclear basins does not coincide with the nuclear position, local atomic dipole moments from the Bader partitioning are quite large, and counteract the molecular dipole moment based solely on the net charges. The total dipole moment of an isolated molecule is, of course, unambiguous, and not dependent on the partitioning scheme.

6.3 Chemical Bonding and the Topology of the Total Electron Density Distribution

6.3.1 The Laplacian of the Electron Density

An important function of the electron density is its *Laplacian*, defined as

$$\nabla^2 \rho(\mathbf{r}) = \partial^2 \rho(\mathbf{r})/\partial x^2 + \partial^2 \rho(\mathbf{r})/\partial y^2 + \partial^2 \rho(\mathbf{r})/\partial z^2 \qquad (6.22)$$

The Laplacian is invariant under a rotation of the coordinate system, and is equal to the trace of the Hessian matrix $\mathbf{H}(\mathbf{r})$ with elements $\partial^2 \rho/\partial x_i \, \partial x_j$. The

---

[1] There are indications that the appearance of non-nuclear attractors in silicon is basis-set dependent.

Laplacian is related to the *electronic energy density* $E(\mathbf{r})$ of the charge distribution, defined as

$$E(\mathbf{r}) = G(\mathbf{r}) + V(\mathbf{r}) \tag{6.23}$$

In this equation, $V(\mathbf{r})$ is the potential energy, including exchange, at the point $\mathbf{r}$. The term $G(\mathbf{r})$ is a local one-electron kinetic energy density, defined as the scalar product of the gradient of the wave function and the gradient of the complex conjugate of the wave function, or (using atomic units) (Bader 1990, p. 147)

$$G(\mathbf{r}) = \tfrac{1}{2}n\, \nabla\psi(\mathbf{r})^* \cdot \nabla\psi(\mathbf{r}) \tag{6.24}$$

where n is the number of electrons.

For a basin Ω defined by the surface of zero flux, and therefore implicitly also for the whole system, integration over $G(\mathbf{r})$ gives the kinetic energy K, defined by

$$K \equiv -\tfrac{1}{2}\langle\psi|\nabla^2|\psi\rangle \tag{6.25}$$

The proof that $\int_\Omega G(\mathbf{r})\, d\mathbf{r} = K(\Omega)$ can be found in work by Bader (1990, p. 148) and Velders (1992, p. 22).

It is common to illustrate the Laplacian by the function $L(\mathbf{r})$, defined as

$$L(\mathbf{r}) \equiv -(\hbar^2/4m)\nabla^2\rho(\mathbf{r}) \tag{6.26a}$$

or, in atomic units,

$$L(\mathbf{r}) = -\tfrac{1}{4}\nabla^2\rho(\mathbf{r}) \tag{6.26b}$$

The relation between $L(\mathbf{r})$ and the components of the local energy density $E(\mathbf{r})$ is given by the equation

$$-L(\mathbf{r}) = (\hbar^2/4m)\nabla^2\rho(\mathbf{r}) = 2G(\mathbf{r}) + V(\mathbf{r}) \tag{6.27}$$

As noted by Bader, this expression is unique in relating a property of the electronic charge density to the local components of the total energy.

It is important that $L(\mathbf{r})$ *vanishes when the integration is performed over the zero-flux surface atomic basin*. This is because the integral over $L(\mathbf{r})$ can be replaced by the surface integral over the flux at the surface (Bader 1990):

$$L(\Omega) = \int_\Omega L(\mathbf{r}) = (\hbar^2/4m)\int_\Omega \nabla^2\rho(\mathbf{r})\, d\mathbf{r}$$

$$= -(\hbar^2/4m)\int_{\text{surface},\,\Omega} \nabla\rho(\mathbf{r})\cdot n(\mathbf{r})\, dS(\Omega, r) = 0 \tag{6.28}$$

in which Ω describes the atomic basin.

Since $L(\Omega)$ vanishes, $2E_{\text{kin}}(\Omega) = -E_{\text{pot}}(\Omega)$, and the *virial theorem*

$$2E_{\text{kin}} = -E_{\text{pot}} \tag{6.29}$$

which is valid for the whole system, is equally obeyed for the atomic basins defined by the zero-flux surface. This is a crucial feature of the *virial partitioning* discovered by Bader.

As a consequence of Eq. (6.27), the sign of the Laplacian at a point determines whether the negative potential energy or the positive kinetic energy is in excess of the virial ratio $|E_{\text{pot}}|/|E_{\text{kin}}| = 2$ at that point. In negative regions of the Laplacian

[$L(\mathbf{r})$ positive], the potential energy dominates the local electronic energy and the local contribution to the virial theorem [Eq. (6.29)]. Conversely, where the Laplacian is positive, the kinetic energy dominates the virial contribution. At the $(3, -3)$ critical points, all curvatures are negative, so the sign of the Laplacian is negative [$L(\mathbf{r})$ is positive], and the potential energy dominates. This is the result of the importance of the electron–nuclear attractions in the regions very close to the nuclei. The spike at the nuclear position in $L(\mathbf{r})$ (Fig. 6.9) is surrounded by a pronounced hole, representing an area where the positive kinetic energy is dominant.

In general, in regions of space where the Laplacian is negative, the electronic charge is concentrated as the negative eigenvalues of the Hessian represent density accumulation. Thus, a plot of $L(\mathbf{r})$ shows *maxima* in regions of density accumulation and minima in regions of depletion.

Relief maps of the charge density and $L(\mathbf{r})$ in the plane of the water molecule are shown in Fig. 6.9. The $L(\mathbf{r})$ at the bond critical point shows an accumulation in the internuclear surface. This is due to the shared electrostatic attraction of the electrons by both nuclei. Such a *shared interaction* is typical for covalent bonds.

6.3.2 Classification of Bonds Based on the Topology of the Electron Density

At the $(3, -1)$ critical points, the charge density along the bond paths connecting bonded atoms attains its minimum, but it is a maximum in the internuclear surfaces containing the $(3, -1)$ critical points. Accordingly, the principal curvature along a bond path, designated λ_3, is positive, while the remaining two, λ_1 and λ_2, are negative. In the plane of λ_1 and λ_2, density is concentrated at the critical point, while density is depleted in the bond path direction.

The value of the density at the critical point ρ_b, the values of the curvature, and the asymmetry of the curvature provide the information for a density-based classification of chemical bonding. Several parameters are used to classify a bond:

1. Bond order
The *bond order* is defined by the value of the charge density at the bond critical point ρ_b. The value of ρ_b increases with the number of assumed electron pair bonds, leading for specific bond types to relations of the form

$$n = \exp\left[A(\rho_b - B)\right] \tag{6.30}$$

where n is the bond order, and the coefficients A and B are constants specific for each bond type. For carbon–carbon bond, for example, B is set to the ρ_b value for ethane, and A is chosen such as to give a bond order of 2 for ethylene, or 3 for acetylene.

2. The value of $\nabla^2\rho$ at the Bond Critical Point
For closed-shell interactions, there will be no density accumulation in the bond. This means a deep minimum along the path connecting the nuclei, that is, a positive value of λ_3, and no contraction perpendicular to the bond and thus no strongly negative values of λ_1 and λ_2. Consequently, a positive value of $\nabla^2\rho$ is typical for a closed-shell interaction.

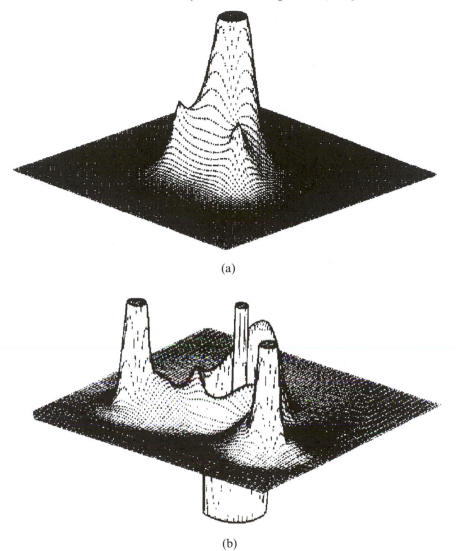

(a)

(b)

FIG. 6.9 Relief maps for (a) ρ and (b) $-\nabla^2\rho$ for the plane of the water molecule. There are spike-like maxima in both functions at the oxygen nucleus (terminated at arbitrary values) and in $-\nabla^2\rho$ at the hydrogen positions. The valence shell of charge concentration (VSCC) of oxygen in the $-\nabla^2\rho$ map exhibits maxima and saddle points in the two-dimensional relief maps. In addition, there are two local maxima, one along each bond path to a proton. *Source*: Bader (1990).

For the covalent bonds in the diatomic molecules studied by Bader and Essén (1984), values of λ_1 and λ_2 vary from -25 to $-45\ e\text{Å}^{-5}$, while λ_3 is positive in the range of 0–$45\ e\text{Å}^{-5}$. The sum of the curvatures, $\nabla^2\rho$, is invariably negative, indicating the concentration of electron density in the internuclear region. But for second-row atoms, the larger positive value of λ_3 may dominate the Laplacian.

For the Si—O bonds in silicates, for example, $\nabla^2\rho$ is as large as $+20$ eÅ^{-5} (chapter 11), though the bond clearly has appreciable covalent character as the charge density is contracted in the plane perpendicular to the bond path.

Closed-shell interactions occur between ions in ionic crystals and between atoms in adjacent molecules in molecular crystals. For X—$H\cdots Y$ hydrogen bonds, the X—H interaction is covalent, but the $H\cdots Y$ region in normal hydrogen bonds shows a small value of ρ_b and a positive value of $\nabla^2\rho$, typical for a closed-shell interaction. The value of $\nabla^2\rho$ in the $O\cdots H$ region of $(H_2O)_2$, for example, is found to be 0.42 eÅ^{-5} (Bader 1990, p. 292). For the stronger hydrogen bond in $(HF)_2$, the value is 0.81 eÅ^{-5} indicating again an absence of covalency. Recent experimental studies, however, indicate a covalent interaction for very short hydrogen bonds, as discussed in chapter 12.

3. The ratio $|\lambda_1|/\lambda_3$

The quantity $|\lambda_1|/\lambda_3$, that is, the ratio between the largest perpendicular contraction at the $(3, -1)$ critical point and the parallel concentration towards the nuclei, is <1 for closed-shell interactions. For shared interactions, its value increases with bond strength and decreasing ionicity of a bond. It decreases, for example, in the sequence ethylene (4.31), benzene (2.64), ethane (1.63).

4. The ellipticity ε of a bond

The *ellipticity* of a bond is defined as

$$\varepsilon = \lambda_1/\lambda_2 - 1 \tag{6.31}$$

As λ_1 represents the contraction of the density perpendicular to the bond path, ε is ≥ 0. For a cylindrically symmetric σ bond, the ellipticity will be zero, while it is different from zero for double bonds which have a π contribution.

The validy of ε in classifying bonds is borne out by the analysis of theoretical densities; the ellipticity of the C—C bonds in the series ethane, benzene, ethylene increases from 0.0 to 0.23 to 0.45 (Bader et al. 1983). For the very long bridgehead bonds in propellanes, which are shared by three rings, the ellipticity can be quite large, as in [2.1.1] propellane,

[2.1.1] propellane

for which a value of 7.21 has been reported (Bader and Laidig 1991). In such bonds, the density at the bond critical point is low and the curvature along the direction connecting the three-membered-ring nuclear attractors is very small. Such bonds are quite likely to rupture, leading to ring opening. The $(3, -1)$ critical point in the bridgehead bonds distinguishes the cyclopropellanes from bicyclic molecules like bicyclopentane and bicyclooctane. As pointed out by Bader and Laidig (1991), the topological analysis of the total density in these bonds has marked advantages over examination of the deformation density, because, in the latter, the density subtracted at the midpoint depends very much on the distance between the proximal atoms.

Since the Laplacian is a second-derivative function, it is very sensitive to small changes in the density. Quantitative results are basis-set dependent, and convergence requires an extensive basis set (Gatti et al. 1992; R. Destro, private communication).

6.3.3 Topological Analysis of Experimental Densities

Topological analysis of the total density has a considerable advantage over the use of the deformation densities in that it is reference-density independent. There is no need to define hybridized atoms to analyze the nature of covalent bonding, and the ambiguity when using the standard deformation density, noted above in the discussion on propellanes, does not occur.

For analysis of experimental results, the static model density must be used to eliminate noise, truncation effects, and thermal smearing. Some caution is called for, because the reciprocal space representation of the Laplacian is a function of $F(\mathbf{H}) \cdot H^2$, and thus has poor convergence properties.[2] This difficulty is only partly circumvented by use of the model density, as high-resolution detail may be quite dependent on the nature of the model functions, as is evident in the experimental study of the quartz polymorph coesite discussed in chapter 11.

The topological analysis is especially informative in the comparison of related molecules and solids. Results for a number of related bonds in the amino acids L-alanine (Gatti et al. 1992) and L-dopa (dihydroxyphenylalanine, Fig. 6.10) (Howard et al. 1995), based on data at 23 K and 173 K, respectively, are compared in Table 6.4. We note quite good agreement between the experiments for ρ_b, the electron density at the bond critical points. But the experimental values tend to be 10–20% higher than the theoretical values. Howard et al. (1995), in the study on L-dopa, performed a multipole refinement on theoretical structure factors, and found that in the corresponding model density the ρ_b values were systematically larger than those from the exact density. This suggests an inadequacy in the multipole model that requires further investigation.

The values of the Laplacian $\nabla^2 \rho_b$ and its components $\lambda_{1,2,3}$ obtained in the

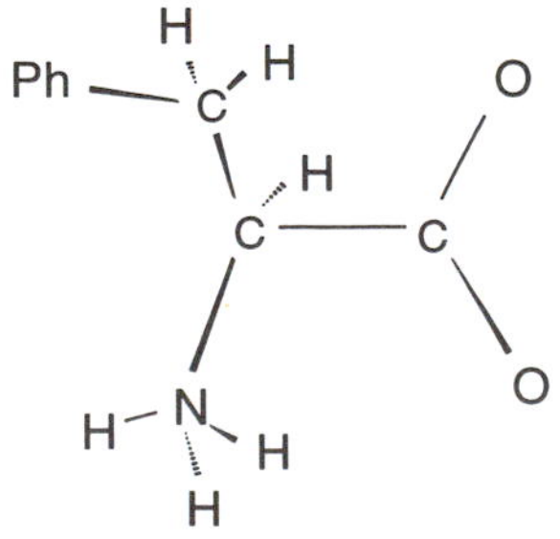

FIG. 6.10 Schematic drawing of the amino acids L-dopa and L-alanine. For L-dopa, Ph = 3,4-dihydroxyphenyl; for L-alanine, Ph = H.

---

[2] See Table 8.1 for the dependence of derived properties on the power of the magnitude of scattering vector $\mathbf{H}$.

TABLE 6.4 Ab-initio RHF (Restricted Hartree–Fock) Bond Critical Point Properties for Corresponding Bonds between C, N, and O in the L-Dopa and L-Alanine Zwitterionic Monomers at the X-ray Experimental Geometry

Bond $(x-y)$	Method	R	R_x	ρ_b	$\nabla^2\rho_b$	λ_1	λ_2	λ_3	ε
C_1-O_1	Exp.[a]	1.251	0.602	2.70	-32.6	-28.3	-22.6	18.4	0.25
	Exp.[b]	1.248	0.517	3.02	-39.0	-31.4	-26.4	18.8	0.19
	IAM[bc]	1.248	0.447	2.05	2.4	-10.8	-10.6	23.8	0.02
	6-31G**[b]		0.405	2.63	-1.8	-26.1	-24.6	49.0	0.06
C_1-O_2	Exp.[a]	1.260	0.646	2.64	-38.8	-28.2	-22.6	12.1	0.25
	Exp.[b]	1.267	0.540	2.86	-29.5	-27.6	-24.4	22.5	0.13
	IAM[b]	1.268	0.463	1.99	-0.8	-10.3	-10.1	19.6	0.02
	6-31G**[b]		0.413	2.57	-8.38	-24.9	-22.9	39.3	0.08
C_1-C_2	Exp.[a]	1.536	0.508	1.71	-12.0	-12.5	-11.0	11.4	0.14
	Exp.[b]	1.535	0.779	1.76	-10.9	-13.5	-11.2	13.8	0.21
	IAM[b]	1.533	0.767	0.12	1.3	-5.3	-5.1	11.7	0.04
	6-31G**[b]		0.697	1.78	-18.7	-13.3	-12.6	7.24	0.06
C_2-C_3	Exp.[a]	1.340	0.510	1.79	-13.1	-13.2	-11.4	11.5	0.16
	Exp.[b]	1.526	0.779	1.67	-10.1	-11.5	-11.1	12.5	0.04
	IAM[b]	1.526	0.763	1.18	1.1	-5.3	-5.3	11.7	0.00
	6-31G**[b]		0.797	1.74	-16.7	-12.0	-11.8	7.24	0.04
C_2-N	Exp.[a]	1.495	0.904	1.62	-8.4	-12.7	-8.6	12.9	0.47
	Exp.[b]	1.488	0.635	1.70	-11.0	-13.9	-10.7	13.6	0.30
	IAM[b]	1.494	0.680	1.35	3.2	-6.7	-6.7	16.6	0.00
	6-31G**[b]		0.464	1.53	-2.7	-8.5	-7.5	13.5	0.14

R: bond length; R_x: distance of bond critical point from first atom; ρ_b density at bond critical point. Units are Å, eÅ^{-3}, and eÅ^{-5}, respectively.
Exp., Experimental.
[a] L-dopa.
[b] L-alanine.
[c] IAM = Independent Atom Model.
Source: Gatti et al. (1992), Howard et al. (1995).

two experiments are in remarkable qualitative agreement. But we note the pronounced discrepancies between theoretical and experimental values for the second derivatives. This is especially evident for $\nabla^2\rho_b$, but also for its components. The agreement among the experiments for the position of the critical point along the bond path, described by the distance R_x of the point from the first atom, is not that satisfactory, perhaps due to differences between the basis functions used in the two analyses.

The promolecule density shows $(3, -1)$ critical points along the bond paths, just like the molecule density. But, as the promolecule is hypothetical and violates the exclusion principle, it would be incorrect to infer that the atoms in the promolecule are chemically bonded. In a series of topological analyses, Stewart (1991) has compared the model densities and promolecule densities of urea, benzene, imidazole, and 9-methyladenine. It is found that the critical points of the promolecule density are generally close to those of the experimental molecular density. As may be expected, the true density at the critical point in the covalent bonds is higher (by about 50% for C—C bonds), and the Laplacian is much more negative than for the promolecule density. The ellipticity of the bonds is essentially zero for the density consisting of a superposition of spherical atoms.

7

The Electrostatic Moments of a Charge Distribution

The moments of a charge distribution provide a concise summary of the nature of that distribution. They are suitable for quantitative comparison of experimental charge densities with theoretical results. As many of the moments can be obtained by spectroscopic and dielectric methods, the comparison between techniques can serve as a calibration of experimental and theoretical charge densities. Conversely, since the full charge density is not accessible by the other experimental methods, the comparison provides an interpretation of the results of the complementary physical techniques. The electrostatic moments are of practical importance, as they occur in the expressions for intermolecular interactions and the lattice energies of crystals.

The first electrostatic moment from X-rays was obtained by Stewart (1970), who calculated the dipole moment of uracil from the least-squares valence-shell populations of each of the constituent atoms of the molecule. Stewart's value of 4.0 ± 1.3 D had a large experimental uncertainty, but is nevertheless close to the later result of 4.16 ± 0.4 D (Kulakowska et al. 1974), obtained from capacitance measurements of a solution in dioxane. The diffraction method has the advantage that it gives not only the magnitude but also the direction of the dipole moment. Gas-phase microwave measurements are also capable of providing all three components of the dipole moment, but only the magnitude is obtained from dielectric solution measurements.

We will use an example as illustration. The dipole moment vector for formamide has been determined both by diffraction and microwave spectroscopy. As the diffraction experiment measures a continuous charge distribution, the moments derived are defined in terms of the method used for space partitioning, and are not necessarily equal. Nevertheless, the results from different techniques (Fig. 7.1) agree quite well.

A comprehensive review on molecular electric moments from X-ray diffraction

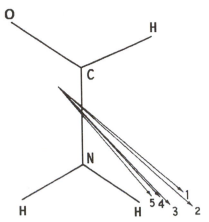

FIG. 7.1 Direction and magnitude of the dipole moment of formamide from various methods; the origin is at the center of mass of the molecule. (1) X-ray spherical atom. (2) X-ray aspherical atom. (3) Theory, double ζ. (4) Theory, extended basis. (5) Microwave (Coppens et al. 1979). For numerical information see Table 7.2.

data has been published by Spackman (1992). Spackman points out that despite a large number of determinations of molecular dipole moments and a few determinations of molecular quadrupole moments, it is not yet widely accepted that diffraction methods lead to valid experimental values of the electrostatic moments. This is quite unwarranted, as is clear from examination of the rapidly increasing number of diffraction results, summarized at the end of this chapter.

7.1 Moments of a Charge Distribution

7.1.1 Definitions

Use of the expectation value expression

$$\langle O \rangle = \int_{V_T} \hat{O}\rho(\mathbf{r})\, d\mathbf{r} \tag{6.5}$$

with the operator $\hat{O} = r_{\alpha_1}r_{\alpha_2}r_{\alpha_3}\ldots r_{\alpha_l}$, gives for the electrostatic moments of a charge distribution $\rho(\mathbf{r})$

$$\mu_{\alpha_1,\alpha_2,\alpha_3\ldots\alpha_l} = \int_{V_T} \rho(\mathbf{r})r_{\alpha_1}r_{\alpha_2}r_{\alpha_3}\ldots r_{\alpha_l}\, d\mathbf{r} \tag{7.1}$$

in which the r_α terms are the three components of the vector $\mathbf{r}$ ($\alpha_i = 1, 2, 3$), and the integral is over the volume V_T of the distribution.

For $l = 0$, Eq. (7.1) simply represents the integral over the charge distribution, which is the total charge—a scalar function described as the *monopole*. The higher moments are, in ascending order of l: the *dipole*, a vector; the *quadrupole*, a second-rank tensor; and the *octupole*, a third-rank tensor. Successively higher moments are named the *hexadecapole* ($l = 4$), the *triacontadipole* ($l = 5$), and the *hexacontatetrapole* ($l = 6$).

While the monopole represents the total charge, the higher poles are a measure of the *charge separation*. Two opposite charges of one electron unit must be separated by 1 Å to give a dipole moment of 4.803 D [1 Debye $= 10^{-18}$ esu $=$ $3.3356 \cdot 10^{-30}$ Cm (Cm $=$ coulomb meter); see Appendix K]. Thus, long molecules may have large dipole moments, especially when they carry groups of opposite polarity at their extremes. The dipeptide glycylglycine, for example, has a zwitterionic structure with NH_3^+ and COO^- groups at the two ends of its backbone. The moment derived from X-ray data is 24 D, compared with ≈ 27 D for the dipole moment in solution (Sakellaridis and Karageorgopolous 1974, Coppens et al. 1979). Similarly, large quadrupole moments occur for molecules with strongly polar, well-separated groups (see Table 7.3).

The moments defined by Eq. (7.1) are referred to as the *unabridged* moments. For moments with $l \geq 2$, an alternative, traceless definition is often used. In the traceless definition, the quadrupole moment $\Theta_{\alpha\beta}$ is given by

$$\Theta_{\alpha\beta} = \tfrac{1}{2} \int \rho(\mathbf{r})[3r_\alpha r_\beta - r^2 \delta_{\alpha\beta}] \, d\mathbf{r} \tag{7.2a}$$

where $\delta_{\alpha\beta}$ is the Kronecker delta function. The term $\int \rho(\mathbf{r})r^2 \, d\mathbf{r}$, which is subtracted from the diagonal elements of the tensor, corresponds to the spherically averaged second moment of the distribution.

The corresponding expression for the octupole moments is

$$\Omega_{\alpha\beta\gamma} = \tfrac{1}{2} \int \rho(\mathbf{r})[5r_\alpha r_\beta r_\gamma - r^2(r_\alpha \delta_{\beta\gamma} + r_\beta \delta_{\alpha\gamma} + r_\gamma \delta_{\alpha\beta})] \, d\mathbf{r} \tag{7.2b}$$

Expressions (7.2a) and (7.2b) follow from the following general expression for the lth-rank traceless tensor elements:

$$M^{(l)}_{\alpha_1\alpha_2\ldots\alpha_l} = \frac{(-1)^l}{l!} \int \rho(\mathbf{r})r^{2l+1} \frac{\partial^l}{\partial r_{\alpha_1} \partial r_{\alpha_2} \ldots \partial r_{\alpha_l}} \left(\frac{1}{r}\right) d\mathbf{r} \tag{7.3}$$

Though the traceless moments can be derived from the unabridged moments, the converse is not the case because the information on the spherically averaged moments is no longer contained in the traceless moments. The general relations between the traceless moments and the unabridged moments follow from Eq. (7.3).

For the quadrupole moments, we obtain with Eq. (7.2):

$$\Theta_{xx} = \tfrac{3}{2}\mu_{xx} - \tfrac{1}{2}(\mu_{xx} + \mu_{yy} + \mu_{zz}) = \mu_{xx} - \tfrac{1}{2}(\mu_{yy} + \mu_{zz}) \tag{7.4a}$$

and

$$\Theta_{xy} = \tfrac{3}{2}\mu_{xy} \tag{7.4b}$$

Expressions for the other elements of the traceless quadrupole tensor are obtained by simple permutation of the indices.

For a site of point symmetry 1, the electrostatic moment $\mu_{\alpha_1, \alpha_2, \alpha_3 \ldots \alpha_l}$ of order

l has $(l + 1)(l + 2)/2$ unique elements. In the traceless definition, the number of independent elements is smaller, because the trace of the tensor has been set to zero. The number of independent elements then is equal to $2l + 1$, that is, equal to the number of spherical harmonic functions of order l, to which the traceless moments are related, as discussed in section 7.2.1.

In a different form, the traceless moment operators can be written as the Cartesian spherical harmonics c_{lmp} multiplied by r^l, which defines the *spherical harmonic electrostatic moments*:

$$\Theta_{lmp} = \int \rho(\mathbf{r}) c_{lmp} r^l \, d\mathbf{r} \tag{7.5}$$

The factor r^l enters because the Cartesian spherical harmonics c_{lmp} are defined in terms of the direction cosines in a Cartesian coordinate system. The expressions for c_{lmp} are listed in appendix D. As an example, the c_{2mp} functions have the form $3z^2 - 1, xz, yz, (x^2 - y^2)/2$ and xy, where x, y and z are the direction cosines of the radial vector from the origin to a point in space.

The linear relationships between the traceless moments $\Theta_{\alpha\beta}$ and the spherical harmonic moments Θ_{lmp} are obtained by use of the definitions of the functions c_{lmp}. For example, for the quadrupolar moment element Θ_{xx}, we obtain the equality $(3x^2 - 1)/2 = a\{(x^2 - y^2)/2\} + b(3z^2 - 1)$. Solution for a and b for this and corresponding equations for the other moments leads to

$$\Theta_{zz} = \tfrac{1}{2}\Theta_{20}$$

$$\Theta_{xx} = \tfrac{1}{2}(3\Theta_{22+} - \tfrac{1}{2}\Theta_{20})$$

$$\Theta_{yy} = \tfrac{1}{2}(-3\Theta_{22+} - \tfrac{1}{2}\Theta_{20})$$

$$\Theta_{xz} = \tfrac{3}{2}\Theta_{21+}$$

$$\Theta_{yz} = \tfrac{3}{2}\Theta_{21-}$$

$$\Theta_{xy} = \tfrac{3}{2}\Theta_{22-} \tag{7.6}$$

The moments discussed in this chapter are sometimes referred to as the *outer moments* of the distribution, in contrast to the *inner moments* for which the powers of r in the operator $\hat{O}$ in Eq. (6.5) are negative. The electric field at the nucleus and the field gradient at the nucleus are examples of inner moments, which will be discussed in chapter 8.

7.1.2 The Origin Dependence of the Electrostatic Moments

With the exception of the charge, the values of the multipole moments, in general, depend on the choice of origin. Let us consider a shift of the origin by $\mathbf{R}$ $(R_\alpha, R_\beta, R_\gamma)$, as depicted in Fig. 7.2. Substitution of $r'_\alpha = r_\alpha - R_\alpha$ in Eq. (7.1) corresponds to a shift of origin by R_α, with components X, Y, Z in the original

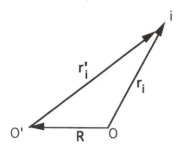

FIG. 7.2 The coordinates of a point i relative to a new origin O' at distance R from O.

coordinate system. For the x component of the first moment,

$$\mu'_x = \int \rho(\mathbf{r})(x - X)\,d\mathbf{r} = \mu_x - X\int \rho(\mathbf{r})\,d\mathbf{r} = \mu_x - Xq \qquad (7.7)$$

Similar algebra shows that the expressions for the transformed first and second moments are

$$\mu'_x = \mu_x - qX;\ \mu'_y = \mu_y - qY;\ \mu'_z = \mu_z - qZ$$
$$\mu'_{\alpha\alpha} = \mu_{\alpha\alpha} - 2\mu_\alpha R_\alpha + qR_\alpha^2$$
$$\mu'_{\alpha\beta} = \mu_{\alpha\beta} - \mu_\alpha R_\beta - \mu_\beta R_\alpha + qR_\alpha R_\beta \qquad (7.8)$$

For the traceless quadrupole moments, the corresponding equations are similarly obtained by substitution of $r'_\alpha = r_\alpha - R_\alpha$ and $\mathbf{r}' = \mathbf{r} - \mathbf{R}$ into Eq. (7.2), which gives

$$\Theta'_{\alpha\beta} = \Theta_{\alpha\beta} + \tfrac{1}{2}(3R_\alpha R_\beta - R^2\delta_{\alpha\beta})q - \tfrac{3}{2}(R_\beta\mu_\alpha + R_\alpha\mu_\beta) + \sum_{\gamma=1}^{3}(R_\gamma\mu_\gamma)\delta_{\alpha\beta} \qquad (7.9)$$

The analogous expressions for the higher moments are reported in the literature (Buckingham 1970).

Expressions (7.8) and (7.9) demonstrate that *the first nonvanishing moment is origin independent*. Thus, the dipole moment of a neutral molecule, but not that of an ion, is independent of origin; the quadrupole moment of a neutral molecule without dipole moment is not dependent on the choice of origin, and so on.

An element of an electrostatic moment tensor can only be nonzero if the distribution has a component of the same symmetry as the corresponding operator. In other words, the integrand in Eq. (7.1) must have a component that is invariant under the symmetry operations of the distribution, namely, it is totally symmetric with respect to the operations of the point group of the distribution. As an example, for the x component of the dipole moment to be nonzero, $\rho(\mathbf{r})x$ must have a totally symmetric component, which will be the case if $\rho(\mathbf{r})$ has a component with the symmetry of x. The symmetry restrictions of the spherical electrostatic moments are those of the spherical harmonics given in appendix section D.4. Restrictions for the other definitions follow directly from those listed in this appendix.

7.2 Electrostatic Moments from Diffraction Data

7.2.1 Atomic Electrostatic Moments in Terms of the Parameters of the Multipole Formalism

The atomic electrostatic moments of an atom are obtained by integration over its charge distribution. As the multipole formalism separates the charge distribution into pseudoatoms, the atomic moments are well defined.

The total charge distribution of atom i consists of the sum of the nuclear and electronic distributions $\rho_{\text{total},i}(\mathbf{r}) = \rho_{\text{nuclear},i} - \rho_{e,i}$. The electrostatic moments follow

$$\mu_{\alpha_1,\alpha_2,\alpha_3\ldots\alpha_l} = \int \rho_{\text{total},i}(\mathbf{r}) r_{\alpha_1} r_{\alpha_2} r_{\alpha_3} \ldots r_{\alpha_l} \, d\mathbf{r} \tag{7.10}$$

If the moments are referred to the nuclear position, only the electronic part of the charge distribution contributes to the integral. According to the multipole formalism of Eq. (3.32),

$$\rho_{e,i}(\mathbf{r}) = P_{i,c}\rho_{i,\text{core}}(r) + P_{i,v}\kappa_i^3 \rho_{i,\text{valence}}(\kappa_i r)$$

$$+ \sum_{l=0}^{l_{\max}} \kappa_i'^3 R_{i,l}(\kappa_i' r) \sum_{m=0}^{l} \sum_p P_{i,lmp} d_{lmp}(\theta, \phi) \tag{7.11}$$

with $p = \pm$ when $m > 0$, and $R_{i,l}(\kappa_i' r)$ representing the radial dependence of the spherical harmonic deformation functions on atom i.

Substitution in Eq. (7.10) gives, for the jth moment of the atomic density with respect to the nuclear position,

$$\mu^j = \mu_{\alpha_1,\alpha_2,\alpha_3\ldots a_j} = -\int \Bigg[P_{i,c}\rho_{\text{core}}(r) + P_{i,v}\kappa_i\rho_{i,\text{valence}}(\kappa_i r)$$

$$+ \sum_{l=0}^{l_{\max}} \kappa_i'^3 R_{i,l}(\kappa_i' r) \sum_{m=0}^{l} \sum_p P_{i,lmp} d_{lmp}(\theta, \phi) \Bigg] r_{\alpha_1} r_{\alpha_2} \ldots r_{\alpha_j} \, d\mathbf{r} \tag{7.12}$$

in which the minus sign arises because of the negative charge of the electrons.

The spherical terms of the charge distribution contribute only to the diagonal elements of the even moments. Evaluation of these contributions is further discussed in section 7.2.3. For the higher-order terms in the summation, we have, using the symbol $\hat{O}_j$ for the jth moment operator:

$$\mu^j = -\kappa_i'^3 \int \hat{O}_j \sum_{l=1}^{l_{\max}} \Bigg[\sum_{m=0}^{l} \sum_p P_{lmp} d_{lmp} R_l \Bigg] d\mathbf{r} \tag{7.13}$$

where, as before, $p = \pm$. The requirement that the integrand be totally symmetric means that only the dipolar terms in the multipole expansion contribute to the dipole moment. In the traceless definitions, this is equally true for all higher moments; for example, only the quadrupolar terms of the multipole expansion will contribute to the quadrupole moment. In general, for the traceless definition, *the lth-order multipoles are the sole contributors to the lth moments*. Accordingly, because of the orthogonality of the spherical harmonics, for $\hat{O} = c_{lmp}r^l$ the sole contributor to the integral of Eq. (7.13) is the density function $d_{l'm'p'}$, for which $l = l'$, $m = m'$, and $p = p'$, or, for the spherical harmonic quadrupole moments

$$\Theta_{lmp} = -P_{lmp} \int \hat{O}_{lmp}[d_{lmp}R_l]\, d\mathbf{r} \tag{7.14}$$

Substituting

$$R_l = \frac{(\kappa'\zeta)^{n(l)+3}}{(n(l)+2)!} r^{n(l)} \exp(-\kappa'\zeta r)$$

and $\hat{O}_{lmp} = c_{lmp}r^l$, and subsequent integration over r gives

$$\Theta_{lmp} = -P_{lmp} \frac{1}{(\kappa'\zeta)^l} \frac{(n(l)+l+2)!}{(n(l)+2)!} \frac{1}{D_{lm}M_{lm}} \int y_{lmp}^2 \sin\theta\, d\theta\, d\phi \tag{7.15}$$

where the definitions (chapter 3 and appendix D)

$$d_{lmp} = L_{lm}c_{lmp} = (L_{lm}/M_{lm})y_{lmp} = (D_{lm})^{-1}y_{lm}, \text{ and } c_{lmp} = (1/M_{lm})y_{lmp} \tag{7.16}$$

have been used.

Since the y_{lmp} functions are wave-function normalized, we get

$$\Theta_{lmp} = -P_{lmp} \frac{1}{(\kappa'\zeta)^l} \frac{(n(l)+l+2)!}{(n(l)+2)!} \frac{L_{lm}}{(M_{lm})^2}$$

$$= P_{lmp} \frac{1}{(\kappa'\zeta)^l} \frac{(n(l)+l+2)!}{(n(l)+2)!} \frac{1}{D_{lm}M_{lm}} \tag{7.17}$$

Application to dipolar terms with $n(l) = 2$, $L_{lmp} = 1/\pi$, and $M_{lm} = (\frac{3}{4}\pi)^{1/2}$, and $D_{lm} = M_{lm}/L_{lm}$ (Fig. 3.5), gives the x component of the atomic dipole moment as

$$\mu_x = -\int P_{11+}d_{11+}R_1 x\, d\mathbf{r} = -\frac{20}{3\kappa'\zeta}P_{11+} \tag{7.18}$$

For the atomic quadrupole moments in *the spherical definition*, we obtain directly, using $n(l) = 2$, $l = 2$ in Eq. (7.17):

$$\Theta_{20} = -\frac{30}{(\kappa'\zeta)^2} \frac{L_{20}}{(M_{20})^2} P_{20} = -\frac{36\sqrt{3}}{(\kappa'\zeta)^2} P_{20} \tag{7.19}$$

and for the other elements

$$\Theta_{2mp} = -\frac{30}{(\kappa'\zeta)^2} \frac{L_{2m}}{(M_{2m})^2} P_{2mp} = -\frac{6\pi}{(\kappa'\zeta)^2} P_{2mp} \tag{7.20}$$

As the traceless quadrupole moments are linear combinations of the spherical harmonic quadrupole moments, the corresponding expressions follow directly from Eqs. (7.19)–(7.20) and (7.6). We obtain, for $n(2) = 2$,

$$\Theta_{zz} = -\frac{18\sqrt{3}}{(\kappa'\zeta)^2} P_{20}$$

$$\Theta_{yy} = +\frac{9}{(\kappa'\zeta)^2} (\sqrt{3}P_{20} + \pi P_{22+})$$

$$\Theta_{xx} = +\frac{9}{(\kappa'\zeta)^2} (\sqrt{3}P_{20} - \pi P_{22+})$$

$$\Theta_{xz} = -\frac{9\pi}{(\kappa'\zeta)^2} P_{21+} \tag{7.21}$$

and analogously for the other off-diagonal elements. Thus, the atomic electrostatic moments are simple functions of the parameters of the multipole formalism.

7.2.2 Molecular Moments as a Sum Over the Pseudoatom Moments

In the multipole-model description, the charge density is a sum of atom-centered density functions. The moments of the entire distribution are obtained as the sum over the individual atomic moments plus contributions due to the shift to a common origin.

If individual atomic coordinate systems are used, as is common when chemical constraints are applied in the least-squares refinement, they must first be rotated to have a common orientation. The transformation of the population parameters under coordinate-system rotation is described in section D.5 of appendix D (Cromer et al. 1976, Su 1993, Su and Coppens 1994).

The transition to a common coordinate origin requires use of the origin-shift expressions (7.7)–(7.9), with $\mathbf{R} = -\mathbf{r}_i$ for an atom at $\mathbf{r}_i$. The first three moments summed over the atoms i located at $\mathbf{r}_i$ become

$$q_{\text{total}} = \sum q_i \tag{7.22}$$

$$\boldsymbol{\mu}_{\text{total}} = \sum_i \boldsymbol{\mu}_i + \sum \mathbf{r}_i q_i \tag{7.23}$$

and

$$\mu_{\alpha\beta,\text{total}} = \sum_i (\mu_{\alpha\beta i} + r'_{\beta i}\mu'_{\alpha i} + r'_{\alpha i}\mu'_{\beta i} + r'_{\alpha i}r'_{\beta i}q_i) \tag{7.24}$$

with $\alpha, \beta = x, y, z$. Analogous expressions for the traceless components $\Theta_{\alpha\beta}$ follow directly from Eq. (7.9).

7.2.3 The Electrostatic Moments of the Deformation Density

When the electrostatic moments are to be obtained by integration over direct space, it is advantageous to use the deformation density rather than the total density in order to reduce truncation effects. It is therefore important to analyze the relation between the moments of the two distributions.

As noted above, in the traceless definition the lth-order multipoles are the sole contributors to the lth electrostatic moments. This implies that the traceless moments derived from the total density $\rho(\mathbf{r})$ and from the deformation density $\Delta\rho(\mathbf{r})$ are identical, that is, $\Theta_{lmp}(\rho) = \Theta_{lmp}(\Delta\rho)$ for $l \geq 2$.

In the nonzero trace definition, this equality is no longer valid. To illustrate the relation for the diagonal elements of the second-moment tensor, we rewrite the xx element as

$$\mu_{xx}(\rho_{\text{total}}) = \int \rho x^2 \, d\mathbf{r} = \int \rho_{\text{promolecule}} x^2 \, d\mathbf{r} + \int \Delta\rho x^2 \, d\mathbf{r} \qquad (7.25)$$

The promolecule is the sum over spherical atom densities, so we may write

$$\int \rho_{\text{promolecule}} x^2 \, d\mathbf{r} = \int \sum_i \rho_{i,\text{spherical atom}} x^2 \, d\mathbf{r} = \sum_i \int \rho_{i,\text{spherical atom}} x^2 \, d\mathbf{r} \qquad (7.26)$$

If $\mathbf{R}_i = (X_i, Y_i, Z_i)$ is the position vector for atom i, the contribution of this atom can be written as

$$\mu_{i,xx,\text{spherical atom}} = \int \rho_{i,\text{spherical atom}} x^2 \, d\mathbf{r} = \int \rho_{i,\text{spherical atom}}(x - X_i)^2 \, d\mathbf{r}$$
$$+ X_i \int 2\rho_{i,\text{spherical atom}}(x - X_i) \, d\mathbf{r} + X_i^2 \int \rho_{i,\text{spherical atom}} \, d\mathbf{r}$$
$$(7.27a)$$

Since the last two integrals are proportional to the atomic dipole moment and its net charge, respectively, they will be zero for neutral spherical atoms, or

$$\mu_{i,xx,\text{spherical atom}} = \int \rho_{i,\text{spherical atom}}(x - X_i)^2 \, d\mathbf{r} \qquad (7.27b)$$

With $\langle (x - X_i)^2 \rangle = \frac{1}{3}\langle r_i^2 \rangle$, $\langle r_i^2 \rangle = \int \rho_i(r) r^2 \, d\mathbf{r}$, we obtain for the promolecule

$$\int \rho_{\text{promolecule}} x^2 \, d\mathbf{r} = \frac{1}{3} \sum_{\text{atoms}} \langle r^2 \rangle_{\text{spherical atom}} \qquad (7.28)$$

and, by substitution in Eq. (7.25),

$$\mu_{xx}(\rho_{\text{tot}}) = \mu_{xx}(\Delta\rho) + \frac{1}{3} \sum_{\text{atoms}} \langle r^2 \rangle_{\text{spherical atom}} \qquad (7.29a)$$

in which the second moment of the molecular deformation density is equal to

$$\mu_{xx}(\Delta\rho) = \sum_i \left(\int \Delta\rho_i x^2 \, d\mathbf{r} + 2X_i\mu_{xi} + X_i^2 q_i \right) \qquad (7.29b)$$

with μ_i and q_i being the atomic dipole moment and the charge on atom i, respectively.

The right-hand side of Eq. (7.28) can be derived readily from analytical expressions for the atomic wave functions. Results for Hartree–Fock wave functions have been tabulated by Boyd (1977). Since the off-diagonal elements of the second-moment tensor vanish for the spherical atom, the second term in Eq. (7.29a) disappears for the off-diagonal elements, and therefore the off-diagonal elements are identical for the total and deformation densities.

The relation between the second moments $\mu_{\alpha\beta}$ of the deformation density and the traceless moments $\Theta_{\alpha\beta}$ can be illustrated as follows. From Eq. (7.2a), we may write

$$\Theta_{\alpha\beta}(\Delta\rho) = \tfrac{3}{2}\mu_{\alpha\beta}(\Delta\rho) - \tfrac{1}{2}\delta_{\alpha\beta}\int \Delta\rho r^2 \, d\mathbf{r} \tag{7.30}$$

Only the spherical and dipolar density terms contribute to the integral on the right. Assuming, for simplicity, that the deformation is represented by the valence-shell distortion (i.e., the second monopole in the aspherical atom expansion is not used), we have, with density functions ρ normalized to 1, for each atom:

$$(\Delta\rho)_{\text{spherical}} = \kappa^3 P_{\text{valence}}\,\rho_{\text{valence}}(\kappa r) - P^0_{\text{valence}}\rho_{\text{valence}}(r) \tag{7.31}$$

and

$$\int \Delta\rho r^2 \, d\mathbf{r} = \int \left[\sum_i (\kappa_i^3 P_{i,\,\text{valence}}\,\rho_{i,\,\text{valence}}(\kappa_i r) - P^0_{i,\,\text{valence}}\,\rho_{i,\,\text{valence}}(r))r^2 \, d\mathbf{r} \right.$$

$$= \sum_i [(P_{i,\,\text{valence}}/\kappa_i^2 - P^0_{i,\,\text{valence}})\langle r_i^2 \rangle_{\text{spherical valence shell}}$$

$$+ R_i^2(P_{i,\,\text{valence}} - P^0_{i,\,\text{valence}})] \tag{7.32}$$

The second term occurs because the integral $X_i^2 \int \rho_{i,\,\text{spherical atom}} \, d\mathbf{r}$ in Eq. (7.27a) is no longer zero for the non-neutral atom. Substitution of Eq. (7.32) into Eq. (7.30) gives the required relation between $\Theta_{\alpha\beta}(\Delta\rho)$ and $\mu_{\alpha\beta}(\Delta\rho)$.

7.2.4 Electrostatic Moments of a Subvolume of Space by Fourier Summation

When space is partitioned with discrete boundaries, as in Eq. (6.7) and in the Bader virial partitioning method, the moments can be derived directly from the structure factors by a modified Fourier summation, as described for the net charge in chapter 6.

For the moments of the distribution within the volume element V_T, expression (7.1) gives

$$\mu_{\alpha_1, \alpha_2 \ldots \alpha_l} = \int_{V_T} \hat{\gamma}_{\alpha_1, \alpha_2 \ldots \alpha_l}\rho(\mathbf{r}) \, d\mathbf{r} \tag{7.33}$$

with $\hat{\gamma}_{\alpha_1\alpha_2\alpha_3 \ldots \alpha_l} = r_{\alpha_1}r_{\alpha_2}r_{\alpha_3} \cdots r_{\alpha_l}$.

Replacing $\rho(\mathbf{r})$ by the Fourier summation over the structure factors, we obtain for the lth moment:

$$\mu^l(V_T) = \frac{1}{V} \int_{V_T} \hat{\gamma}_l \sum_H F(\mathbf{H}) \exp\left(-2\pi i \mathbf{H} \cdot \mathbf{r}\right) d\mathbf{r} = \frac{1}{V} \sum_H F(\mathbf{H}) \int_{V_T} \hat{\gamma}_l \exp\left(-2\pi i \mathbf{H} \cdot \mathbf{r}\right) d\mathbf{r}$$

(7.34)

which is equivalent to Eq. (6.8).

In the case of the net charge, the integral becomes $\int_{V_T} \exp\left(-2\pi i \mathbf{H} \cdot \mathbf{r}\right) d\mathbf{r}$, which is the shape transform S_T of the volume V_T described in chapter 6. For higher moments, the volume integrals are given by $S_T^l = \int_{V_T} \hat{\gamma}_l \exp\left(-2\pi i \mathbf{H} \cdot \mathbf{r}\right) d\mathbf{r}$. Because the operator γ contains the position vector, the integrals are no longer identical for different volumes of integration, even though their shapes may be identical.

For complicated volumes of integration, the parallelepiped divisioning can again be used. We write

$$\mu^l(V_T) = \frac{1}{V} \sum_H F(\mathbf{H}) \sum_i \left\{ \int_{v_i} \left[\hat{\gamma}_l(\mathbf{r}) \exp 2\pi i \mathbf{H}(r - r_i)\, d\mathbf{r}\right] \exp\left(2\pi i \mathbf{H} \cdot \mathbf{r}_i\right) \right\}$$

(7.35)

in which v_i is the subunit volume.

As before, we write

$$\mu^l(V_T) = \frac{1}{V} \sum F(\mathbf{H}) S_T^l(\mathbf{H})$$

with

$$S_T^l = \sum_i s_i^l \exp 2\pi i \mathbf{H} \cdot \mathbf{r}_i$$

and

$$s_i^l(\mathbf{H}) = \int_{v_i} \hat{\gamma}_l(\mathbf{r}) \exp 2\pi i \mathbf{H} \cdot (\mathbf{r} - \mathbf{r}_i)\, d\mathbf{r}$$

(6.13)

and analogously for the deformation density:

$$\mu^l(V_T) = \frac{1}{V} \sum \Delta F(\mathbf{H}) S_T^l(\mathbf{H})$$

(7.36)

For identical volumes of integration, $s_i^l(\mathbf{H})$ can be written as the sum of a position-independent and a position-dependent term. When γ is the dipole-moment operator, this is accomplished as follows.

For the dipole moments,

$$\hat{\gamma}_1(\mathbf{r}) = \mathbf{r} = (\mathbf{r} - \mathbf{r}_i) + \mathbf{r}_i$$

(7.37)

Substitution in Eq. (6.13) gives

$$s_i^1(\mathbf{H}) = \int_{v_i} (\mathbf{r} - \mathbf{r}_i) \exp 2\pi i \mathbf{H} \cdot (\mathbf{r} - \mathbf{r}_i)\, d\mathbf{r} + \mathbf{r}_i \int_{v_i} \exp 2\pi i \mathbf{H} \cdot (\mathbf{r} - \mathbf{r}_i)\, d\mathbf{r}$$

$$= s_0^1(\mathbf{H}) + \mathbf{r}_i \int_{v_i} \exp 2\pi i \mathbf{H} \cdot (\mathbf{r} - \mathbf{r}_i)\, d\mathbf{r}$$

(7.38)

in which $s_0^1(\mathbf{H})$ is no longer dependent on the position of the integration volume.

TABLE 7.1 Expressions for the Shape Factors $s_0^l(\mathbf{H})$ for a Parallelepiped with Edges $2\delta_x$, $2\delta_y$, and $2\delta_z$

$\hat{\gamma}$	Property	$s_0^l(\mathbf{H})$
1	Charge	$V_T j_0(2\pi H_x \delta_x) j_0(2\pi H_\beta \delta_\beta) j_0(2\pi H_\gamma \delta_\gamma)$
r_α	Dipole μ_α	$-i V_T \delta_\alpha j_1(2\pi H_\alpha \delta_\alpha)$ $j_0(2\pi H_\beta \delta_\beta) j_0(2\pi H_\gamma \delta_\gamma)$
$r_\alpha r_\beta$	Second-moment $\mu_{\alpha\beta}$ off-diagonal	$-V_T \delta_\alpha \delta_\beta j_1(2\pi H_\alpha \delta_\alpha) j_1$ $(2\pi H_\beta \delta_\beta) j_0(2\pi H_\gamma \delta_\gamma)$
$r_\alpha r_\alpha$	Second-moment $\mu_{\alpha\alpha}$ diagonal	$V_T \delta_\alpha^2 \left\{ \dfrac{j_1(2\pi H_\alpha \delta_\alpha)}{\pi h_\alpha \delta_\alpha} - j_0(2\pi H_\alpha \delta_\alpha) \right\}$ $\cdot j_0(2\pi H_\beta \delta_\beta) j_0(2\pi H_\gamma \delta_\gamma)$

Terms j_0 and j_1 are the zero- and first-order spherical Bessel functions: $j_0(x) = (\sin x)/x$, $j_1(x) = (\sin x)/x^2 - (\cos x)/x$; V_T is the volume of the parallelepiped.
Source: Moss and Coppens (1981).

With Eq. (7.35) we obtain, for the dipole moment,

$$\mu^1(V_T) = \sum_i \{\mu^1(t_i) + \mathbf{r}_i q_i\} \tag{7.39}$$

The first term in Eq. (7.39) represents the sum over the dipole moments of the individual subunits, each referred to its own origin; the second term represents the effect of the origin shift. As expected, Eq. (7.39) for the sum over subunit dipole moments is identical in form to Eq. (7.23) for the sum over pseudoatom dipole moments.

Expressions for $s_0^l(\mathbf{H})$ for $l \le 2$ and a subvolume of parallelepipedal shape are given in Table 7.1. Though the shape factor for the dipole moment is imaginary, combination of the Friedel pairs $F(\mathbf{H})$ and $F(\bar{\mathbf{H}})$ in the summation

$$\mu^l(V_T) = \frac{1}{V} \sum F(\mathbf{H}) S_T^l(\mathbf{H})$$

leads to a result that is real, as required for an observable physical property. We obtain

$$\mu_\alpha^1 = \frac{2v\delta_\alpha}{V} \sum_{1/2} \sum_i \{(A \sin 2\pi\mathbf{H} \cdot \mathbf{r}_i + B \cos 2\pi\mathbf{H} \cdot \mathbf{r}_i) j_1(2\pi H_\alpha \delta_\alpha) j_0(2\pi H_\beta \delta_\beta) j_0(2\pi H_\gamma \delta_\gamma)\} \tag{7.40}$$

in which the first summation is over a hemisphere in reciprocal space and v is the subunit volume.

As noted in chapter 6, since the spherical Bessel functions $j_n(x)$ generally decrease with increasing x, the moments are less dependent on the high-order reflections in a data set than the electron density itself.

Performing the summation over ΔF_{obs} further reduces the possibility of series termination effects. It also causes the spherical-atom density of the nuclei within the volume of integration to be assigned to that volume, whereas in the F_{obs} summation the promolecule density may extend into adjacent regions.

The properties obtained by Fourier summation over ΔF_{obs} or F_{obs} are those of the thermally averaged density. But because of the decreased dependence on the high-order reflections, the effect of thermal vibrations is not pronounced, especially for large volumes of integration. In other words, as long as a density unit vibrates harmonically within the volume of integration, neither the charge nor the dipole moment components will be affected.

7.2.5 Error Analysis of Diffraction Moments

The accuracy of the electrostatic moments based on the multipole parameters is a function of the errors in both the population coefficients P_{val} and the atomic parameters P_{lmp}. Let $\mathbf{M}_x$ represent the $m \times m$ variance–covariance matrix for these parameters, as in chapter 4. Let $\mathbf{D}$ be the derivative matrix with elements

$$D_{ij} = \frac{\partial \mu_i}{\partial x_j} \tag{7.41}$$

in which μ_i is an element of the moment tensor, and x_j is a least-squares variable. In the atomic case, $\mathbf{D}$ is a $3 \times m$ or $6 \times m$ matrix for the first and second moments, respectively. According to Eq. (4.31), the variances and covariances of the elements of μ are given by

$$\mathbf{M}_\mu = \mathbf{D}\mathbf{M}_x\mathbf{D}^T \tag{7.42}$$

As mentioned before, when the population parameters have been defined with respect to local atomic coordinate systems, the moments must be transferred to a common coordinate system for the calculation of molecular properties. The matrix $\mathbf{D}$ will have to be modified accordingly. Analogous to Eq. (7.41), the elements of $\mathbf{D}'$ are given by

$$D'_{ij} = \frac{\partial \mu'_i}{\partial x_j} \tag{7.43}$$

in which the primed quantities refer to the molecular coordinate system.

We will illustrate the equations with an example. As discussed above, the molecular dipole moment vector is obtained from the atomic dipole moments μ_i and the atomic net charges q_i by

$$\mu_{total} = \sum_i \mu_i + \sum \mathbf{r}_i q_i \tag{7.23}$$

The atomic dipole moment is dependent on eight variables: the net charge of the atom, derived from P_{val}, the κ parameter, the atomic coordinates, and the three population parameters P_{10}, P_{11+} and P_{11-}. If we are interested in the error in the magnitude of the molecular dipole moment $|\mu_{total}|$, and we omit columns of $\mathbf{D}'$ which contain only zero's, $\mathbf{D}'$ will be a $1 \times 8N$ matrix, where N is the number of atoms in the molecule. To evaluate the error in $|\mu_{total}|$, the elements of $\mathbf{D}'$ are obtained as

$$D'_j = \frac{\partial |\mu_{total}|}{\partial x_j} \tag{7.44}$$

and Eq. (7.42) gives a single number, which is the variance of $|\mu_{total}|$.

In the case that the three components of the molecular dipole moment are to be evaluated for subsequent calculation of its direction, $\mathbf{D}'$ becomes a $3 \times 8N$ matrix, and the result of Eq. (7.42) is the 3×3 variance–covariance matrix $\mathbf{M}_\mu$ of the components.

The contribution of the errors in the positional parameters is usually much smaller than that of the population parameters, except perhaps for the hydrogen atom contributions.

The expression for the errors in the moments obtained by direct integration of the density follows from Eq. (6.17):

$$\sigma^2(\langle O \rangle) = \frac{1}{V^2} \sum_H S(\mathbf{H})^2 \sigma^2[F(\mathbf{H})] \tag{6.17}$$

or

$$\sigma^2(\mu^l) = \frac{1}{V^2} \sum_H S^l_T(\mathbf{H})^2 \sigma^2[F(\mathbf{H})] \tag{7.45a}$$

and

$$\sigma^2(\mu^l) = \frac{1}{V^2} \sum_H S^l_T(\mathbf{H})^2 \sigma^2[\Delta F(\mathbf{H})] \tag{7.45b}$$

in case the deformation density has been used. The ΔF values will also be subject to errors in the structural parameters.

7.3 Discussion of Experimental Results

7.3.1 A Compilation of X-ray Electrostatic Moments

A critical discussion of molecular dipole and quadrupole moments from X-ray diffraction data, with comprehensive coverage of the pre-1992 literature, has been given by Spackman (1992). The numerical values from this survey are listed in Tables 7.2–7.4. Spackman uses the units 10^{-30} Cm for the dipole moments, and 10^{-40} Cm^2 for the second and quadrupole moments. They are SI units and have a simple relation to the Debye (D) and to the Buckingham (B), which are both based on the esu (1 D = $3.336 \cdot 10^{-30}$ Cm, 1 B = $3.336 \cdot 10^{-40}$ Cm^2; see appendix K for conversion factors). The dipole moments listed in Table 7.2 are for neutral species, and are therefore origin independent. The quadrupole and second moments in Tables 7.3 and 7.4 are given with respect to the center of mass, and coordinate axes are chosen to maximize the use of molecular symmetry. This means that the z axis is taken along the molecular axis for linear molecules, along the two-fold axis for molecules with C_{2v} or D_{2h} symmetry, and perpendicular to the molecular plane for less symmetric planar molecules.

TABLE 7.2 Diffraction Estimates of Dipole Moments Compared with Other
Experimental Results, Where Available, and with ab-initio Theoretical Results[a]

Molecule	Diffraction Results		Other Experiment or Theory
H_2O	7.7 (10)	Multipole[1]	6.186 (1)/gas[2]
Water	5.3	Multipole[3]	7.29/6-31G**
	6.40 (17)	DI–DB[4]	
	7.0 (7)	Multipole[4]	
	7.2 (5)	Multipole[5]	
	7.9 (5)	Multipole[5]	
	7.6–8.3 (5)	Monopole[6]	
	7.0–7.6 (5)	Monopole[6]	
	3.9–6.5	DI–DB[6]	
	8.13 (3)	Multipole[7]	
	6.5–8.9	Monopole[8]	
H_3NO_3S	32.7 (20)	Monopole[9]	34.0–44.4/solution[10]
Sulfamic acid (1)	33.0 (20)	Multipole[11]	29.97/6-31G**
$H_{12}N_3B_3$	13.7 (13)	Monopole[12]	9.0 (4)/solution[13]
Cyclotriborazene (2)			12.98/6-31G**
CH_2N_2	18.5	Multipole[14]	13.3–15.1/solution[15]
Cyanamide (3)			16.26/6-31G**
CH_3NO	14.6 (17)	Monopole[9]	12.4 (2)/gas[16]
Formamide (4)	16.1 (17)	Multipole[9]	12.8 (1)/solution[17]
	16.5	DI–DB[18]	14.18-31G**
	13.4	DI–FB[18]	
CH_4N_2O	18.0 (17)	Multipole[19]	12.8 (1)/gas[20]
Urea (5)	19.0 (17)	Multipole[19]	15.2/solution[21]
			15.5/crystal[22]
			17.06/6-31G**
CH_4N_2S	18.0 (83)	Multipole[23]	16.3/solution[21]
Thiourea			22.31/6-31G**
$C_2H_4N_4$	22.6	Multipole[24]	27.2/solution[15]
2-Cyanoguanidine (6)	37.5	Multipole[24]	30.27/6-31G**
C_2H_5NO	16.5	Multipole[25]	12.3 (1)/gas[26]
Acetamide (7)			12.9 (1)/solution[17]
			15.16/6-31G**
$C_2H_8NO_4P$	43 (7)	Multipole[27]	72.82/6-31G**
Phosphorylethanol amine (8)			
$C_3H_2N_2O_2S$	7 (3)	Multipole[28]	9.61 (7)/solution[29]
2,5-Diaza-1,6-dioxa-6a-thiapentaline (9)	44.7	Monopole[7]	
$C_3H_2N_2O_3$	7.7 (10)	Multipole[30]	8.92/6-31G**
Parabanic acid (10)	11.7 (60)	Multipole[31]	
	15.7 (70)	Multipole[23]	
$C_3H_4N_2$	16.0 (20)	Multipole[32]	12.2 (2)/gas[33]
Imidazole (12)			13.2 (1)/solution[34]
$C_3H_7NO_2$	43.0 (23)	Multipole[35]	41.0/solution[36]
L-Alanine (13)			41.44/6-31G**[35]

(continued)

Table 7.2 (*continued*)

Molecule	Diffraction Results		Other Experiment or Theory
$C_4H_2N_2O_4$ Alloxan (14)	0.7 (33)	Multipole[37]	9.93/6-31G**
$C_4H_4N_2O_2$ Uracil (16)	13.3 (43) 14.7 (43)	Monopole[38–40] Monopole[42]	13.9 (1)/solution[41] 16.22/6-31G**
$C_4H_5N_3O$ Cytosine (17)	26.7 (47) 19.5	Multipole[43] Multipole[2]	23.3 (est)/solution[44] 27.47/6-31G**
$C_4H_8N_2O_3$ Glycylglycine (18)	80.4 (28)	Monopole[9]	91–93/solution[45]
$C_4H_9NO_2$ γ-Aminobutyric acid (19)	43.3 (33)	Multipole[46]	55.7–67.4/solution[47,48] 67.00/6-31G**
$C_5H_4N_2O_3$ p-Nitropyridine-N-oxide (20)	1.3 (33)	Monopole[9]	2.3 (1)/solution[49] 1.00/6-31G**
$C_5H_6N_2O_2$ 1-Methyluracil	14.7 (73) 21.3 (90)	Multipole[50] Multipole[50]	13.9 (1)/solution[42] 18.27/6-31G**
$C_6H_6N_2O_3$ 3-Methyl-p-nitropyrine-N-oxide	5.3 (43) 11.9 (47)	Monopole[51] Monopole[51]	
$C_6H_7N_5$ 9-Methyladenine (21)	6.0 (33) 8.0	Multipole[52] Multipole[2]	10.8 (7)/solution[53]
$C_8H_5N_3$ Pyridinium dicyanomethylide (22)	62.7 32.7 20.0 34.0 27.6	Multipole[54] Multipole[54] Multipole[54] DI–FB DI–DB	30.7/solution[55] 36.17/6-31G**
$C_8H_{12}N_2O_3$ Barbital (23)	2.3 (40)	Multipole[56]	3.77 (3)/solution[57]
$C_9H_6CrO_3$ Benzenechromium tricarbonyl	18.3 (83)	Monopole[58]	16.8/solution[59]
$C_9H_{13}N_3O_5$ Cytidine	52.0	Monopole[7]	
$C_9H_{14}N_3O_7P$ Deoxycytidine 5'-mono-phosphate	50.7	Monopole[7]	
$C_{10}H_{13}N_5O_3$ Deoxyadenosine			
$C_{10}H_{13}N_5O_4$ Adenosine	25.7 (170) 8.0	Multipole[60] Monopole[7]	
$C_{10}H_{14}N_2O_5$ Deoxythymidine	49.7	Monopole[7]	

[a] All quantities are given in units of 10^{-30} Cm (see Appendix K for conversion factors). The 6-31G** ab-initio results have been obtained at the SCF level, generally at the neutron crystal geometry. DI: direct integration, DB: discrete boundary, FB: fuzzy boundary, i.e., stockholder concept.

(*continued*)

TABLE 7.3 Diffraction Estimates of Quadrupole Moments Compared with Other Experimental Results, Where Available, and with ab-initio Theory[a]

Molecule	Quadrupole Moment	Diffraction Results		Other Experiment or Theory
Cl_2	Θ_{zz}	+11.0 (20)	Multipole[1]	+10.8 (5)/gas[2]
Chlorine		+7.7 (20)	Multipole[1]	+16.5 (17)/gas[3]
				+8.91/6-31G**
H_2O	Θ_{xx}	+11.0	Multipole[4]	+8.77 (7)/gas[5]
Water		+3.3	Multipole[6]	+7.93/6-31G**
	Θ_{yy}	−13.0	Multipole	−8.34 (7)/gas
		−2.9	Multipole	−7.59/6-31G**
	Θ_{zz}	+2.0	Multiple	−0.43 (10)/gas
		−0.4	Multipole	−0.33/6-31G**
CH_3NO	Θ_{xx}	−0.6 (26)	Monopole[7]	−1.0 (7)/gas[8]
Formamide (4)				−4.44/6-31G**
	Θ_{yy}	+9.1 (18)	Monopole	−11.3 (13)/gas
				+12.58/6-31G**
	Θ_{zz}	−8.5 (14)	Monopole	−10.3 (27)/gas
				−8.14/6-31G**
	Θ_{xy}	−5.5 (12)		−4.87/6-31G**
CH_4N_2O	Θ_{xx}	+38.7 (60)	Multipole[9]	+30.03/6-31G**
Urea (5)		+36.4 (60)	Multipole[9]	
	Θ_{yy}	−35.4 (40)	Multipole	−19.00/6-31G**
		−30.0 (40)	Multipole	
	Θ_{zz}	−3.3 (43)	Multipole	−11.04/6-31G**
		−6.3 (40)	Multipole	
C_2H_2	Θ_{zz}	+24.3 (58)	Multipole[10]	+20.1 (6)/gas[11]
Acetylene		+19.0 (36)	Multipole[10]	+23.23/6-31G**
		+25.9 (38)	Monopole[10]	
		+21.3 (53)	Monopole[10]	
		+20.3	DI-FB[12]	
C_2H_4	Θ_{xx}	+5.7	Multipole[10]	+4.7/gas[13]
Ethylene				+5.4 (3)/gas[14]
				+4.99/6-31G**

(continued)

Table 7.2 (*continued*)

References: [1] Weber and Craven (1990), [2] Dyke and Muenter (1973), [3] Eisenstein (1988), [4] Stevens and Coppens (1980), [5] Bats et al. (1986), [6] Bats and Fuess (1986), [7] Delaplane et al. (1990), [8] Pearlman and Kim (1990), [9] Coppens et al. (1979), [10] Sears et al. (1966), [11] Coppens et al. (1980), [12] Corfield and Shore (1973), [13] Leavers and Taylor (1977), [14] Koritsanszky et al. (1991), [15] Schneider (1950), [16] Kurland and Wilson (1957), [17] Aroney et al. (1965), [18] Moss and Coppens (1980), [19] Swaminathan et al. (1984), [20] Brown et al. (1975), [21] Kumler and Fohlen (1942), [22] Lefebvre (1973), [23] Weber and Craven (1987), [24] Hirshfeld and Hope (1980), [25] Berkovitch-Yellin and Leiserowitz (1980), [26] Kojima et al. (1987), [27] Swaminathan and Craven (1984), [28] Fabius et al. (1989), [29] Larsen et al. (1984), [30] He et al. (1988), [31] Craven and McMullan (1979), [32] Epstein et al. (1982), [33] Christen et al. (1982), [34] Calderbank et al. (1981), [35] Destro et al. (1989), [36] Khanarian and Moore (1980), [37] Swaminathan et al. (1985), [38] Stewart (1970), [39] Stewart (1974), [40] Stewart (1980), [41] Kulakowska et al. (1974), [42] Yanez and Stewart (1978), [43] Weber et al. (1980), [44] Palmer et al. (1983), [45] Sakellaridis and Karageorgopolous (1974), [46] Craven and Weber (1983), [47] Edward et al. (1973), [48] Pottel et al. (1975), [49] Katritzky et al. (1957), [50] Klooster et al. (1992), [51] Baert et al. (1988), [52] Craven and Benci (1981), [53] Bergmann et al. (1970), [54] Baert et al. (1982), [55] Treiner et al. (1964), [56] Craven et al. (1982), [57] Soundararajan (1958), [58] Rees and Coppens (1973), [59] Lumbroso et al. (1973), [60] Klooster and Craven (1992).

Table 7.3 (*continued*)

Molecule	Quadrupole Moment	Diffraction Results		Other Experiment or Theory
	Θ_{yy}	−7.8	Multipole	−12.0/gas −10.8 (7)/gas −11.04/6-31G**
	Θ_{zz}	+2.1	Multipole	+7.3 (10)/gas +5.4 (3)/gas +6.05/6-31G**
$C_2H_4N_4$ 2-Cyanoguanidine (6)	Θ_{11}	+16.2 +27.2	Multipole[15] Multipole[15]	+27.23/6-31G**
	Θ_{22}	+0.3 −4.0	Multipole Multipole	−12.95/6-31G**
	Θ_{33}	−16.5 −23.2	Multipole Multipole	−14.28/6-31G**
$C_3H_3N_3$ s-Triazine (11)	Θ_{zz}	−20 (13) −24 (16) −2.7 (12) −3.5 (6) −1.2 (4)	Multipole[16] Multipole[16] Monopole[16] Monopole[16] Monopole[18]	−2.8 (31)/solution[17] +2.03/6-31G**
$C_3H_4N_2$ Imidazole (12)	Θ_{xx}	+19.4	Multipole[19]	−3.1 (9)/gas[20] −1.58/6-31G**
	Θ_{yy}	+3.3	Multipole	+22.6 (11)/gas +17.43/6-31G**
	Θ_{zz}	−22.7	Multipole	−19.6 (18)/gas −15.84/6-31G**
	Θ_{xy}	−19.4	Multipole	−9.61/6-31G**
$C_4H_4N_2$ Pyrazine (15)	Θ_{xx}	+37.5	Multipole[12]	+40.47/6-31G**
	Θ_{yy}	−9.8	Multipole	−8.83/6-31G**
	Θ_{zz}	−27.7	Multipole	−31.64/6-31G**
	Θ_{xy}	+55.0	Multipole[21]	+8.56/6-31G**
$C_4H_5N_3O$ Cytosine (17)	Θ_{yy}	+6.8	Multipole	+8.56/6-31G**
	Θ_{xx}	−61.8	Multipole	−17.11/6-31G**
	Θ_{xy}	+66.2	Multipole	+45.4/6-31G**
	Θ_{zz}	−40.3 (35) −32.3 (35) −28.7 (35)	Multipole[22] Multipole Monopole	−29.0 (17)/gas[23] −28.3 (12)/solution[24] −28.30/6-31G**
C_6H_6 Benzene	Θ_{xx}	−4.7 (50) −6.9	Multipole[25] Multipole	+34.61/6-31G**
$C_8N_2F_4$ p-Dicyanotetra-fluorobenzene	Θ_{yy}	+72.5 (46) +61.5	Multipole Multipole	+72.61/6-31G** −107.22/6-31G**
	Θ_{zz}	−68.8 (88) −54.6	Multipole Multipole	

[a] All quantities are given in units of 10^{-40} Cm^2 (see Appendix K for conversion factors). Other experimental results are labeled as either "gas" (i.e., microwave Zeeman or induced birefringence), or "solution" (i.e., induced birefringence in a nonpolar solvent). The 6-31G** ab-initio results have been obtained at the SCF level, generally at the neutron crystal geometry. DI: direct integration, FB: Fuzzy boundary, i.e., Stockholder concept.

(*continued*)

Since space partitioning introduces an ambiguity, there will be differences between the moments obtained by the various methods. In accordance with Kurki-Suonio's requirement of locality (Kurki-Suonio 1971, Kurki-Suonio and Salmo 1971) (chapter 6), excellent dipole and quadrupole moments are often obtained with the simple spherical-atom κ-formalism. The neglect of higher-order atomic deformations in the κ-formalism is apparently compensated by a subtle bias in the spherical charge density parameters, such that a sum over all atoms in the molecule still gives reliable estimates of the molecular properties.

7.3.2 Dipole Moments

X-ray dipole moments of formamide, of sulfamic acid, of benzene chromium tricarbonyl, and of water, obtained from κ-refinements, are in good agreement with those from other physical techniques. When hydrogen-atom positions are of crucial importance, as in the case of the water molecule, the availability of positional information from neutron diffraction becomes essential if accurate moments are to be obtained. In other cases, extension of the $X-H$ bond to accepted values provides a reasonable alternative.

Not surprisingly, formalisms with very diffuse density functions tend to yield large electrostatic moments. This appears, in particular, to be true for the Hirshfeld formalism, in which each $\cos^n$ term in the expansion (3.48) includes diffuse spherical harmonic functions with $l = n, n - 2, n - 4, \ldots (0, 1)$ with the radial factor r^n. For instance when the refinement includes $\cos^4$ terms, monopoles and quadrupoles with radial functions containing a factor r^4 are present. For pyridinium dicyanomethylide (Fig. 7.3), the dipole moment obtained with the coefficients from the Hirshfeld-type refinement is $62.7 \cdot 10^{-30}$ Cm (18.8 D), whereas the dipole moments from the spherical harmonic refinement, from integration in direct space, and the solution value (in dioxane), all cluster around $31 \cdot 10^{-30}$ Cm (9.4 D) (Baert et al. 1982).

On the other hand, dipole moments obtained with stockholder partitioning tend to be systematically low; for pyridinium dicyanomethylide, the value is

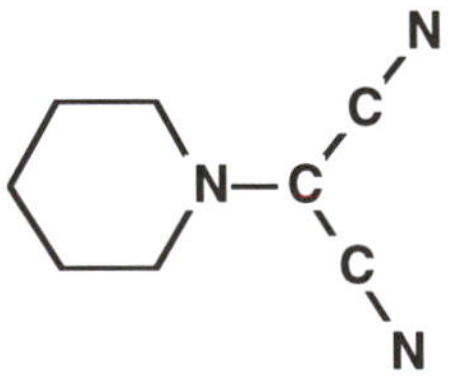

FIG. 7.3 The pyridinium dicyanomethylide molecule.

Table 7.3 (continued)

References: [1] Stevens (1979), [2] Buckingham et al. (1983), [3] Emrich and Steele (1980), [4] Weber and Craven (1990), [5] Verhoeven and Dymanus (1970), [6] Eisenstein (1988), [7] Stevens (1978), [8] Tigelaar and Flygare (1972), [9] Swaminathan et al. (1984), [10] van Nes and van Bolhuis (1979), [11] Coonan and Ritchie (1993), [12] Moss and Feil (1981), [13] Kukolich et al. (1983), [14] Dagg et al. (1982), [15] Hirshfeld and Hope (1980), [16] Price et al. (1978), [17] Dennis (1986), [18] Stewart (1970), [19] Epstein et al. (1982), [20] Stolze and Sutter (1987), [21] Weber et al. (1980), [22] Spackman (1991), [23] Battaglia et al. (1981), [24] Dennis and Ritchie (1991), [25] Hirshfeld (1984).

TABLE 7.4 Diffraction Estimates of Second Moments Compared with ab-initio 6-31G** Results Obtained at the Same Geometry and with Respect to the Same Coordinate System

Molecule	Method	μ_{xx}	μ_{yy}	μ_{zz}	μ_{xy}	μ_{xz}	μ_{yz}
Water[b]	Multipole	−15.4 (24)	−21.4 (23)	−14.6 (41)	−0.4 (15)	−2.5 (16)	−6.7 (15)
	6-31G**	−18.93	−20.67	−17.13	+0.35	−1.09	−4.75
Urea[c]	Multipole	−64.6 (37)	−86.1 (24)	−36.7 (70)			
	Multipole	−68.9 (29)	−84.7 (24)	−40.5 (54)			
	6-31G**	−79.87	−85.17	−52.49			
2-Cyanoguanidine[d]	Multipole	−95.3	−116.2	−129.0			
	Multipole	−104.9	−115.5	−126.7			
	6-31G**	−96.45	−123.2	−124.2			
Imidazole[b]	Multipole	−100.8 (33)	−95.8 (44)	−67.5 (60)	+0.3 (23)	−10.3 (21)	+13.7 (27)
	6-31G**	−101.45	−85.06	−89.96	+5.46	−1.40	+13.97
Cytosine[b]	Multipole	−129.7 (109)	−60.1 (141)	−149.8 (71)	+28.6 (86)	−9.4 (46)	−18.7 (51)
	6-31G**	−173.94	−100.45	−160.75	+16.33	+2.38	−5.25
Benzene[b]	Multipole	−126.4 (34)	−111.0 (51)	−126.5 (39)	−7.5 (23)	−19.2 (25)	−6.2 (23)
	Multipole	−116.2 (27)	−105.7 (41)	−116.0 (32)	−6.7 (18)	−15.6 (20)	−5.1 (18)
	6-31G**	−117.72	−106.70	−117.87	−4.85	−13.34	−4.94
p-Dicyanotetrafluorobenzene[e]	Multipole	−329.7 (87)	−287.6 (24)	−236.2 (8)			
	Multipole	−315.8	−284.0	−238.4			
	6-31G**	−359.50	−264.95	−239.62			

[a] The x and y axes are defined in terms of the crystallographic axes; in all cases, z completes a right-handed set of Cartesian axes. All quantities are in units of 10^{-40} Cm^2.
[b] $x \parallel a$, $y \parallel b^*$.
[c] x along two-fold axis, z in molecular plane.
[d] Principal directions; x and z in molecular plane.
[e] x along two-fold axis, y in molecular plane.

Source: Spackman 1992.

161

$20.0 \cdot 10^{-30}$ Cm (6.0 D); for water, $5.3 \cdot 10^{-30}$ Cm, versus $7-8 \cdot 10^{-30}$ Cm by most other methods in agreement with theoretical values; and $22.6 \cdot 10^{-30}$ Cm for 2-cyanoguanidine versus $27-30 \cdot 10^{-30}$ Cm from solution measurements and theoretical calculations. Hirshfeld and Hope (1980) have interpreted the *lower* values obtained with the stockholder recipe as resulting from a mutual cancellation of positive and negative deformation functions centered on neighboring molecules. A deficiency in density in an interatomic or intermolecular region is proportionally distributed among the contributing atoms, but there may be situations in which the deficiency should be assigned entirely to one of the contributors, and in which the other participants may, in fact, contribute excess density. The stockholder recipe partitions the density according to each atom's contribution to the promolecule density, and the partitioned fragments therefore tend towards the distribution of the promolecule.

The dipole moments are generally affected by intermolecular electrostatic interactions. In a crystal, molecules tend to line up such that opposite charges are proximal, in order to maximize the electrostatic attractions. Induced polarization therefore tends to enhance the electrostatic moments.

A simple calculation for urea by Spackman is instructive. Urea crystallizes in an acentric space group (it is a well-known nonlinear optical material), in which the symmetry axes of the molecules coincide with the two-fold axes of the space group. All molecules are lined up parallel to the tetragonal c axis. If the electric field is given by $\mathbf{E}$, and the principal element of the diagonalized molecular polarizability tensor along the c axis by α_{zz}, the induced moment along the polar c axis is

$$\mu_z^{ind} = \alpha_{zz} E_z \qquad (7.46)$$

A dipole of magnitude μ_z causes an electric field along z, which, at a point described by the polar coordinates (r, θ) (with the origin at the center of the dipole), is equal to

$$E_z = \frac{\mu_z}{4\pi\varepsilon_0} \frac{3\cos^2\theta - 1}{r^3} \qquad (7.47)$$

The dipole induced by a single urea molecule at (r, θ) is then

$$\mu_z^{ind} = \frac{\alpha_{zz}\mu_z}{4\pi\varepsilon_0} \frac{3\cos^2\theta - 1}{r^3} \qquad (7.48)$$

As all molecules are oriented parallel to the tetragonal c axis, the summation over the neighboring unit cells is relatively simple. Using the gas phase value of $12.8 \cdot 10^{-30}$ Cm for the dipole moment, and a theoretical value of $6.54 \cdot 10^{-40}$ Cm^2 V^{-1} for α_{zz}, the induced dipole moment is calculated as $2.57 \cdot 10^{-30}$ Cm, in excellent agreement with the size of the discrepancy of $2.4 \cdot 10^{-30}$ Cm between the gas- and condensed-phase experimental results of 12.8 and $15.2 \cdot 10^{-30}$ Cm, respectively (Table 7.2).

The solid-state molecular dipole moment has been evaluated in a recent HF calculation on the crystal of urea. The result of $23.5 \cdot 10^{-30}$ Cm, compared with $17.2 \cdot 10^{-30}$ Cm for the molecule in the observed geometry in the crystal, indicates a considerably larger enhancement (Gatti et al. 1994). Part of the underestimate obtained with the simple interaction model will be due to its lack of self-consistency. The more polar molecules in the crystal induce a larger moment in their neighbors, so the calculation should be repeated until consistency is obtained. Other effects, such as dipole–quadrupole interactions, may also contribute.

The evidence summarized in Table 7.2, and that obtained since the table was compiled, indicate polarization effects in the crystal to exceed the errors of the diffraction method. Additional examples of enhanced dipole moments in crystals are mentioned in chapter 12.

7.3.3 Second Moments and Quadrupole Moments

In the derivation of the traceless quadrupole moments from the electrostatic moments, the spherical components are subtracted. Thus, the quadrupole moments can be derived from the second moments, but the opposite is not the case. Spackman (1992) notes that the subtraction introduces an ambiguity in the comparison of quadrupole moments from theory and experiment. The spherical component subtracted is not that of the promolecule, but is based on the distribution itself. It is therefore generally not the same in the two densities being compared. On the other hand, the moments as defined by Eq. (7.1) are based on the total density without the intrusion of a reference state.

It is found that comparison based on the second moments often gives the better agreement. This is the case for the in-plane moments of benzene. For cytosine and imidazole, the agreement with theoretical values is poor for the quadrupole moments (Table 7.3). But comparison based on the second moments (Table 7.4) reveals much better agreement, though the diagonal elements of the moments appear systematically lowered relative to the theoretical values, especially for cytosine. This is interpreted as being due to the strong hydrogen-bonding in the crystals, which causes a contraction of the density, in accordance with conclusions based on model studies on oxalic acid dihydrate by Krijn (1988) and Krijn and Feil (1988).

Nevertheless, the diffraction quadrupole moments frequently compare well with values from techniques such as induced birefringence and the Zeeman effect. The Θ_{zz} value from a multipole analysis of Stevens' diffraction data on chlorine, for example, is in good agreement with a later birefringence study (Table 7.3). The κ-refinement values often show good agreement with results from other experiments and from theory, as is the case for formamide, discussed at the beginning of this chapter. For the Θ_{zz} (out-of-plane) component of benzene, the agreement is excellent, provided neutron diffraction thermal parameters are used. For acetylene, the multipole values of Θ_{zz} of $+25.9$ (38)$\cdot 10^{-40}$ Cm and $+21.3$ (53)$\cdot 10^{-40}$ Cm, as well as the direct space integration value of $20.3 \cdot 10^{-40}$ Cm, agree well with an experimental gas-phase value of $+20.1$ (6)$\cdot 10^{-40}$ Cm and with an ab-initio calculation that gives $\Theta_{zz} = 23.2 \cdot 10^{-40}$ Cm.

For further discussion of the already extensive experimental information, the reader is referred to the review by Spackman (1992). Spackman concludes that, while the diffraction method may never become the routine method of choice for the determination of electrostatic moments, the diffraction data give a much more detailed description of the distribution on which the quantitative values are based, and contain a vast amount of information on intermolecular interactions which must be accessible for work in other fields.

8

X-ray Diffraction and the Electrostatic Potential

The distribution of positive and negative charge in a crystal fully defines physical properties like the electrostatic potential and its derivatives, the electric field, and the gradient of the electric field.

The electrostatic potential at a point in space, defined as the energy required to bring a positive unit of charge from infinite distance to that point, is an important function in the study of chemical reactivity. As electrostatic forces are relatively long-range forces, they determine the path along which an approaching reactant will travel towards a molecule. A nucleophilic reagent will first be attracted to the regions where the potential is positive, while an electrophilic reagent will approach the negative regions of the molecule.

As the electrostatic potential is of importance in the study of intermolecular interactions, it has received considerable attention during the past two decades (see, e.g., articles on the molecular potential of biomolecules in Politzer and Truhlar 1981). It plays a key role in the process of molecular recognition, including drug-receptor interactions, and is an important function in the evaluation of the lattice energy, not only of ionic crystals.

This chapter deals with the evaluation of the electrostatic potential and its derivatives by X-ray diffraction. This may be achieved either directly from the structure factors, or indirectly from the experimental electron density as described by the multipole formalism. The former method evaluates the properties in the crystal as a whole, while the latter gives the values for a molecule or fragment "lifted" out of the crystal.

Like other properties derived from the charge distribution, the experimental electrostatic potential will be affected by the finite resolution of the experimental data set. But as the contribution of a structure factor $F(\mathbf{H})$ to the potential is proportional to H^{-2}, as shown below, convergence is readily achieved. A summary of the dependence of electrostatic properties of the magnitude of the scattering

TABLE 8.1 Dependence of the Electrostatic Properties and Other Quantities Derived from the Experimental Structure Factors $F(\mathbf{H})$ on $|\mathbf{H}|^n$

Property	Type	n in $|\mathbf{r}|^n$	n in $|\mathbf{H}|^n$
Electrostatic potential	Scalar	-1	-2
Diamagnetic shielding tensor	Second rank tensor	-1	-2
Electrostatic energies	Scalar	-1	-2
Electric field	Vector	-2	-1
Electric field gradient	Traceless Second-rank tensor	-3	0
Charge density	Scalar	-3	0
Diamagnetic current density	Vector	-3	0
Gradient of field gradient	Third rank tensor	-4	1
Gradient of charge density	Vector	-4	1
Grad–grad of field gradient	Fourth rank tensor	-5	2
Hessian of charge density	Second rank tensor	-5	2
Laplacian of charge density	Scalar	-5	2

Source: Stewart (1991, p. 68).

vector $\mathbf{H}$ is given in Table 8.1, which shows that the electrostatic potential is among the most accessible of the properties listed.

8.1 Definitions and Units

8.1.1 Definition of the Electrostatic Potential

The *electrostatic potential* at $\mathbf{r}'$, $\Phi(\mathbf{r}')$, due to a charge Q at $\mathbf{r}$ is defined by the Coulomb equation (Jackson 1975)

$$\Phi(\mathbf{r}') = \frac{Q}{4\pi\varepsilon_0|\mathbf{r} - \mathbf{r}'|} \tag{8.1}$$

The factor $4\pi\varepsilon_0$ occurs when the quantities are expressed as SI units; ε_0 is the "permittivity of free space," equal to $8.854\,187\,82\cdot10^{-12}\ \mathrm{C}^2\,\mathrm{N}^{-1}\,\mathrm{m}^{-2}$. The factor $4\pi\varepsilon_0$ ($=1.112\,626\,5\cdot10^{-10}\ \mathrm{C}^2\,\mathrm{N}^{-1}\,\mathrm{m}^{-2}$) disappears when either atomic or cgs units are used, and we shall omit it in the expressions given. However, its inadvertent omission in numerical calculations will lead to meaningless results.

The difference between the electrostatic potential at two points is equal to the work required to bring a unit charge from one point to the other. The choice of zero potential is arbitrary, but the potential is commonly defined as zero when the particles are at infinite distance. Thus, the electrostatic potential at a point is the work required to bring a unit of charge from infinity to that point.

For a continuous charge distribution, the potential is obtained by integration over the space containing the distribution. At a point defined by $\mathbf{r}'$, the potential is given by

$$\Phi(\mathbf{r}') = \int \frac{\rho_{\text{total}}(\mathbf{r})}{|\mathbf{r} - \mathbf{r}'|}\,d\mathbf{r} \tag{8.2}$$

in which ρ_{total} represents both the nuclear and the electronic charge.

For an assembly of positive point nuclei and a continuous distribution of negative electronic charge, we obtain

$$\Phi(\mathbf{r}') = \sum_M \frac{Z_M}{|\mathbf{R}_M - \mathbf{r}'|} - \int \frac{\rho(\mathbf{r})}{|\mathbf{r} - \mathbf{r}'|} \, d\mathbf{r} \tag{8.3}$$

in which Z_M is the charge of nucleus M located at $\mathbf{R}_M$.

Since the first contribution is positive, while the second is negative, the sign of the potential at a point depends on whether the nuclei or the electrons dominate at that particular point. An electrophilic reagent approaching this distribution will be attracted by the electrons, but repelled by the nuclei.

8.1.2 The Electric Field

The *electric field* vector $\mathbf{E}$ at a point in space is the negative gradient of the electrostatic potential at that point:

$$\mathbf{E}(\mathbf{r}) = -\nabla_{\mathbf{r}}\Phi(\mathbf{r}) = -\mathbf{i}\frac{\partial\Phi(\mathbf{r})}{\partial x} - \mathbf{j}\frac{\partial\Phi(\mathbf{r})}{\partial y} - \mathbf{k}\frac{\partial\Phi(\mathbf{r})}{\partial z} \tag{8.4}$$

As $\mathbf{E}$ is the *negative* gradient vector of the potential, the electric force is directed "downhill" and is proportional to the slope of the potential function. The explicit expression for E is obtained by differentiation of the operator $|\mathbf{r} - \mathbf{r}'|^{-1}$ in Eq. (8.2) towards x, y, z, and subsequent addition of the vector components. With $\partial|\mathbf{r} - \mathbf{r}'|^{-1}/\partial x = |\mathbf{r} - \mathbf{r}'|^{-2} \partial\mathbf{r}'/\partial x$, we obtain, for the negative slope of the potential in the x direction,

$$E_x(\mathbf{r}') = \int \frac{\rho_{\text{total}}(\mathbf{r})}{|\mathbf{r} - \mathbf{r}'|^2} \frac{(\mathbf{r}' - \mathbf{r})_x}{|\mathbf{r} - \mathbf{r}'|} \, d\mathbf{r} = \int \frac{\rho_{\text{total}}(\mathbf{r})}{|\mathbf{r} - \mathbf{r}'|^3} (\mathbf{r}' - \mathbf{r})_x \, d\mathbf{r} \tag{8.5}$$

which leads, after addition of all three components, to the expression

$$\mathbf{E}(\mathbf{r}') = -\nabla_{\mathbf{r}}\Phi(\mathbf{r}') = \int \frac{\rho_{\text{total}}(\mathbf{r})(\mathbf{r}' - \mathbf{r})}{|\mathbf{r} - \mathbf{r}'|^3} \, d\mathbf{r} \tag{8.6}$$

8.1.3 The Electric Field Gradient

The electric field gradient (EFG) is the tensor product of the gradient operator $\nabla = \mathbf{i}\,\partial/\partial x + \mathbf{j}\,\partial/\partial y + \mathbf{k}\,\partial/\partial z$ and the electric field vector $\mathbf{E}$:

$$\nabla E = \nabla_{\mathbf{r}}:E = -\nabla_{\mathbf{r}}:\nabla\Phi \tag{8.7}$$

As in a Cartesian coordinate system the tensor product $u:v$, of the vectors $\mathbf{u}$ and $\mathbf{v}$, has the elements $u_\alpha v_\beta$, the EFG tensor is a symmetric tensor with elements

$$\nabla E_{\alpha\beta} = -\frac{\partial^2\Phi}{\partial r_\alpha \, \partial r_\beta} \tag{8.8}$$

The EFG tensor elements can be obtained by differentiation of the operator in expression (8.5) for E_α to each of the three directions β. This procedure removes the spherical component, which does not affect the electric field, and yields the traceless result

$$\nabla E_{\alpha\beta}(\mathbf{r}') = \frac{\partial E_\alpha}{\partial (r_\beta - r_\beta')} = -\int \frac{1}{|\mathbf{r} - \mathbf{r}'|^5} \{3(r_\alpha - r_\alpha')(r_\beta - r_\beta') - |\mathbf{r} - \mathbf{r}'|^2\, \delta_{\alpha\beta}\} \rho_t(\mathbf{r})\, d(\mathbf{r})$$

(8.9)

Including the spherical component, the electric field gradient can be interpreted as the second-moment tensor of the distribution $\rho(\mathbf{r})/|\mathbf{r} - \mathbf{r}'|^5$.

The definition of Eq. (8.8) and the result of Eq. (8.9) differ in that Eq. (8.8) does not correspond to a zero-trace tensor. The situation is analogous to the two definitions of the second moments, discussed in the preceding chapter. The trace of the tensor defined by Eq. (8.8) is given by

$$-\nabla^2 \Phi = -\nabla \cdot \nabla \Phi = -\left(\frac{\partial^2 \Phi}{\partial x^2} + \frac{\partial^2 \Phi}{\partial y^2} + \frac{\partial^2 \Phi}{\partial z^2}\right)$$

(8.10)

An important equation of electrostatics, which follows directly from Maxwell's equations (Jackson 1975) is Poisson's equation. It relates the divergence of the gradient of the potential $\Phi(\mathbf{r})$ to the charge density at that point:

$$\nabla^2 \Phi(\mathbf{r}) = -4\pi \rho_{total}(\mathbf{r})$$

(8.11)

Thus, the trace of the EFG tensor is only equal to zero if the charge density at $\mathbf{r}$ is zero.

The potential and its derivatives are sometimes referred to as *inner* moments of the charge distribution, since the operators in expressions (8.2), (8.4), and (8.6) contain the negative power of the position vector. Using the same terminology, the electrostatic moments discussed in the previous chapter are described as the *outer* moments.

8.1.4 Units

As $\rho(\mathbf{r})$ is expressed in $e\text{Å}^{-3}$, and *dr* and $|\mathbf{r} - \mathbf{r}'|^{-1}$ have dimensions of Å^3 and Å^{-1}, respectively, expression (8.3) gives the potential in units of $e\text{Å}^{-1}$, which equals $1.602 \cdot 10^{-9}$ Cm^{-1}. For conversion to SI energy units, the value of $4\pi\varepsilon_0$ is needed. Since $4\pi\varepsilon_0$ is in C^2 N^{-1} m^{-2}, the resulting unit has the dimension Nm C^{-1}, or J C^{-1}. A more common unit in molecular studies is the kJ per electron per mole: $1\,e\text{Å}^{-1} = 1389$ kJ e^{-1} mol^{-1}.[1] The electric field from Eq. (8.5) is expressed in $e\text{Å}^{-2}$, while Eq. (8.6) gives the electric field gradient in $e\text{Å}^{-3}$, like $\rho(\mathbf{r})$. Conversion factors to other units are given in appendix K.

8.2 Evaluation of the Electrostatic Potential and its Derivatives in Reciprocal Space

8.2.1 The Electrostatic Potential by Fourier Summation of the Structure Factors

Since the electrostatic potential and its derivatives are directly related to the charge density, it is not surprising that they also can be obtained by Fourier summation

---

[1] $1\,e\text{Å}^{-1} = 1.602 \cdot 10^{-9}$ Cm$^{-1} = (1.602 \cdot 10^{-9})(6.022 \cdot 10^{23})(1.602 \cdot 10^{-19})/1.112\,626 \cdot 10^{-10}$ kJ (e mol^{-1}) $=$ 1389 kJ (e mol)$^{-1} = 322$ Kcal (e mol)$^{-1}$.

over the structure factors. We will give two different derivations of the relations involved.

The electrostatic potential is defined in reciprocal space via the Fourier transform expression

$$\Phi(\mathbf{H}) = \int \Phi(\mathbf{r}) \exp{(2\pi i \mathbf{H} \cdot \mathbf{r})} \, d\mathbf{r} \qquad (8.12a)$$

and the inverse relation

$$\Phi(\mathbf{r}) = \int \Phi(\mathbf{H}) \exp{(-2\pi i \mathbf{H} \cdot \mathbf{r})} \, d\mathbf{r} \qquad (8.12b)$$

Differentiation of both sides of Eq. (8.12b) gives

$$\nabla_{\mathbf{r}} \Phi(\mathbf{r}) = -2\pi i \int \mathbf{H} \Phi(\mathbf{H}) \exp{(-2\pi i \mathbf{H} \cdot \mathbf{r})} \, d\mathbf{H} \qquad (8.13a)$$

and

$$\nabla_{\mathbf{r}}^2 \Phi(\mathbf{r}) = -4\pi^2 \int H^2 \Phi(\mathbf{H}) \exp{(-2\pi i \mathbf{H} \cdot \mathbf{r})} \, d\mathbf{H} \qquad (8.13b)$$

Substitution of the Poisson equation (8.11) into Eq. (8.13b) now leads to

$$-4\pi \rho_{\text{total}} = -4\pi^2 \int H^2 \Phi(\mathbf{H}) \exp{(-2\pi i \mathbf{H} \cdot \mathbf{r})} \, d\mathbf{H} \qquad (8.14)$$

In analogy to the definition of the total charge density, we define the *total structure factor* $F_{\text{total}}(\mathbf{H})$, which includes both the nuclei and the electrons, and is, excluding thermal effects, given by

$$F_{\text{total}}(\mathbf{H}) = \sum_A Z_A \exp{2\pi i \mathbf{H} \cdot \mathbf{R}_A} - F_{\text{electronic}}(\mathbf{H}) \qquad (8.15a)$$

and

$$\rho_{\text{total}}(\mathbf{r}) = \int F_{\text{total}}(\mathbf{H}) \exp{(-2\pi i \mathbf{H} \cdot \mathbf{r})} \, d\mathbf{H} \qquad (8.15b)$$

Combination of Eqs. (8.14) and (8.15a, b) gives (Bertaut 1952)

$$\Phi(\mathbf{H}) = + F_{\text{total}}(\mathbf{H})/\pi H^2 \qquad (8.16)$$

or, by inverse Fourier transformation,

$$\Phi(\mathbf{r}) = \frac{1}{\pi V} \sum_{\mathbf{H}} \frac{F(\mathbf{H})}{H^2} \exp{(-2\pi i \mathbf{H} \cdot \mathbf{r})} \qquad (8.17)$$

which expresses the electrostatic potential as a Fourier summation with coefficients $F_{\text{total}}(\mathbf{H})/H^2$.

It is of interest to note a second, quite different, derivation of Eq. (8.17) (Su 1993). The electrostatic potential of Eq. (8.2) can be written as a convolution:

$$\Phi(\mathbf{r}') = \int \frac{\rho_{\text{total}}(\mathbf{r})}{|\mathbf{r} - \mathbf{r}'|} \, d\mathbf{r} = \frac{1}{r} * \rho_{\text{total}}(\mathbf{r}) \qquad (8.18)$$

As the Fourier transform of $1/r$ is given by

$$\int \frac{1}{r} \exp{(2\pi i \mathbf{S} \cdot \mathbf{r})} \, d\mathbf{r} = \frac{1}{\pi S^2} \tag{8.19}$$

and the Fourier transformation of $\rho_{total}(\mathbf{r})$ is $F_{total}(\mathbf{S})$, the Fourier convolution theorem implies that the potential $\Phi(\mathbf{r}')$ is given by the inverse Fourier transform (p. 93) of the products of $1/(\pi S^2)$ and $F_{total}(\mathbf{S})$, or

$$\Phi(\mathbf{r}') = \int \frac{F_{total}(\mathbf{S})}{\pi S^2} \exp{(-2\pi i \mathbf{S} \cdot \mathbf{r}')} \, d\mathbf{S} = \frac{1}{\pi V} \sum_{\mathbf{H}} \frac{F_{total}(\mathbf{H})}{H^2} \exp{(-2\pi i \mathbf{H} \cdot \mathbf{r}')}$$

which is identical to Eq. (8.17).

As the Fourier coefficients in Eq. (8.17) contain the factor $1/H^2$, the high-order structure factors are of decreasing importance in the potential summation. The emphasis on the low-order structure factors is less pronounced for the higher-order electrostatic functions, such as the electric field and the electric field gradient, as summarized in Table 8.1.

In analogy with the electron density summation of the structure factors, the potential summation of Eq. (8.17) can be simplified using $F(\mathbf{H}) = A(\mathbf{H}) + iB(\mathbf{H})$, and Friedel's law, $F(\mathbf{H}) = F(\bar{\mathbf{H}})^*$. The result is

$$\Phi(\mathbf{r}) = \frac{2}{\pi V} \sum_{1/2} \left[\frac{A_{total}(\mathbf{H})}{H^2} \cos{(2\pi \mathbf{H} \cdot \mathbf{r})} + \frac{B_{total}(\mathbf{H})}{H^2} \sin{(2\pi \mathbf{H} \cdot \mathbf{r})} \right] \tag{8.20}$$

8.2.2 The $\Phi(0)$ Term in the Potential Summation

The summation in expression (8.17) contains a term with $H = 0$, representing the average potential in the unit cell. This origin term is singular, and its evaluation has been discussed extensively in the literature (see Su 1993, p. 9ff. for a critical analysis).

We consider the behavior of the origin term when the continuous variable S ($= H$ at the reciprocal lattice points) approaches zero:

$$\Phi(0) = \lim{(S \to 0)} \frac{\langle F_{total}(\mathbf{S}) \rangle}{\pi V S^2} \tag{8.21}$$

in which $\langle F_{total}(\mathbf{S}) \rangle$ is the angular average of $F(\mathbf{S})$. If the nuclear contribution is not included in the structure factor $F(\mathbf{S})$, the numerator of Eq. (8.21) will not go to zero in the limit $(S \to 0)$, and $\Phi(0)$ is infinite. Thus, the Fourier summation of Eq. (8.20) requires the use of the total structure factor.

Starting with

$$F_{total}(\mathbf{S}) = \int_{cell} \rho_{total}(\mathbf{r}) \exp{(2\pi i \mathbf{H} \cdot \mathbf{r})} \, d\mathbf{r} \tag{8.22}$$

we obtain, retaining the first two terms of the Taylor expansion of $\exp{(2\pi i \mathbf{S} \cdot \mathbf{r})}$,

$$\langle F_{total}(\mathbf{S}) \rangle = 2\pi i \int_{cell} \langle \mathbf{S} \cdot \mathbf{r} \rangle \rho_{total}(\mathbf{r}) \, d\mathbf{r} - 2\pi^2 \int_{cell} \langle \mathbf{S} \cdot \mathbf{r} \rangle^2 \rho_{total}(\mathbf{r}) \, d\mathbf{r} \tag{8.23}$$

in which the averages are over the reciprocal space angular coordinates. The first term in this expression is zero, while the average in the second term equals $S^2/3$. Substitution in Eq. (8.21) gives the expression (Becker and Coppens 1990)

$$\Phi(0) = \frac{2\pi}{3V} \int_{\text{cell}} r^2 \rho_{\text{total}}(\mathbf{r}) \, d\mathbf{r} \tag{8.24}$$

where the integration is over the volume of the unit cell.

If the unit cell has a dipole moment, as is the case for polar crystals, the macroscopic polarization must be taken into account. At the point $\mathbf{r}$, it is given by $(4\pi/3V)\mu_{\text{unit cell}} \cdot \mathbf{r}$ (P. Becker 1990, unpublished results), or

$$\Phi(0) = -\frac{2\pi}{3V} \int_{\text{cell}} r^2 \rho_{\text{total}}(\mathbf{r}) \, d\mathbf{r} + \frac{4\pi}{3V} \mu_{\text{unit cell}} \cdot \mathbf{r} \tag{8.25}$$

in which the origin of $\mathbf{r}$ is located at the center of the crystal and $\mu_{\text{unit cell}}$, the unit-cell dipole moment, is related to the macroscopic polarization $\mathbf{P}$ per unit volume by $\mu_{\text{unit cell}} = \mathbf{P} V_{\text{unit cell}}$. The macroscopic polarization contribution varies linearly with distance $\mathbf{r}$ from the center of the crystal. Its existence introduces an ambiguity in the comparison of potentials in different polar crystals, as it depends on the position of the unit cell in the crystal. For most chemical applications, this term can be omitted. It is possible to select the unit cell such that it has a vanishing dipole moment (Spackman and Stewart 1981, Avery et al. 1984), but this creates a large surface charge, with an identical contribution to Eq. (8.25).

For the static density, the zero term in the potential can be expressed in terms of the multipole coefficients of the aspherical-atom formalism. Substituting for atom j at $\mathbf{R}_j$:

$$\mathbf{r} = \mathbf{r} - \mathbf{R}_j + \mathbf{R}_j = \mathbf{r}' + \mathbf{R}_j$$

and integrating, we obtain, from Eq. (8.25) for the contribution to $\Phi(0)$ due to pseudoatom j,

$$\Phi_j(0) = -\frac{2\pi}{3V} (R_j^2 q_j + 2\mathbf{R}_j \cdot \boldsymbol{\mu}_j + \omega_j) \tag{8.26}$$

and, for the total distribution (excluding the macroscopic polarization term),

$$\Phi(0) = -\frac{2\pi}{3V} \sum_j^{\text{cell}} (R_j^2 q_j + 2\mathbf{R}_j \cdot \boldsymbol{\mu}_j + \omega_j) \tag{8.27}$$

where q_j, μ_j, and ω_j are the net charge, the dipole-moment vector, and the spherically averaged second moment of atom j, respectively, all with the sign convention of electronic charge being negative. All these quantities can be expressed in terms of the population parameters as described in chapter 7. The important first two terms in the summation of Eq. (8.27) are frequently omitted in calculations, but cannot be ignored.

For the procrystal, composed of neutral spherical atoms, the first two terms are zero, and we obtain

$$\Phi(0)_{\text{procrystal}} = -\frac{2\pi}{3V} \sum_j Z_j \langle r_j^2 \rangle \tag{8.28}$$

with

$$\langle r_j^2 \rangle = \frac{\int \rho_{e,j}(\mathbf{r}) r^2 \, d\mathbf{r}}{\int \rho_{e,j}(\mathbf{r}) \, d\mathbf{r}} = \int \rho_{e,j}(\mathbf{r}) r^2 \, d\mathbf{r}/Z_j \tag{8.29}$$

Equations (8.27) and (8.28) indicate that $\Phi(0)$ will tend to be more positive for a crystal containing heavier atoms. This is confirmed by experimental measurements of $\Phi(0)$ using electron-beam techniques. Measurements by electron holography, for example, give the following values for a number of crystals: Si, 9.26 (8); MgO, 13.01 (8); GaAs, 14.53 (17); PbS, 17.19 (12) V (Gajdardziska-Josifoska et al., as quoted in O'Keefe and Spence 1994). Thus, $\Phi(0)$ must be taken into account when different solids are compared, as in the studies of zeolites described in chapter 11.

Unlike Eq. (8.27), the Fourier summation method for evaluation of the electrostatic potential and its derivatives leads to the potential of the thermally averaged distribution. However, the differences are likely to be small relative to other uncertainties, especially when low-temperature data are used in the analysis.

8.2.3 The Electric Field and the Electric Field Gradient by Direct Fourier Summation of the Structure Factors

Like the potential, other electrostatic functions can be expressed as Fourier summations over the structure factors (Stewart 1979). The electric field, being the (negative) gradient of the potential, is a Fourier series in which the power of the magnitude of H increases from -2 to -1, as expected from the reciprocal relationship between direct space and Fourier space. Starting with

$$\mathbf{E}(\mathbf{r}) = -\nabla_r \Phi(\mathbf{r}) = -2\pi i \int \mathbf{H}\Phi(\mathbf{H}) \exp(-2\pi i \mathbf{H} \cdot \mathbf{r}) \, d\mathbf{H} \tag{8.13a}$$

and substituting

$$\Phi(\mathbf{H}) = +F_{\text{total}}(\mathbf{H})/\pi H^2 \tag{8.16}$$

the Fourier summation of the electric field vector is obtained as

$$\mathbf{E}(\mathbf{r}) = -\frac{2\pi i}{V} \Sigma(F_{\text{total}}(\mathbf{H})/H^2)\mathbf{H} \exp(-2\pi i \mathbf{H} \cdot \mathbf{r}) \tag{8.30}$$

The electric field gradient tensor has a trace equal to $-4\pi\rho_{\text{total}}(\mathbf{r})$ [Eq. (8.11)]. According to Eq. (8.7), its Fourier component $\mathbf{H}$ is obtained by taking the tensor product of the divergence vector ∇_r, with $\mathbf{E}(\mathbf{H})$. Since $\mathbf{E}(\mathbf{H}) = -2\pi i \mathbf{H}\Phi(\mathbf{H})$ [Eq. (8.13a)] and

$$\Phi(\mathbf{H}) = \int \Phi(\mathbf{r}) \exp(2\pi i \mathbf{H} \cdot \mathbf{r}) \, d\mathbf{r} \tag{8.12a}$$

we obtain

$$[\nabla_r : \mathbf{E}](\mathbf{H}) = +4\pi^2 \mathbf{H} : \mathbf{H}\Phi(\mathbf{H}) = +4\pi \mathbf{H} : \mathbf{H} F_{\text{total}}(\mathbf{H})/H^2 \tag{8.31}$$

in which $\mathbf{i} : \mathbf{k}$ represents the 3×3 tensor product of two vectors.

The expression for the electric field gradient in direct space then becomes

$$[\nabla:\mathbf{E}](\mathbf{r}) = \frac{4\pi}{V} \sum \mathbf{H}:\mathbf{H} F_{\text{total}}(\mathbf{H})/H^2 \exp(-2\pi i \mathbf{H} \cdot \mathbf{r}) \tag{8.32}$$

Both the components of $\mathbf{E}$ and the elements of the electric field gradient as given by Eqs. (8.30) and (8.32) are with respect to the reciprocal-lattice coordinate system. A transformation is required if the values in the direct-space coordinate systems are needed. To obtain the elements of the traceless ∇E tensor, the quantity $-(4\pi/3)\rho_e(\mathbf{r}) = -(4\pi/3V) \sum F(\mathbf{H}) \exp(-2\pi i \mathbf{H} \cdot \mathbf{r})$ must be subtracted from each of the diagonal elements ∇E_{ii}.

8.2.3.1 Application of the Expressions; Use of ΔF Series

As noted previously, the order of the derivative of the potential in real space increases on going from the potential to the electric field to the field gradient. Thus, the summation in reciprocal space becomes increasingly dependent on the high-order structure factors (Table 8.1). So, though Eqs. (8.28) and (8.30) would appear to provide a straightforward recipe for evaluating the corresponding physical properties, their application is hampered by series termination effects. Spackman and Stewart (1984) report that a data set on the mineral stishovite (SiO_2) with $\sin \theta/\lambda = 1.2 \, \text{Å}^{-1}$, gave electric field maps with large termination ripples, while even the potential maps were not entirely ripple free. Such effects will, of course, be less serious for softer crystals with larger thermal motion, which suggests the introduction of an artificial thermal motion as a remedy, though this again raises the question of the equivalence of the properties of the average distribution and the average of the properties of the instantaneous distributions.

A better alternative is to use the difference structure factor ΔF in the summations. The electrostatic properties of the procrystal are rapidly convergent and can therefore be easily evaluated in direct space. Stewart (1991) describes a series of model calculations on the diatomic molecules N_2, CO, and SiO, placed in cubic crystal lattices and assigned realistic mean-square amplitudes of vibration. He reports that for an error tolerance level of 1%, $(\sin \theta/\lambda)_{\text{max}} = 1-1.1 \, \text{Å}^{-1}$ is adequate for the *deformation* electrostatic potential, $\approx 1.5 \, \text{Å}^{-1}$ for the electric field, and $\approx 2.0 \, \text{Å}^{-1}$ for the deformation density and the deformation electric field gradient (which both have Fourier coefficients proportional to H^0).

Stewart's conclusion underscores the need for short-wavelength, low-temperature studies, if very high accuracy electrostatic properties are to be evaluated by Fourier summation. But, as pointed out by Hansen (1993), the convergence can be improved if the spherical atoms subtracted out are modified by the κ values obtained with the multipole model. Failure to do this causes pronounced oscillations in the deformation density near the nuclei. For the binuclear manganese complex (μ-dioxo)Mn(III)Mn(IV)(2,2'-bipyridyl)$_4$, convergence of the electrostatic potential at the Mn nucleus is reached at $0.7 \, \text{Å}^{-1}$, as checked by the inclusion of higher-order data (Frost-Jensen et al. 1995).

8.2.4 Application of the ΔF Summation to the Electrostatic Potential in L-alanine

When the electrostatic properties are evaluated by ΔF summation, the effect of the spherical-atom molecule must be evaluated separately. According to electrostatic theory, on the surface of any *spherical* charge distribution, the distribution acts as if concentrated at its center. Thus, outside the spherical-atom molecule's density, the potential due to this density is zero. At a point inside the distribution the nuclei are incompletely screened, and the potential will be repulsive, that is, positive. Since the spherical atom potential converges rapidly, it can be evaluated in real space, while the deformation potential $\Delta\Phi(\mathbf{r})$ is evaluated in reciprocal space. When the promolecule density, rather than the superposition of κ-modified non-neutral spherical-atom densities advocated by Hansen (1993), is evaluated in direct space, the pertinent expressions are given by (Destro et al. 1989)

$$\Phi(\mathbf{r}) = \Phi_{\text{spherical-atom crystal}}(\mathbf{r}) - \langle\Phi_{\text{spherical-atom crystal}}(\mathbf{r})\rangle + \Delta\Phi(\mathbf{r}) \quad (8.33)$$

with, analogous to Eq. (8.17).

$$\Delta\Phi(\mathbf{r}) = \frac{1}{\pi V}\sum_{\mathbf{H}\neq 0}\frac{\Delta F(\mathbf{H})}{H^2}\exp\left(-2\pi i\mathbf{H}\cdot\mathbf{r}\right) \quad (8.34)$$

The term $\langle\Phi_{\text{spherical-atom crystal}}(\mathbf{r})\rangle$ in Eq. (8.33) is the average potential in the unit cell of the promolecule crystal, equal to $\Phi(0)$ for the promolecule crystal. Expression (8.33) thus gives the *deviation* from the average promolecule potential in the crystal. Modification of Eq. (8.33) for the direct space evaluation of the κ-modified non-neutral spherical atom densities is straightforward.

The crystal potential for L-alanine calculated with Eq. (8.34) is shown in Fig. 8.1(a). The term $\Phi_{\text{spherical-atom}}(\mathbf{r})$ can be evaluated in direct space by the methods described in the following section. The term $\Phi(0)$ for the independent-atom model [not exactly equal to the true $\Phi(0)$] was evaluated by a summation of the IAM potential over the unit cell.

Figure 8.1(b) shows the electrostatic potential for the isolated molecule consisting of pseudoatoms, calculated as described below. In both maps, the zwitterionic nature of L-alanine is evident, with pronounced negative and positive potential regions occurring near the opposing ends of the molecule near the COO^- and NH_3^+ groups, respectively. The crystal potential shows a saddle point in the $O\cdots H$ hydrogen-bond regions, which is absent for the molecule lifted out of the crystal.

8.3 Evaluation of the Electrostatic Functions in Direct Space

8.3.1 Basic Expressions

The multipole description of the charge density makes it possible to identify a *pseudomolecule* in a crystal. The pseudomolecule is distinct from the molecules in the *procrystal*, composed of noninteracting molecules, in the same sense as

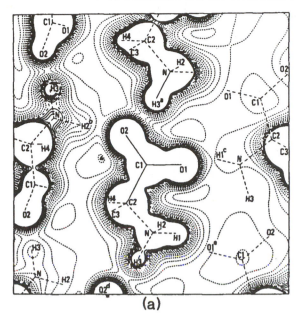

(a)

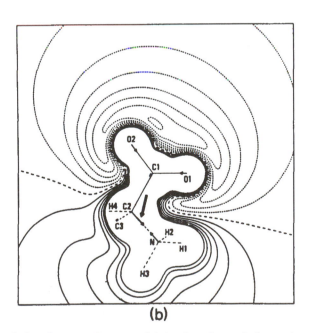

(b)

FIG. 8.1 Maps of the electrostatic potential in the plane of the carboxylate group of L-alanine from 23 K X-ray diffraction data. Atoms not close to the plane are connected by broken lines. Contour levels at 0.05 eÅ^{-1}. (a) By Fourier summation, Eqs. (8.33) and (8.34). (b) From the multipole coefficients for the isolated molecule. *Source*: Destro et al. (1989).

pseudoatoms are different from isolated atoms. The electrostatic properties of the pseudomolecule can be derived directly from the multipole population coefficients. The electrostatic quantities at the periphery of the pseudomolecule will include the effect of induced polarization by neighboring entities, and are thus relevant for the analysis of intermolecular interactions. The $\Phi(0)$ term encountered in the reciprocal space summation of the potential is absent in the direct space analysis.

To proceed, we rewrite Eqs. (8.3), (8.6), and (8.9) for the electrostatic properties at point P as a sum over atomic contributions:

$$\Phi(\mathbf{R}_P) = \sum_{M \neq P} \frac{Z_M}{|\mathbf{R}_{MP}|} - \sum_{M} \int \frac{\rho_{e,M}(\mathbf{r}_M)}{|\mathbf{r}_P|} \, d\mathbf{r} \qquad (8.35)$$

$$\mathbf{E}(\mathbf{R}_P) = -\sum_{M \neq P} \frac{Z_M \mathbf{R}_{MP}}{|\mathbf{R}_{MP}|^3} + \sum_{M} \int \frac{\mathbf{r}_P \rho_{e,M}(\mathbf{r}_M)}{|\mathbf{r}_P|^3} \, d\mathbf{r} \qquad (8.36)$$

$$\nabla \mathbf{E}_{\alpha\beta}(\mathbf{R}_P) = -\sum_{M \neq P} \frac{Z_M(3R_\alpha R_\beta - \delta_{\alpha\beta}|\mathbf{R}_{MP}|^2)}{|\mathbf{R}_{MP}|^5} + \sum_{M} \int \frac{\rho_{e,M}(\mathbf{r}_M)(3r_\alpha r_\beta - \delta_{\alpha\beta}|\mathbf{r}_P|^2)}{|\mathbf{r}_P|^5} \, d\mathbf{r}$$

$$(8.37)$$

where the index M represents the atomic centers, a density function $\rho_{e,M}$ being centered at nucleus M. The label P refers to the *field point* at which the property is being evaluated. The exclusion $M \neq P$ applies only when P coincides with a nucleus. Terms Z_M and $\mathbf{R}_M$ are the nuclear charge and the position vector of atom M, respectively, while $\mathbf{r}_M$ is the vector from a point $\mathbf{r}$ to the nucleus, and $\mathbf{r}_P$ and $\mathbf{R}_{MP}$ are, respectively, the vectors from P to a point $\mathbf{r}$ and to the nucleus $\mathbf{M}$, such that $\mathbf{r}_P = \mathbf{r} - \mathbf{R}_P$ and $\mathbf{R}_{MP} = \mathbf{R}_m - \mathbf{R}_P$, as illustrated in Fig. 8.2.

The derivation of the electrostatic properties from the multipole coefficients given below follows the method of Su and Coppens (1992). It employs the Fourier convolution theorem used by Epstein and Swanton (1982) to evaluate the electric field gradient at the atomic nuclei. A direct-space method based on the Laplace expansion of $1/|\mathbf{R}_P - \mathbf{r}|$ has been described by Bentley (1981).

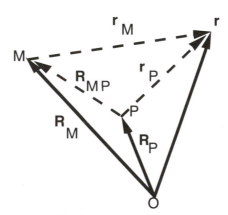

FIG. 8.2 Definition of vectors used in direct-space evaluation of the electrostatic properties. P is the field point, M the nuclear position. The electron position is defined by the vector $\mathbf{r}$.

8.3.2 Analytical Evaluation of the Electrostatic Properties Due to the Spherical Components of the Charge Density

Since the spherical core- and valence-scattering factors in the multipole expansion are based on theoretical wave functions, expressions for the corresponding density functions are needed in the analytical evaluation of the integrals in Eqs. (8.35)–(8.37). The expansion of the Roothaan–Hartree–Fock wave functions for the ground-state atoms tabulated by Clementi and Roetti (1974) can be used for this purpose.

In the Clementi and Roetti tables, the radial wave function of all orbitals in each electron subshell j is described as a sum of Slater-type functions:

$$\psi_j(r) = \sum_i^{v_j} C_{j,i} r^{n_j} e^{-\zeta_{j,i} r} \tag{8.38}$$

where v_j is the number of basis-set functions for the orbitals in the subshell j. The orbital wave functions are obtained by multiplication of the radial functions with the appropriate spherical harmonics. The expansion coefficients $C_{j,i}$ and the exponents $\zeta_{j,i}$ are, for basis functions with principal quantum number n, given by Clementi and Roetti (1974). The power of r, $n_j = n - 1$.

The spherically averaged atomic core and valence densities are obtained as the sum over products of the radial orbital functions, or, including normalization,

$$N_c \rho_c(r) = \frac{1}{4\pi} \sum_{j=1}^{\mu_c} N_j \sum_{i=1}^{v_j} \sum_{k=1}^{v_j} C_{j,i} C_{j,k} \sqrt{\frac{(2\zeta_{j,i})^{(2n_{j,i}+3)}(2\zeta_{j,k})^{(2n_{j,k}+3)}}{(2n_{j,i}+2)!(2n_{j,k}+2)!}}$$
$$\times r^{(n_{j,i}+n_{j,k})} e^{-(\zeta_{j,i}+\zeta_{j,k})r} \tag{8.39}$$

and

$$N_v \rho_v(r) = \frac{1}{4\pi} \sum_{j=1}^{\mu_v} N_j \sum_{i=1}^{v_j} \sum_{k=1}^{v_j} C_{j,i} C_{j,k} \sqrt{\frac{(2\zeta_{j,i})^{(2n_{j,i}+3)}(2\zeta_{j,k})^{(2n_{j,k}+3)}}{(2n_{j,i}+2)!(2n_{j,k}+2)}}$$
$$\times r^{(n_{j,i}+n_{j,k})} e^{-(\zeta_{j,i}+\zeta_{j,k})r} \tag{8.40}$$

where μ_c and μ_v are the number of atomic orbitals in the core shells and valence shells, respectively, N_j is the occupancy of the jth subshell and $N_c = \sum_{j=1}^{\mu_c} N_j$ and $N_v = \sum_{j=1}^{\mu_v} N_j$. The terms $\rho_c(r)$ and $\rho_v(r)$ are normalized to one electron. For a ^3P ground-state carbon atom, for example, with valence configuration $2s^2 2p^2$, the number of subshells in the valence shell μ_v equals 2, and $N_v = 4$, $N_1 = 2$, and $N_2 = 2$. For carbon, the subshells are expansions of 6, respectively, 4, Slater-type functions , that is, $v_1 = 6$, $v_2 = 4$. Because of the spherical averaging of ρ_c and ρ_v, the occupancies of orbitals with the same n and l values are the same, regardless of their m values. In other words, the electrons in a subshell are evenly distributed among the orbitals with different values of the magnetic quantum number m.

8.3.3 Electrostatic Properties and the Multipole Expansion

While the calculation of the electrostatic functions from the multipole parameters parallels that of the calculation of the atomic electrostatic moments, there is an important distinction. The atomic moments are local, that is, they depend only on the density at one atom, but the electrostatic properties depend *both* on the local and on the peripheral electron distribution. When the properties are evaluated at the nuclear position, we will distinguish between *central* contributions originating from the charge density centered at that nucleus, and *peripheral* contributions by all other atomic densities. For a field point not coinciding with a nucleus, all contributions are by definition peripheral.

8.3.3.1 Central Contributions

The electron density centered at M is the only central contributor at the nuclear position M, as in this case the nucleus coincides with the field point P, which is excluded from the integrals. For transition metal atoms, the central contributions are the largest contributors to the properties at the nuclear position, which can be compared directly with results from other experimental methods. The electric field gradient at the nucleus, for instance, can be measured very accurately for certain nuclei with nuclear quadrupole resonance and/or Mössbauer spectroscopic methods, while the electrostatic potential at the nucleus is related to the inner-shell ionization energies of atoms, which are accessible by photoelectron and X-ray spectroscopic methods.

The operators for the potential, the electric field, and the electric field gradient have the same symmetry, respectively, as those for the atomic charge, the dipole moment, and the quadrupole moment discussed in chapter 7. In analogy with the moments, only the spherical components on the density give a central contribution to the electrostatic potential, while the dipolar components are the sole central contributors to the electric field, and only quadrupolar components contribute to the electric field gradient in its traceless definition.

We start with the spherical terms. Substitution of the density expressions (8.39) and (8.40) into the second term of Eq. (8.35), and integration over the coordinates, gives for the central electrostatic potential:

$$
\Phi^{\text{central}}(\mathbf{R}_M) = -\frac{P_{M,c}}{N_c} \sum_{j=1}^{\mu_c} N_j \sum_{i=1}^{v_j} \sum_{k=1}^{v_j} C_{j,i} C_{j,k} \sqrt{\frac{(2\zeta_{j,i})^{(2n_{j,i}+3)}(2\zeta_{j,k})^{(2n_{j,k}+3)}}{(2n_{j,i}+2)!(2n_{j,k}+2)!}}
$$

$$
\times \frac{(n_{j,i}+n_{j,k}+1)!}{(\zeta_{j,i}+\zeta_{j,k})^{(n_{j,i}+n_{j,k}+2)}}
$$

$$
-\frac{P_{M,v}}{N_v} \sum_{j=1}^{\mu_v} N_j \sum_{i=1}^{v_j} \sum_{k=1}^{v_j} C_{j,i} C_{j,k} \sqrt{\frac{(2\kappa\zeta_{j,i})^{(2n_{j,i}+3)}(2\kappa\zeta_{j,k})^{(2n_{j,k}+3)}}{(2n_{j,i}+2)!(2n_{j,k}+2)!}}
$$

$$
\times \frac{(n_{j,i}+n_{j,k}+1)!}{[\kappa(\zeta_{j,i}+\zeta_{j,k})]^{(n_{j,i}+n_{j,k}+2)}} - P_{M,00}\frac{\kappa'\zeta_0}{(n_0+2)} \tag{8.41}
$$

The last term in Eq. (8.41) originates from the Slater-type monopole in the scattering formalism, with n_0 as power of r, and ζ_0 as exponent.

The dipolar terms contribute to the electric field. With the density deformation functions of the multipole model (chapter 3) and Eq. (8.36), one obtains

$$E_x^{\text{central}}(\mathbf{R}_M) = \frac{4}{3} P_{M,11+} \frac{(\kappa'\zeta_1)^2}{(n_1 + 1)(n_1 + 2)}$$

$$E_y^{\text{central}}(\mathbf{R}_M) = \frac{4}{3} P_{M,11-} \frac{(\kappa'\zeta_1)^2}{(n_1 + 1)(n_1 + 2)} \qquad (8.42)$$

$$E_z^{\text{central}}(\mathbf{R}_M) = \frac{4}{3} P_{M,10} \frac{(\kappa'\zeta_1)^2}{(n_1 + 1)(n_1 + 2)}$$

The electric field gradient elements $\nabla E_{\alpha\beta}(\mathbf{r})$, according to the traceless definition of Eq. (8.9), are, according to Eq. (8.37), the expectation values of the operator

$$\left\langle \frac{1}{|\mathbf{r} - \mathbf{r}'|^5} \{3(r_\alpha - r_\alpha')(r_\beta - r_\beta') - |\mathbf{r} - \mathbf{r}'|^2 \delta_{\alpha\beta}\} \right\rangle$$

which differs from the traceless quadrupole moment operator in Eq. (7.2) only by the factor of $2/|\mathbf{r} - \mathbf{r}'|^5$. The expressions for the central contributions therefore contain the same linear combinations of the population parameters P_{2mp} as the quadrupole moment expressions of Eq. (7.21). The results are:

$$\nabla E_{xx}^{\text{central}}(\mathbf{R}_M) = \frac{3}{5} \frac{(\kappa'\zeta_2)^3}{n_2(n_2 + 1)(n_2 + 2)} (\pi P_{M,22+} - \sqrt{3} P_{M,20})$$

$$\nabla E_{yy}^{\text{central}}(\mathbf{R}_M) = -\frac{3}{5} \frac{(\kappa'\zeta_2)^3}{n_2(n_2 + 1)(n_2 + 2)} (\pi P_{M,22+} + \sqrt{3} P_{M,20})$$

$$\nabla E_{zz}^{\text{central}}(\mathbf{R}_M) = \frac{6}{5} \frac{(\kappa'\zeta_2)^3}{n_2(n_2 + 1)(n_2 + 2)} \sqrt{3} P_{M,20}$$

$$\nabla E_{xy}^{\text{central}}(\mathbf{R}_M) = \nabla E_{yx}^{\text{central}}(\mathbf{R}_M) = \frac{3}{5} \frac{(\kappa'\zeta_2)^3}{n_2(n_2 + 1)(n_2 + 2)} \pi P_{M,22-} \qquad (8.43)$$

$$\nabla E_{xz}^{\text{central}}(\mathbf{R}_M) = \nabla E_{zx}^{\text{central}}(\mathbf{R}_M) = \frac{3}{5} \frac{(\kappa'\zeta_2)^3}{n_2(n_2 + 1)(n_2 + 2)} \pi P_{M,21+}$$

$$\nabla E_{yz}^{\text{central}}(\mathbf{R}_M) = \nabla E_{zy}^{\text{central}}(\mathbf{R}_M) = \frac{3}{5} \frac{(\kappa'\zeta_2)^3}{n_2(n_2 + 1)(n_2 + 2)} \pi P_{M,21-}$$

in which ζ_1, ζ_2, n_1, and n_2 are defined by Eqs. (8.38) and (3.35).

8.3.3.2 Peripheral Contributions

In Eq. (8.18), we wrote the potential as a convolution of the total density and the operator $1/r$. Similarly, the integrals encountered in the evaluation of the peripheral electronic contributions to Eqs. (8.35)–(8.37) are convolutions of the electron density $\rho(\mathbf{r})$ and the pertinent operator. They can be evaluated with the Fourier convolution theorem (Prosser and Blanchard 1962), which implies that the convolution of $f(\mathbf{r})$ and $g(\mathbf{r})$ is the inverse transform of the product of their

Fourier transforms $F(\mathbf{S})$ and $G(\mathbf{S})$, or

$$\int f(\mathbf{r})g^*(\mathbf{r} - \mathbf{R})\, d\mathbf{r} = \int F(\mathbf{S})G^*(\mathbf{S})\exp\left(-2\pi i\mathbf{S}\cdot\mathbf{R}\right) d\mathbf{S} \tag{8.44}$$

where * indicates the complex conjugate.

The calculation thus consists of three steps: (1) calculating the scattering factors of the analytical charge density functions (see appendix G for closed-form expressions), (2) Fourier transformation of the electrostatic operator, and (3) back transformation of the product of two Fourier transforms.

Using the appropriate operators, the back transforms are given by

$$\int \frac{\rho_{e,M}(\mathbf{r}_M)}{|\mathbf{r}_P|}\, d\mathbf{r} = \frac{1}{\pi} \int \frac{f_{e,M}(\mathbf{S})}{|\mathbf{S}|^2}\exp\left(2\pi i\mathbf{S}\cdot\mathbf{R}_{MP}\right) d\mathbf{S} \tag{8.45}$$

$$\int \frac{\mathbf{r}_P\rho_{e,M}(\mathbf{r}_M)}{|\mathbf{r}_P|^3}\, d\mathbf{r} = 2i \int \frac{\mathbf{S}f_{e,M}(\mathbf{S})}{|\mathbf{S}|^2}\exp\left(2\pi i\mathbf{S}\cdot\mathbf{R}_{MP}\right) d\mathbf{S} \tag{8.46}$$

$$\int \frac{\rho_{e,M}(\mathbf{r}_M)(3x_{m,P}x_{n,P} - \delta_{mn}|\mathbf{r}_P|^2)}{|\mathbf{r}_P|^5}\, d\mathbf{r}$$

$$= -\frac{4\pi}{3} \int \frac{(3S_m S_n - \delta_{mn}|\mathbf{S}|^2)f_{e,M}(\mathbf{S})}{|\mathbf{S}|^2}\exp\left(2\pi i\mathbf{S}\cdot\mathbf{R}_{MP}\right) d\mathbf{S} \tag{8.47}$$

In order to separate the integration into a radial and an angular part, the exponential $\exp(i2\pi\mathbf{S}\cdot\mathbf{r})$ may be expressed as the expansion (Cohen-Tannoudji et al. 1977, Arfken 1970)

$$\exp(i2\pi\mathbf{S}\cdot\mathbf{r}) = \sum_{l=0}^{\infty} i^l(2l + 1)j_l(2\pi Sr)P_l(\cos\gamma) \tag{8.48}$$

where $P_l(x)$ is the lth order Legendre polynomial and γ is the angle between $\mathbf{S}$ and $\mathbf{r}$. It can be shown that Eq. (8.48) is equivalent to the plane wave expansion in terms of the complex spherical harmonics, used in chapter 3 for the evaluation of scattering factors.

Assuming that all vectors on atom M are referred to the same local Cartesian coordinate system of that atom, the results for the peripheral contributions to the potential are[2]

$$\Phi^{\mathrm{per}}(\mathbf{R}_P) = \sum_{M \neq P} \frac{Z_M}{|\mathbf{R}_{MP}|}$$

$$- \sum_{M \neq P}'' \left\{ \frac{2}{\pi}\frac{P_{M,c}}{N_c} \sum_{j=1}^{\mu_c} N_j \sum_{i=1}^{v_j} \sum_{k=1}^{v_j} C_{j,i}C_{j,k}\sqrt{\frac{(2\zeta_{j,i})^{(2n_{j,i}+3)}(2\zeta_{j,k})^{(2n_{j,k}+3)}}{(2n_{j,i}+2)!(2n_{j,k}+2)!}} \right.$$

$$\left. + A_{n_i+n_k,0,0,0}(\zeta_{j,i} + \zeta_{j,k}, |\mathbf{R}_{MP}|) \right.$$

---

[2] The symbol $\sum''$ indicates that the contributions of the individual atoms must be referred to the same coordinate system.

$$+ \frac{2}{\pi} \frac{P_{M,v}}{N_v} \sum_{j=1}^{\mu_v} N_j \sum_{i=1}^{v_j} \sum_{k=1}^{v_j} C_{j,i} C_{j,k}$$

$$\times \sqrt{\frac{(2\kappa\zeta_{j,i})^{(2n_{j,i}+3)} (2\kappa\zeta_{j,k})^{(2n_{j,k}+3)}}{(2n_{j,1}+2)!(2n_{j,k}+2)!}}$$

$$\times A_{n_{j,i}+n_{j,k}, 0, 0, 0}(\kappa(\zeta_{j,i}+\zeta_{j,k}), |\mathbf{R}_{MP}|)$$

$$+ 8 \sum_{l_1=0}^{l_{1,max}} \sum_{m_1=0}^{l_1} \sum_{p_1} (-1)^{l_1} \frac{\zeta_{l_1}^{(n_{l_1}+3)}}{(n_{l_1}+2)!}$$

$$\left. \times A_{n_{l_1}, l_1, l_1, 0}(\kappa'\zeta_{l_1}, R_{MP}) P_{l_1 m_1 p_1} d_{l_1 m_1 p_1}(\theta_{\mathbf{R}_{MP}}, \phi_{\mathbf{R}_{MP}}) \right\} \qquad (8.49)$$

As before, the exclusion $M \neq P$ applies when the point P coincides with a nucleus.

For compactness, the subscript M for the electronic density parameters has been omitted in Eq. (8.49). The polar coordinate system has the z axis of the local Cartesian coordinate system as the polar axis, and the vector $\mathbf{R}_{MP}$ is referred to this local coordinate system.

$$A_{N, l_1, l_2, k}(Z, R) = \int_0^\infty G_{N+2, l_1}(Z, S) j_{l_2}(SR) S^k \, dS \qquad (8.50)$$

is described in appendix H. Corresponding expressions for the peripheral contributions to the electric field and the electrostatic potential from the multipole parameters are given in Su and Coppens (1992).

The expressions given here are valid for the multipole formalism of Hansen and Coppens, as described by Eq. (3.35). With other formalisms, similar expressions are used. Experimental molecular potentials reported in the literature include those of imidazole (Spackman and Stewart 1984, pp. 302–320), phosphorylethanolamine (Swaminathan and Craven 1984), alloxan (Swaminathan and Craven 1985), and parabanic acid (He et al. 1988).

8.4 Comparison of Diffraction Results with Theory and Other Experimental Values

8.4.1 The Comparison of Experimental and Theoretical Potentials

Comparison of the experimental potential in a crystal and the theoretical potential for an isolated molecule is an excellent test for the transferability of theoretical isolated molecule densities to problems such as molecular packing and protein folding. A systematic study of this kind was done on L-alanine. Figure 8.3 shows a comparison between theory and experiment for a plane containing the C—N bond in this molecule. The comparison is with the 6-31G** basis set of double-zeta-plus-polarization quality. The agreement of experiment with more modest basis-set calculations was found to be inferior, which gives confidence in the experimental results. Both in the plane shown, and in the plane of the carboxyl

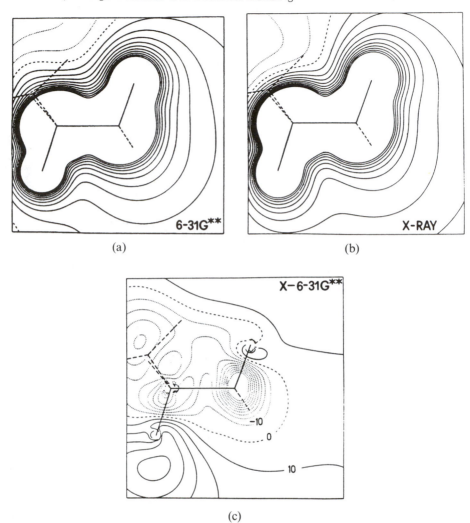

FIG. 8.3 (a) Theoretical, (b) experimental, and (c) difference between theoretical and experimental potential in a plane containing the C—N bond of the L-alanine molecule. The NH_3^+ group is on the right, with one hydrogen atom in the plane, and the others above and below the plane, respectively. *Source*: Destro et al. (1989).

group, the experimental potential is more diffuse than the theoretical result, the positive areas near the nuclei being lower, and the negative areas in remote regions being less negative. An anomaly is found near the H(4) hydrogen atom bound to the α-carbon atom, where a large residual feature occurs in the difference potential [Fig. 8.3(c)]. Both oxygen-atom regions show a similar potential, notwithstanding the different environment of the two atoms. The overall pattern is very similar in theory and experiment.

Further systematic studies are required to establish overall patterns suitable for application in molecular modeling calculations.

8.4.2 The Relation Between the Electrostatic Potential at the Nuclear Position, the 1s Binding Energy and the Net Atomic Charge

The electrostatic potential at the nuclear position correlates qualitatively with the 1s-electron binding energy ε. The correlation is negative: when the electrostatic potential is more positive, the negative electron will be more tightly bound.

For first-row atoms, the quantitative relation between the *variation* in binding energy ε_{1s} and the potential at the nuclear positions due to the valence electrons, Φ_{val}, is well established (Basch 1970). Early SCF calculations showed that for C, N, O, and F atoms in a series of small molecules, the ratio $\Delta(-\varepsilon_{1s})/\Delta\Phi_{val}$ is very nearly constant, as illustrated in Fig. 8.4 (Schwartz 1970). $\Delta(-\varepsilon_{1s})$ is the difference between the 1s binding energy in a molecule and that in the corresponding hydride, and $\Delta\Phi_{val}$ is the corresponding difference in Φ_{val}. Here, $\Phi_{val} = \Phi_{total} - \Phi_{1s}$. The

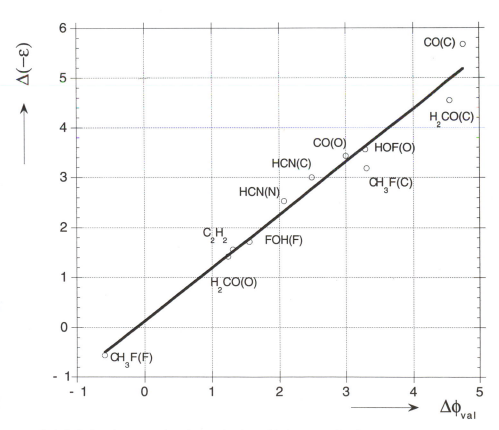

FIG. 8.4 Relation between the change in 1s orbital energy for first-row atoms and the potential at the nuclear position due to the valence electrons. Both are expressed relative to the corresponding hydride (i.e., CH_4, NH_3, H_2O, and HF for C, N, O, and F, respectively). The label in the brackets specifies the nucleus being considered. *Source*: Data in Schwartz (1970).

potential at the nucleus due to the $1s$ electrons, Φ_{1s}, is for a given element virtually unaffected by chemical substitution, but increases with increasing nuclear charge. The variations in the potential at the nucleus for a given element are due to bonding-induced changes in the valence-shell distribution.

The binding energy can be derived experimentally through its relation with the ionization energy I:

$$I = \varepsilon - R \qquad (8.51)$$

where R represents the reorganization energy released when the ionized atom relaxes to a lower energy state. In the much-used approximation of Koopman's theorem (1934), the relaxation energy is neglected, and the ionization energy becomes equal to the binding energy. But even when Koopmans' theorem is not valid, the correlation between ionization energy and the binding energy holds approximately, as long as the relaxation energies do not vary much among the species being compared. Saethre et al. (1985) have derived the reorganization energies upon $1s$ ionization for a number of diatomic halides by combining core-ionization energies and Auger kinetic energies. Results on F_2, HF, and ClF, and corresponding analogues for other halogen atoms, show non-negligble variations in the reorganization energy R, especially for the fluorine-containing species.

When experimental values of ΔR are available, the variation in the ionization energy among species can be related to the change in the potential at the nuclear position by

$$\Delta\Phi_{nucleus} \approx \Delta\varepsilon = \Delta I + \Delta R \qquad (8.52)$$

The approximate relation between the electrostatic potential $\Phi_{nucleus}$ and the net charge on the atom q is frequently being used for the derivation of net atomic charges in molecules from ionization energies (Siegbahn et al. 1967). As pointed out by Saethre et al. (1991), the assumption that $q_i \approx \Delta\Phi_{nucleus,i}$ is too simplistic and ignores the effect of the detailed charge distribution, which is implicit in Eq. (8.52).

It is exactly in the evaluation of the detailed charge distribution that X-ray methods provide the needed information. The comparison of diffraction-derived electrostatic potentials with binding energies determined by either photoelectron spectroscopy or X-ray resonance scattering is a field that merits further attention. The comparison of resonance-edge shifts and X-ray electrostatic potentials for a mixed valence complex $(\mu$-dioxo)Mn(III)Mn(IV)(2,2'-bipyridyl)$_4$ (BF$_4$)$_3$ gives encouraging results, which are in agreement with theoretical values for the binding energy (Table 8.2) (Gao et al. 1992, Frost-Jensen et al. 1995).

8.4.3 The Electric Field Gradient at the Nuclear Position

For nuclei possessing an electric quadrupole moment, the electric field gradient at the atomic nuclei can be measured accurately by techniques such as nuclear quadrupole resonance, Mössbauer spectroscopy, nuclear magnetic resonance, and, for gaseous species, by microwave spectroscopy. The diffraction data permit an

TABLE 8.2 Differences in Ionization Energy I, $1s$ Binding Energy ε_{1s}, and Electrostatic Potential Φ at the Mn Nuclei in (μ-dioxo)Mn(III)Mn(IV) $(2,2'\text{-bipyridyl})_4$ $(BF_4)_3$

		Mn(III)–Mn(IV) (eV)
ΔI	Resonance scattering	3.7–4·0 (1.0)
$\Delta \varepsilon_{1s}$	Theoretical calculation	3.17
$\Delta \Phi$	Diffraction:	
	Direct space	3.6 (2.0)
	Reciprocal space	2.5 (>0.2)
ΔI	Diffraction, corrected for reorganization energy:	
	Direct space	4.0 (2.0)
	Reciprocal space	2.9 (>0.4)

Source: Gao et al. (1992), Frost-Jensen et al. (1995).

interpretation of the spectroscopic results in terms of the detailed charge distribution, and can provide the signs, which are not generally available via the other methods.

For transition metal atoms, the dominant contribution to the electric field gradient at a nucleus originates in the valence shell centered on that nucleus. The application of Eq. (8.41) to transition metal complexes will be discussed in chapter 10.

Peripheral contributions become important when short interatomic distances are involved, as, for example, for the EFG at nitrogen nuclei and especially at nuclei of hydrogen atoms. Since hydrogen has only one electron, the electric field gradient is mainly due to the density farther from the nucleus, and has therefore been described as less sensitive to the precise charge distribution (Tegenfeldt and Hermansson 1985).

This is borne out by calculations, and by the good agreement between X-ray, and NMR, and theoretical values, illustrated for the molecule of benzene in Table 8.3. Unless the hydrogen atoms participate in strong hydrogen-bonds, such as in

TABLE 8.3 Experimental and Theoretical Values of the Electric Field Gradient at the Deuterons in Deuterobenzene (e au^{-3})

	X-ray[a]	NMR[b]	Theory (6-31G**)
∇E_{xx}	0.16 (2)	0.143	0.181
∇E_{yy}	0.13 (2)	0.132	0.157
∇E_{zz}	−0.29 (2)	−0.275	−0.338
η	0.10 (1)	0.04	0.07

Source: [a] Z. Su and P. Coppens, unpublished, based on X-ray and neutron data of Jeffrey et al. (1987) and private communication.
[b] Millet and Dailey (1972).

$LiOH \cdot 2H_2O$, the unperturbed spherical atom contribution to the EFG dominates, while the contribution from the deformation density is relatively small.

The EFG data from the multipole parameters are, in principle, for the static crystal; while the spectroscopic data are affected by vibrations. There may therefore by a systematic difference between the two sets of values, which is evident for a number of hydrogen-bonded hydroxyl groups and water molecules studied by Tegenfeldt and Hermansson (1985), but is not apparent in the data in Table 8.3. The EFG values for H atoms in hydrogen-bonds is further discussed in chapter 12.

8.5 The Electrostatic Potential Outside a Charge Distribution

8.5.1 The Electrostatic Potential Outside a Charge Distribution in Terms of the Multipole Moments

The potential outside a charge distribution can be expressed in terms of a finite series of the outer moments of the distribution. The expression is obtained through a power series expansion of r^{-1}, where r is the distance from the field point to the origin of the distribution, and subsequent integration (Hirshfelder et al. 1954, Buckingham 1978). At a point $\mathbf{r}_i$, with components $\mathbf{r}_\alpha$, for unit value of $4\pi\varepsilon_0$, one obtains

$$\Phi(\mathbf{r}_i) = \frac{q}{r_i} + \frac{\mu_\alpha r_\alpha}{r_i^3} + \frac{1}{3}[3r_\alpha r_\beta - r^2\delta_{\alpha\beta}]\frac{\Theta_{\alpha\beta}}{r_i^5} + \text{higher order terms} \quad (8.53)$$

where the Einstein convention implying summation over repeated indices is used. In this expression, the moments $\Theta_{\alpha\beta}$ and $\Omega_{\alpha\beta\gamma}$ of the total charge density are traceless, as defined in Eq. (7.2). Since the traceless electrostatic moments are not dependent on the spherical components of the neutral atom, they can equally well be calculated from the deformation density.

The summation in Eq. (8.53) is slowly converging if a molecular charge distribution is represented by a single set of moments. However, the expression can be written as the summation over the distributed moments, centered at the nuclei j, which is precisely the information available from the multipole analysis:

$$\Phi(\mathbf{r}_i) = \sum_j \left[\frac{q_j}{r_{ij}} + \frac{\mu_\alpha r_{\alpha j}}{r_{ij}^3} + \frac{1}{3}[3r_{\alpha j}r_{\beta j} - r_j^2\delta_{\alpha\beta}]\frac{\Theta_{\alpha\beta}}{r_{ij}^5} \right.$$
$$\left. + [5r_{\alpha j}r_{\beta j}r_{\gamma j} - r^2(r_{\alpha j}\delta_{\beta\gamma} + r_{\beta j}\delta_{\gamma\alpha} + r_{\gamma j}\delta_{\alpha\beta})]\frac{\Omega_{\alpha\beta\gamma j}}{5r_{ij}^7} + \cdots \right] \quad (8.54)$$

in which r_{ij} measures the distance from each center j of the expansion to the field point i. Equation (8.54) is the limiting case of Eq. (8.49) for very large values of R_{MP}, or $\exp(-R_{MP}Z) \approx 1$ in the integrals $A_{N,l_1,l_2,\kappa}(Z, R)$ occurring in Eq. (8.49).

8.5.2 Net Atomic Charges Reproducing the Electrostatic Potential

For a study of nonbonded interactions, it is of practical importance to identify the net atomic charges that best reproduce the potential on the periphery of the

TABLE 8.4 Dipole Moments (in Debye), According to the SCF Wave Function, from Potential-Adjusted Charges and from Mulliken Charges Derived from the Same Wave Function, Compared with Experimental Values

	SCF (6-31G**)	Potential Charges	Mulliken Charges	Experimental Values (Gas Phase)
H_2O	2.196	2.258	1.011	1.850
NH_3	1.886	1.850	0.762	1.470
N_2CO	2.749	2.695	1.644	2.330
CH_3OH	1.914	1.864	1.336	1.700
H_2NCHO	4.486	4.458	2.412	3.730

Source: Chirlian and Francl (1987).

molecule. The idea of fitting the charges to the theoretical potential around the molecule was first proposed by Momany (1978), and elaborated by Chirlian and Francl (1987). The resulting net charges are reasonable and reproduce the molecular dipole moments from the same calculations much better than Mulliken charges, as illustrated in Table 8.4.

The methods adjust the atomic net charges q_i in a least-squares minimization with a discrepancy function equal to the sum of the potential differences over all n sampling points:

$$\Delta = \sum_{k=1}^{n} \left(\Phi_{\text{exact}}(k) - \sum_{i} \frac{q_i}{R_{ik}} \right)^2 \qquad (8.55)$$

The location of the sampling points used in Eq. (8.55) is crucial. Typically, points are selected on and just outside the van der Waals envelope of the molecule. Woods et al. (1990) selected 100 points per atom randomly within a spherical shell around each atom, but outside the van der Waals envelope of the molecule. Ghermani et al. (1993), in an analysis of experimental charge densities of peptides and pseudopeptides (modified peptide molecules), place the sample points on the surfaces of atom-centered spheres. In the application to the pseudopeptide N-acetyl-α,β-dehydrophenylalanine methylamide, the sampling points are selected to be equidistant and are located on composite spherical surfaces with radii 2–8 Å around the nuclei in the molecule (Fig. 8.5). The potential used as the starting point in the fitting was calculated for a molecule composed of spherical atoms, with experimental charges and contraction parameters from the κ-refinement. For a sphere radius of 2 Å, the fitted charges differ by several percent from those obtained with larger radii, but convergence is reached rapidly when the sphere radius is increased, or more spheres with large radii are added. The charges obtained are within 0.06 e of the values from the κ-refinement providing the starting density for the minimalization. No molecular neutrality constraint was used; a total charge of −0.19 e resulted.

The observed and fitted potentials in the region of a peptide link are compared in Fig. 8.6. Except for the inner regions, the agreement is satisfactory. In particular, the pronounced negative areas in the proximity of the oxygen atom are well reproduced.

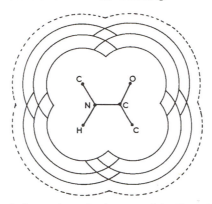

FIG. 8.5 Sampling-point shells used to obtain potential-adjusted charges of N-acetyl-α,β-dehydrophenylalanine. *Source*: Ghermani et al. (1993).

A geodesic scheme for the selection of sampling points has been proposed by Spackman (1996). The points are arranged on a series of fused-spherical van der Waals surfaces, and are arranged in a regular pattern, based on the tesselation of the icosahedron. The resulting charges are reported to have less dependence on molecular conformation than those from a number of other sampling techniques.

Francl et al. (1996) examined the conditioning of the least squares matrix in the fitting procedure, and conclude that the method cannot be used to assign statistically valid charges to all atoms in a given molecule. This problem cannot be alleviated by the selection of more sampling points, and thus may require the introduction of chemical constraints to reduce the number of charges to be determined.

Rather than fit the potential at the periphery of the molecule, Su (1993) has fitted Φ at the nuclear positions. The potential at the nuclear positions can be directly related to the total energy of a molecule (chapter 9) (Polizer 1981), as well as to the atomic ionization energies (as discussed in section 8.4).

TABLE 8.5 Comparison of Equivalent-Potential and Refinement Charges for Atoms of COO$^-$ Groups in L-Alanine (top line) and D,L-Histidine (bottom line)

	Potential		Refinement	
	Periphery	At nuclei	κ	Multipole[a]
C	+0.55	−0.06	+0.49 (4)	−0.06 (6)
	+0.11	−0.08	+0.51 (5)	−0.16 (22)
O(1)	−0.65	−0.33	−0.58 (3)	−0.41 (3)
	−0.68	−0.34	−0.65 (4)	−0.45 (8)
O(2)	−0.76	−0.40	−0.69 (3)	−0.45 (3)
	−0.65	−0.35	−0.68 (4)	−0.48 (9)

Source: Su (1993).

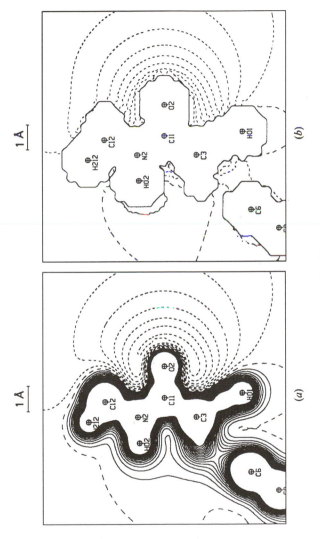

FIG. 8.6 Electrostatic potential maps in the region of one of the peptide links in N-acetyl-α,β-dehydrophenylalanine methylamide. (a) Observed. (b) From net charges fitted to the potential. Contours are at 0.05 eÅ^{-1} (1 eÅ^{-1} = 332.1 kcal mol^{-1}). Zero and negative contours are dashed lines. *Source:* Ghermani et al. (1993).

TABLE 8.6 Molecular Dipole Moments (D) from Fit to the Potential at the Nuclei, and Directly from X-ray Refinements

| | Φ at nucleus | | | | | Solution | |
| | $\kappa_H^a =$ | | κ | Multipole | | Value | |
	1.0	1.4	Refinement	Refinement	Reference		Reference
L-Alanine	8.6	12.7	12.5 (2)	13.3	Li (1989)	12.3–17.0[c]	McClellan (1974)
D,L-Histidine	13.9	16.0	16.9 (7)	23 (1)	Li (1989)	13.5[c]	Khanarian and Moore (1980)
MNA[b]	19.2	23.4	25 (6)	25 (9)	Howards et al. (1992)	6.98[d]	McClellan (1963)
PDM[b]	7.0	8.5	11 (2)	9.8	Baert et al. (1982)	9.2[d]	Treiner et al. (1964)

[a] κ Expansion–contraction parameter in definition of hydrogen density function.
[b] MNA = 2-methyl-4-nitroaniline, PDM = pyridium dicyanomethylide.
[c] In H_2O.
[d] In dioxane.
Source: Su (1993).

The charges obtained with nuclear sampling points are very different from those derived with the fit at the periphery. In Table 8.5, the results of the two methods at the carbon and oxygen atoms in the carboxyl groups in alanine and histidine are compared with the primary X-ray charges. Interestingly, for this example, the periphery-adjusted charges agree well with the results of the κ-refinement, while the nucleus-adjusted charges are closer to those from the multipole refinement. The molecular dipole moments of a number of molecules are quite well reproduced by the nucleus-adjusted charges, particularly when the fixed κ-value of 1.4 is used for the hydrogen valence shell in the fitting procedure (Table 8.6).

In all fitting procedures, electroneutrality and electrostatic-moment constraints may be introduced to provide additional observational equations.

9

The Electron Density and the Lattice Energy of Crystals

9.1 The Total Energy of a System

The total energy of a quantum-mechanical system can be written as the sum of its kinetic energy T, Coulombic energy E_{Coul}, and exchange and electron correlation contributions E_x and E_{corr}, respectively:

$$E = T + E_{\text{Coul}} + E_x + E_{\text{corr}} \tag{9.1}$$

The only term in this expression that can be derived directly from the charge distribution is the Coulombic energy. It consists of nucleus–nucleus repulsion, nucleus–electron attraction, and electron–electron repulsion terms. For a medium of unit dielectric constant,

$$E_{\text{Coul}} = \tfrac{1}{2} \sum_{i \neq j} \frac{Z_i Z_j}{R_{ij}} + \sum_i Z_i \Phi(\mathbf{R}_i) + \tfrac{1}{2} \int \int \frac{\rho(\mathbf{r})\rho(\mathbf{r}')}{(\mathbf{r} - \mathbf{r}')} \, d\mathbf{r} \, d\mathbf{r}' \tag{9.2}$$

where Z_i is the nuclear charge for an atom at position $\mathbf{R}_i$, and $\mathbf{R}_{ij} = \mathbf{R}_j - \mathbf{R}_i$. Even though the other terms in Eq. (9.1) cannot be directly calculated from the electron distribution, they can be related to it through the expressions of density functional theory, as discussed in the following sections.

9.2 Density Functional Expressions for the Energy

9.2.1 The Hohenberg–Kohn Theorem

A theorem due to Hohenberg and Kohn points to the central role of the electron density in representing the properties of a system. In 1964, Hohenberg and Kohn (1964) proved that the properties of a system with a nondegenerate ground state are unique functionals of the electron density.

The proof of the Hohenberg–Kohn theorem is quite straightforward. Excluding nucleus–nucleus interactions and the nuclear kinetic energy, the Hamiltonian may be written as

$$\hat{H} = -\tfrac{1}{2}\sum_i \nabla_i^2 + \sum_i V(\mathbf{r}_i) + \sum_{i>j}\frac{1}{r_{ij}} \qquad (9.3)$$

where the sum is over all the electrons in the system. The term $V(\mathbf{r}_i)$ represents the electrostatic electron–nucleus interaction operator, while the first and the last terms are the electronic kinetic energy and electron–electron repulsion operators, respectively.

Suppose a Hamiltonian $\hat{H}$ has an exact nondegenerate ground state Ψ, with energy E, and a second Hamiltonian $\hat{H}'$ has a nondegenerate ground state Ψ' with energy E', where $\hat{H}$ and $\hat{H}'$ differ by their local potentials $V(\mathbf{r})$ and $V'(\mathbf{r})$, respectively. Then Ψ' will not be an eigenfunction of $\hat{H}$ as long as $V(\mathbf{r}) - V'(\mathbf{r})$ is not a constant. So, if Ψ' is used as a trial wave function for $\hat{H}$, the corresponding energy will, according to the variational theorem be larger than the true energy. Thus, from $\hat{H}\psi' > E\psi'$

$$E' + \int (V - V')\rho'(\mathbf{r})\,dr > E \qquad (9.4)$$

Similarly, if Ψ is used as a trial function of $\hat{H}'$,

$$E + \int (V' - V)\rho(\mathbf{r})\,dr > E' \qquad (9.5)$$

In the case that $\rho(\mathbf{r}) = \rho'(\mathbf{r})$, addition of the two equations gives

$$E' + E > E' + E \qquad (9.6)$$

Since this is a contradiction, it follows that $\rho(\mathbf{r}) \neq \rho'(\mathbf{r})$. Thus, there is a one-to-one correspondence between the local potential $V(\mathbf{r})$ and the electron density $\rho(\mathbf{r})$. This implies that V, Ψ, and E are uniquely determined by the electron density, and therefore are *functionals* of the electron density. If E is a unique functional of the electron density ρ, the kinetic and exchange-correlation energies, T and E_{xc} must be functionals of ρ also.

The Hohenberg–Kohn theorem does not go beyond this point; it offers no guidance on the nature of the functionals that it shows must exist.

9.2.2 Density Functional Expressions

Density functionals are discussed extensively in the literature (Dahl and Avery 1984, Parr and Yang 1989, Ziegler 1991), and their development is an active field of research.

In the simplest form, the Thomas–Fermi–Dirac model, the functionals are those which are valid for an electronic gas with slow spatial variations (the "nearly free electron gas"). In this approximation, the kinetic energy T is given by

$$T = c_k \int \rho(\mathbf{r})t(\rho)\,d\mathbf{r} \qquad (9.7)$$

where $c_k = \frac{3}{10}(3\pi^2)^{2/3}$, and the function $t(\rho) = \rho(\mathbf{r})^{2/3}$. The exchange-correlation energy E_{xc} is also a functional of ρ:

$$E_{xc} = -c_x \int \rho(\mathbf{r}) e_{xc}(\rho) \, d\mathbf{r} \tag{9.8}$$

where $c_x = \frac{3}{4}(3/\pi)^{1/3}$ and $e_{xc}(\rho) = \rho(\mathbf{r})^{1/3}$.

There are extensive discussions in the literature concerning to what extent Eq. (9.8) includes at least part of the correlation energy. A critical examination has been made by Sabin and Trickey (1984). Expressions (9.7) and (9.8) are referred to as *local density approximations* (LDA), as the functionals depend on the local density only. Nonlocal functionals include the gradient of $\rho(\mathbf{r})$ and generally give improved agreement with exact values.

Density functional calculations of molecules, using a Hamiltonian including density functionals, frequently reproduce observed properties, such as bond and excitation energies, reaction profiles, and ionization energies (Ziegler 1991). For tetrafluoroterephthalonitrile (1,4-dicyano-2,3,5,6 tetrafluorobenzene), there is excellent agreement between the electron density from a density functional calculation (Delley 1986) and the X-ray diffraction results (Hirshfeld 1992) (see chapter 5). Avery et al. (1984) have proposed the use of experimental densities in crystals as a basis for band structure calculations.

9.2.3 The Total Energy as a Function of the Electrostatic Potential at the Nuclear Position

In the Thomas–Fermi theory (March 1957), the electrostatic potential at $\mathbf{r}$ is related to the electron density of a neutral atom by the density functional

$$\Phi(\mathbf{r}) = c\rho(\mathbf{r})^{2/3} \tag{9.9}$$

where $c = 4.7854$.

This expression has been used to derive an approximate value for the total energy of an atom in terms of the potential at the nuclear position

$$E_{atom} = kZ\Phi_{nucleus} \tag{9.10}$$

in which k is $3/7 = 0.4286$, compared with the exact value for the hydrogen atom of 0.5. In an extension of this expression to molecules, due to Politzer (1979), the total electronic energy of a molecule is expressed as

$$E_{molecule,\, electronic} = \sum_{all\, atoms} k_i Z_i \Phi_i(0) \tag{9.11}$$

in which $\Phi(0)$ is the potential at the nucleus i due to all electrons, and k_i is equal to $3/7$, as it is for atoms, or selected such that either free atoms or a number of small molecules have the exact Hartree–Fock energy (Politzer 1981). The values of k from the best energy fit differ only a little from 0.4286; they vary between 0.4379 and 0.4159, from which it is concluded that $k = 3/7$ will lead to energies within 2–3% of the correct values.

TABLE 9.1 Politzer Molecular Energies (au) Derived from the X-ray Charge Density

	E_{pol}	6-31G**	MP2[a]	Σ (isolated atoms)
benzene	− 237.5 (6)	− 230.7	− 231.5	− 229.1
D,L-Histidine	− 557 (3)	− 538.4[b]		− 543.5
L-Alanine	− 329 (3)	− 321.8		− 320.1

[a] Möller-Plesset-2 calculation, including correlation.
[b] STO-3G.
Source: Su and Coppens (1993).

Politzer energies from the X-ray charge densities of a number of molecules are given in Table 9.1. They are within several percent of the best theoretical values. The difference between the isolated atom energy (last column of the table) and the total energy gives an estimate for the binding energy of the system. But the uncertainties in the density-functional energies are of the same order as the binding energies; thus, the utility of the method, at present, appears limited.

Bentley (1979) has used experimental data on beryllium and diamond to obtain values for the binding energy in the Politzer approximation. Theoretical atomic densities are projected into density functions as used in the experimental analysis, and the atomic energies are subsequently obtained with Eq. (9.11) and compared with isolated atom energies in the same approximation. Bentley reports reasonable agreement within 0.02 H ($\sim 10\%$) for diamond, but a large discrepancy for the beryllium binding energy.

9.3 The Cohesive Energy of Ionic Crystals

9.3.1 The Point-Charge Model

Cohesive energies are defined as the difference between the total energy of a system and the sum of the energies of its components. If there is a rearrangement of the separate component densities when the components are brought together, this distortion must be taken into account.

It is quite remarkable that electrostatic calculations based on a simple model of integral point charges at the nuclear positions of ionic crystals have produced good agreement with values of the cohesive energy as determined experimentally with use of the Born–Haber cycle. The point-charge model is a purely electrostatic model, which expresses the energy of a crystal relative to the assembly of isolated ions in terms of the Coulombic interactions between the ions.

The geometry of the crystal introduces a factor multiplying the pairwise ionic interaction, which is the Madelung constant μ. It is a dimensionless constant, dependent on the geometry of the crystal under consideration. For an ionic binary crystal, consisting of N each positive and negative ions, μ is defined by

$$E_{\text{electrostatic}} = \frac{1}{2} \sum_{j \neq k} \sum_{k} \frac{q_j q_k}{r_{jk}} \equiv \frac{N q^+ q^- \mu}{r} \tag{9.12}$$

where r is the nearest-neighbor distance, and the summation is over all ion pairs in the crystal.

Equation (9.12) implies the assumption that the kinetic energy and exchange-correlation terms in Eq. (9.1) are the same for the crystal and the assembly of isolated ions.

9.3.2 Fourier Series for the Total Electrostatic Energy

The Coulombic electronic energy of a continuous charge distribution is defined as

$$E_{Coul} = \frac{1}{2} \int \int \frac{\rho(\mathbf{r})\rho(\mathbf{r}')}{(\mathbf{r} - \mathbf{r}')} \, d\mathbf{r} \, d\mathbf{r}' = \frac{1}{2} \int \Phi(\mathbf{r})\rho(\mathbf{r}) \, d\mathbf{r} \tag{9.13}$$

The integration can be performed in reciprocal space, like the reciprocal-space evaluation of the electrostatic potential discussed in chapter 8. According to Parseval's rule (discussed in chapter 5),

$$\int \Phi(\mathbf{r})\rho(\mathbf{r}) \, d\mathbf{r} = \int \Phi(\mathbf{H})F(\mathbf{H}) \, d\mathbf{H} \tag{9.14}$$

For the periodic crystal, the integral is replaced by a summation. With Eq. (8.16) for $\Phi(\mathbf{H})$, we obtain

$$E = \frac{1}{2\pi V} \sum F_{total}^2(\mathbf{H})/H^2 \tag{9.15}$$

where E_{total}, the total structure factor as defined in chapter 8 [Eq. (8.15)], includes the nuclear contribution.

Expression (9.15) gives the total electrostatic energy and not the cohesive energy of a molecular crystal. It ignores the quantum-mechanical nature of the charge distribution; an electron cannot interact with itself, but just such a "self-energy" is included in the expression.

9.3.3 The Accelerated Convergence Method and the Electrostatic Potential in a Point-Charge Crystal

The summation of the electrostatic interaction over a crystal, according to Eq. (9.12), converges poorly because of the increasing number of neighbors at large distances. Ewald's (1921) method of accelerated convergence circumvents this problem.

The electrostatic properties of a point-charge crystal are given by the direct space sum

$$s_n = \frac{1}{2} \sum_{j \neq k} \sum_k q_j q_k r_{jk}^{-n} \tag{9.16}$$

where the sum over j is over one unit cell and the sum over k is over the lattice, and excludes $j = k$ in the origin cell; the q_i are generalized coefficients. The term s_1, for example, is the electrostatic interaction energy defined in Eq. (9.12), and

s_6, with the proper choice of coefficients, equals the London dispersion energy, to be discussed later.

Expression (9.16) may be divided into two parts:

$$S_n = \tfrac{1}{2} \sum_{j \neq k} \sum_k q_j q_k r_{jk}^{-n} \varphi(r_{jk}) + \tfrac{1}{2} \sum_{j \neq k} \sum_k q_j q_k r_{jk}^{-n} [1 - \varphi(r_{jk})] \tag{9.17}$$

The second term in this expression is evaluated in reciprocal space using Eq. (9.15); $\varphi(r_{jk})$ is the convergence function, given by (Nijboer and De Wette 1957)

$$\varphi(r) = \frac{\Gamma(n/2, K^2 \pi r^2)}{\Gamma(n/2)} = \frac{1}{\Gamma(n/2)} \int_{K^2 \pi r^2}^{\infty} t^{n/2 - 1} \exp(-t) \, dt \tag{9.18}$$

The value K determines the relative importance of the direct- and reciprocal-space summations. Like S, K is in units of Å^{-1}.

The Fourier transform of $[1 - \varphi(r)]/r$ is given by

$$\hat{F}\left[\frac{1 - \varphi(r)}{r^n} \right] = \pi^{n - 3/2} S^{n - 3} \Gamma\left(-\frac{n}{2} + \frac{3}{2}, \frac{\pi S^2}{K^2} \right) \bigg/ \Gamma\left(\frac{n}{2} \right) \tag{9.19}$$

in which S is the reciprocal space coordinate, and the Γ function is defined by

$$\Gamma(x, y) = \int_y^{\infty} t^{x - 1} \exp(-t) \, dt \tag{9.20}$$

where $\Gamma(x) = \Gamma(x, 0)$.

The final result is (Williams 1971, 1981)

$$
\begin{aligned}
S_n = \frac{1}{2\Gamma(n/2)} &\left[\sum_{j \neq k} \sum_k q_j q_k r_{jk}^{-n} \Gamma(n/2, a^2) \right. \\
&+ V^{-1} \pi^{n - 3/2} \sum_{H \neq 0} |F_{\text{total}}(\mathbf{H})|^2 |\mathbf{H}|^{n - 3} \Gamma(-n/2 + 3/2, b^2) \\
&\left. + V^{-1} \pi^{n/2} K^{n - 3} \frac{2}{n - 3} \left[\sum_{\text{cell}} q_j \right]^2 - 2n^{-1} \pi^{n/2} K^n \left\{ \sum_{\text{cell}} q_j^2 \right\} \right]
\end{aligned}
\tag{9.21}
$$

where $a^2 = \pi K^2 r_{jk}^2$ and $b^2 = \pi H^2 / K^2$.

In Eq. (9.21), the second summation is over lattice vectors $\mathbf{H}$. The last two terms of this equation represent the (000) term in the Fourier summation and the self-energy correction. The latter describes the interaction of the point charge with itself, which, as noted above, is included in the reciprocal space summation and must therefore be subtracted.

Equation (9.21) can be written in terms of the complementary error function $ERFC(x)$, using $\Gamma(\tfrac{1}{2}) = \pi^{1/2}$, $\Gamma(1, b^2) = \exp(-b^2)$, and $\Gamma(\tfrac{1}{2}, a^2) = \sqrt{\pi} ERFC(a)$, where $ERFC(x)$ is defined as

$$ERFC(x) = \frac{2}{\sqrt{\pi}} \int_x^{\infty} \exp(-t^2) \, dt \tag{9.22}$$

9.3.4 The Lattice Energy of a Crystal Consisting of Spherical Ions

The electrostatic interaction energy between two spherical atoms or ions located at A and B is the sum of the internuclear repulsions, the nucleus–electron attractions, and the electron–electron repulsions (Su and Coppens 1995):

$$E_{AB} = E_{nn} - Z_A \Phi^B(A) - Z_B \Phi^A(B) + E_{ee} \qquad (9.23)$$

where

$$E_{nn} = \frac{Z_A Z_B}{R} \qquad (9.24)$$

and R is the length of the vector $\mathbf{R}$, connecting A and B; $\Phi^B(A)$ is the electrostatic potential at the nucleus A due to the electronic charge distribution B:

$$\Phi^B(A) = \int \frac{\rho^B(r)}{|\mathbf{R} - \mathbf{r}|} \, d\mathbf{r} \qquad (9.25)$$

and, similarly,

$$\Phi^A(B) = \int \frac{\rho^A(r)}{|\mathbf{R} - \mathbf{r}|} \, d\mathbf{r} \qquad (9.26)$$

while

$$E_{ee} = \int \int \frac{\rho^A(r_1)\rho^B(r_2)}{|r_{12}|} \, d\mathbf{r}_1 \, d\mathbf{r}_2 \qquad (9.27)$$

9.3.4.1 Evaluation of the Internuclear Repulsion and the Nuclear–Electronic Attraction

For a nucleus at R_i, the peripheral contribution to the potential Φ, due to a spherical density component centered at R_j, consists of a point-charge term and a *penetration term*. The point-charge term is due to the nuclear charge at R_j and the electronic density within the sphere with radius $|\mathbf{R}_i - \mathbf{R}_j|$, centered on R_j, which passes through the nucleus i (Fig. 9.1). The penetration terms are due to the electronic charge outside that sphere. They decay exponentially as the distance $R_{ij} = |\mathbf{R}_i - \mathbf{R}_j|$ increases (Hirshfeld and Rzotkiewicz 1974).

The electrostatic potential in a crystal of spherical atoms or ions is therefore the sum of the electrostatic potential of a point-charge crystal and a penetration correction. Only atoms for which the product of $R_{ij}\zeta_j$ is small contribute to the

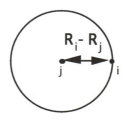

FIG. 9.1 The sphere with radius $|\mathbf{R}_i - \mathbf{R}_j|$ centered on nucleus j.

penetration term, where ζ_j is the radial exponent of the monopolar density centered at R_j.

For the point-charge contributions, the accelerated convergence expression, Eq. (9.21), is used with the substitution, Eq. (9.22). The explicit expression for the point-charge contribution is then (Bertaut 1952)

$$\Phi^{\text{peripheral, point charge}}(\mathbf{R}_i) = \sum_{j \neq i} q_j |R_{ij}|^{-1} ERFC(\sqrt{\pi K R_{ij}})$$

$$+ \pi^{-1} V^{-1} \sum_{\mathbf{H} \neq 0} |F_{\text{total}}^{\text{point charge}}(\mathbf{H}) H^{-2}$$

$$\times \exp\left(\frac{-\pi H^2}{K^2}\right) \exp(-2\pi i \mathbf{H} \cdot \mathbf{R}_i) - 2K q_i \quad (9.28)$$

where $F_{\text{total}}^{\text{point charge}}(\mathbf{H}) = \sum_k q_k \exp(2\pi i \mathbf{H} \cdot \mathbf{R}_k)$, and q_k is the total charge in the sphere with radius $R_{ij} = |\mathbf{R}_i - \mathbf{R}_j|$.

The rates at which the direct-space and the reciprocal-space parts of the lattice sums converge are a function of the value of K. According to Williams (1981), the choice of $K = 0.3/a$ minimizes the total computation time in the case of NaCl. With a lower K value of $\approx 0.2/a$, the reciprocal sum can be neglected completely because of the rapid decay of the exponential factor in the Fourier summation. Generally, K can be chosen to be of the order of 0.1 Å^{-1}.

The penetration contribution to the electrostatic potential at R_i is evaluated by application of the general expression of Eq. (8.49) for Φ^{per} for the spherical density ($l_1 = m_1 = 0$). The point-charge term, proportional to $1/R_{ij}$, must subsequently be subtracted. Due to the rapid decrease of the penetration terms with increasing R_{ij}, convergence is quickly achieved. For spherically averaged Hartree–Fock atom densities, inclusion of penetration terms for atoms within 10 Å of the point under consideration is more than adequate.

9.3.4.2 Evaluation of the Electron–Electron Repulsion

The electron–electron repulsion

$$E_{\text{ee}} = \int \int \frac{\rho^A(r_1) \rho^B(r_2)}{|r_{12}|} d\mathbf{r}_1 \, d\mathbf{r}_2 \quad (9.29)$$

can be evaluated with the help of the Fourier convolution theorem, as defined by expression (8.44).[1] For spherical densities, separated by a distance R, the result is particularly simple, and given by (Spackman and Maslen 1986)

$$E_{\text{ee}}^{\text{spherical}} = 4 \int_0^\infty f_A(S) f_B(S) j_0(2\pi S R) \, dS \quad (9.30)$$

where $S = 2 \sin \theta / \lambda$, and the spherical-atom scattering factor f_A of the radial density $\rho_A(r)$ is defined as

$$f_A(S) = 4\pi \int r^2 \rho_A(r) j_0(2\pi S r) \, dr \quad (9.31)$$

---

[1] In this case the Fourier convolution theorem is applied to a convolution of three functions.

As discussed in Chapter 1 [Eq. (1.28)], the term $V_{ec}^{spherical}$ may be evaluated numerically with Eq. (9.30) using Gaussian quadrature (Press et al. 1986).

9.3.5 Application to the Lattice Energy of Alkali Halides

9.3.5.1 The Electrostatic Energy for the Point-Charge and Overlapping-Ion Models

Sodium fluoride, NaF, is a favorable choice for X-ray analysis of the lattice energy of an ionic crystal. Both Na and F are relatively light atoms, and the Na 3s-radial distribution, though diffuse, is not quite as spread out as the Li 2s shell (single-ζ values are 0.8358 and 0.6396 au^{-1}, respectively; see appendix F), and therefore contributes to a larger number of reflections.

With K in the convergence function of Eq. (9.18) equal to 0.4 Å^{-1}, the accelerated convergence method reaches an accuracy of nine significant figures with inclusion of only 1000 unit cells. By comparison, about 50,000 unit cells would be required for a summation performed exclusively in direct space.

The point-charge model gives the electrostatic energy per NaF unit as $-0.402\,950$ au. With $E_{electrostatic} = -\mu/a$,[2] where a is the unit-cell edge at 0 K, and the value of $a_{NaF} = 4.590$ Å $= 8.674$ au, this gives, as reported in the literature (Glasser and Zucker 1980),

$$\mu_{NaF} = 3.495\,129 \tag{9.32}$$

For the unit-point-charge crystal, the absolute value of the electrostatic energy is equal to the potential at the nuclear position. This potential will be equal for both ions in the alkali halide structure, as their positions are equivalent; that is,

$$\Phi(Na^+) = -\Phi(F^-) = -0.4030 \text{ au} \tag{9.33}$$

For the crystal composed of ions, the equality given in Eq. (9.33) is no longer valid, because the two ionic charge distributions, which partly shield the nuclear charges, are different. For the free-ion crystal, the values of the potential at the nuclear positions in NaF, evaluated according to Section 9.3.2, are

$$\Phi^{per}(Na^+) = -0.3958 \text{ au}$$

$$\Phi^{per}(F^-) = +0.4034 \text{ au} \tag{9.34}$$

These values indicate that in the ionic crystal the *effective charges*, which act upon immediate neighbors and some distance beyond, are larger than $+1$ for the Na, and less negative than -1 for the F ions. At large separations, the effective charges will revert to $+1$ and -1.

---

[2] Note that in part of the literature and in Eq. (9.12), the Madelung potential is defined such that $V_{total,\,unit\,cell} = -\mu'/r$, where r is the nearest distance between the cation and the anion. For the rock-salt lattice, $\mu' = \mu/2$.

The electrostatic energy for the free-ion crystal is -1120.5 kJ mol^{-1}, compared with -1058.0 kJ mol^{-1} for the point-charge crystal. But the ions in the crystal are affected by their environment. According to the κ-refinement of the single-crystal AgKα data on NaF (Howard and Jones 1977), both the Na^{+} and F^{-} ions in the crystal are more contracted than the free ions, and the net charges have magnitudes of 0.95 (1) e.[3] The electrostatic energy based on this distribution is -1045.6 kJ mol^{-1}, remarkably close to that of the crude point-charge model.

The success of the point-charge model is to be attributed to the cancellation of the effects of incomplete charge transfer and the interpenetration of the electron shells of adjacent atoms. The electrostatic energy is increased by a factor of $(100/95)^2$, or about 10%, by the assumption of full charge transfer. But the neglect of the spatial distribution of the electrons gives a larger electron–electron repulsion for ten point-electrons on each nucleus than is the case for the spread-out real density. The calculation shows that this destabilization is about 0.0247 au, or ≈ 65 kJ mole^{-1}, for the interaction between first neighbors alone, relative to the experimental distribution in the crystal.

9.3.5.2 The Lattice Energy of NaF

A second major contributor or the lattice energy of an ionic crystal is the repulsive energy. Following Born and Huang (1954), the repulsive energy per mole may be written as

$$E_{\text{rep}} = N_A B'/r^n \tag{9.35}$$

The term B' is a repulsive coefficient, which is evaluated from the equilibrium condition

$$\frac{\partial(E_{\text{electrostatic}} + E_{\text{rep}})}{\partial r} = 0 \tag{9.36}$$

where $r = a/2$ is the nearest separation between the Na and F ions.

Using the point-charge expression $E_{\text{electrostatic}} = \mu/2r$, and $E_{\text{repulsive}} = B'/r^n$, one obtains $B' = \mu r^{n-1}/2n$. Substitution in Eq. (9.35) gives, for the repulsive energy per mole,

$$E_{\text{rep}} = \frac{N_A \mu}{na} \tag{9.37}$$

The lattice energy is the sum of the electrostatic and repulsive energies, and per mole is given by

$$U = E_{\text{electrostatic}} + E_{\text{rep}} = -N_A \mu/a(1 - 1/n) \tag{9.38}$$

where N_A is Avogadro's number.

---

[3] To reduce the effect of extinction and to eliminate possible correlation between the charge density and extinction parameters, the six strongest reflections were eliminated from the refinement. Because the 3s shell of Na is barely populated, its κ parameter cannot be refined separately in such a refinement. In the NaF study, this problem was circumvented by assignment of a common κ value to the L and M shells of Na.

TABLE 9.2 The Effective Madelung Constant μ^{eff}, the Effective Born Coefficient n^{eff} and the Lattice Energy U (kJ mol^{-1}) for NaF According to Point-charge and Ionic Models

	μ^{eff}	μ^{eff}/μ	n^{eff}	U	$\dfrac{U - U^{exp}}{U^{exp}}$ (%)
± 1 Point-charge	3.495 129	1.0000	7.370	-914.2	-1.0
Free Hartree–Fock ions	3.701 7	1.0591	7.015	-960.6	-6.1
± 0.95 Point-charge	3.154 4	0.9025	8.058	-836.4	7.6
X-ray density	3.454 3	0.9883	7.445	-913.4	-0.92

The Born coefficient n can be derived from the experiment bulk modulus $B = -V(dP/dV)$, where V is the volume and P is the pressure (Kittel 1966). With the value $B_{NaF} = 0.5143 \cdot 10^{12}$ dyn cm^{-2} (Sangster and Atwood 1978), n_{NaF} is obtained as

$$n_{NaF} = \frac{18 B r_0^4}{\mu q^2} + 1 = 7.37 \tag{9.39}$$

Substitution in Eq. (9.38) gives, for the energy of the point-charge NaF lattice,

$$U_{\text{point charge, NaF}} = -218.5 \text{ kcal mol}^{-1} \tag{9.40}$$

as compared to the experimental value of -216.3 kcal mol^{-1} (Sangster et al. 1978).

The lattice energy is defined relative to the unperturbed components of the crystal. For an alkali halide crystal such as NaF, it is the difference between the energy of the solid and the energy of the ions Na$^+$(g) + F$^-$(g). Because the real crystal contains incompletely charge-transferred ions, a correction term is required in the calculation of the lattice energy based on the κ-refinement results. The correction is the difference between the energy of the 0.95 e$^-$ charged ions and the energy of fully charge-transferred particles. It equals -0.05 ($I_{Na} + E_F$), where I_{Na} and E_F are the ionization energy of sodium and the electron affinity of fluorine, respectively. With $I_{Na} = 495.85$ kJ mol^{-1} and $E_F = -328.16$ kJ mol^{-1} (Lide 1993), this amounts to -8.38 kJ mol^{-1}.

The results for the three models are summarized in Table 9.2, which lists the effective Madelung constant μ^{eff}, defined by $E_{\text{electrostatic}} = -\mu^{eff}/a$; the effective Born coefficient, defined by $U = -N_A \mu^{eff}/a(1 - 1/n^{eff})$; and the lattice energy U.

The discrepancy between the κ-refinement results and the experimental value for U is 0.92%. The exact value of the lattice energy is very sensitive to the amount of charge transfer. If the charge transfer is reduced by only one standard deviation to 0.94 e, the agreement with the experimental lattice energy is within 0.12%. Conversely, an increase by one standard deviation worsens the discrepancy. Nevertheless, the agreement with the calorimetric lattice energy is satisfying, given the simplicity of the treatment of the nonelectrostatic interactions.

9.4 The Cohesive Energy of Molecular Crystals

9.4.1 Molecular Interactions

Short-range repulsive forces are a direct result of the Pauli exclusion principle and are thus quantum mechanical in nature. Kitaigorodskii (1961) has emphasized that such short-range repulsive forces play a major role in determining the packing in molecular crystals. The size and shape of molecules is determined by the repulsive forces, and the molecules pack as closely as is permitted by these forces.

Repulsive forces determine, for example, the melting point of a solid. Whenever the packing is efficient, the melting point tends to be high. The attractive forces, on the other hand, govern the heat of vaporization and therefore the boiling point. Trouton's rule, which relates the normal boiling point of a liquid to its heat of vaporization, is a manifestation of this relation.

Like the Coulombic forces, the van der Waals interactions decrease less rapidly with increasing distance than the repulsive forces. They include interactions that arise from the dipole moments induced by nearby charges and permanent dipoles, as well as interactions between instantaneous dipole moments, referred to as *dispersion* forces (Israelachvili 1992). Instantaneous dipole moments can be thought of as arising from the motions of the electrons. Even though the electron probability distribution of a spherical atom has its center of gravity at the nuclear position, at any very short instance the electron positions will generally not be centered on the nucleus.

Quantitative treatment of the interaction between two identical Bohr atoms, consisting of point electrons and nuclei, leads to an expression which, apart from a numerical factor, is the same as that derived quantum-mechanically by London (1937):

$$E_{\mathrm{disp}}(r) = -\frac{C_{\mathrm{disp}}}{r^6} = \tfrac{3}{4}\alpha_0^2 I/(4\pi\varepsilon_0)^2 \times \frac{1}{r^6} \qquad (9.41)$$

Here, α_0 and I are the polarizability and the ionization energy of the atom, respectively. The r^6 dependence is also encountered in the so-called Debye interaction between a permanent dipole and an induced dipole, given by

$$E_{\mathrm{Debye}}(r) = -\mu^2\alpha_0/(4\pi\varepsilon_0)^2 \times \frac{1}{r^6} \qquad (9.42)$$

The dipole–dipole interactions, frequently referred to as Keesom interactions, are historically included in the van der Waals interactions, even though they are purely electrostatic. For molecules that are free to orient themselves, the dipole–dipole interactions must be averaged over the molecular orientations, as the angular dependence of the interaction energy is comparable to the Boltzmann energy $k_B T$ (Israelachvili 1992, p. 62). With the averaging of the Keesom interactions, the total van der Waals energy for two dissimilar bodies is given by

$$E_{VDW} = -\frac{C_{VDW}}{r^6} = -\left[\frac{C_{Debye}}{r^6} + \frac{C_{Keesom}}{r^6} + \frac{C_{disp}}{r^6}\right]$$

$$= -\left[(\mu_1^2\alpha_2 + \mu_2^2\alpha_1) + \frac{\mu_1^2\mu_2^2}{3k_BT} + \frac{3\alpha_1\alpha_2 I_1 I_2}{2(I_1 + I_2)}\right] \Big/ (4\pi\varepsilon_0)^2 r^6 \qquad (9.43)$$

When the interaction is expressed in terms of pairwise atom–atom potential functions, all three components of the van der Waals interactions are grouped together because of their common r^{-6} distance dependence. A repulsive term is added, while Coulombic interactions may be accounted for separately. In the expression due to Lennard–Jones, the repulsion has an r^{-12} dependence, to give the pairwise potential function

$$E(r) = Ar^{-12} - Br^{-6} \qquad (9.44)$$

in which A and B are specific for the type of atoms.

Alternatively, the repulsion is frequently described by an exponential term, or

$$E(r) = Ae^{-Br} - Cr^{-6} \qquad (9.45)$$

in which A, B, and C are, again, element specific.

For interactions between unlike atoms i and j, the coefficients A, B, and C can be derived from the coefficients of the homoatomic interactions by approximately valid combining rules, defined as

$$E_{ij}(r) = \sqrt{(A_i A_j)} \exp[-\sqrt{(B_i B_j)}r] - \sqrt{(C_i C_j)}r^{-6} \qquad (9.46)$$

The interaction energy between two molecules or molecular fragments is obtained as a sum over all pairwise atom–atom interactions. The atom–atom potential expressions implicitly assume that the interactions are two-body interactions, undisturbed by other bodies in the vicinity, and that they are isotropic about the atomic centers.

The pairwise interaction may be obtained using density functional theory. Let ρ_A and ρ_B be the density for individual A and B subsystems. In terms of density functional theory, the interaction energy E_{AB} consists of correlation, kinetic energy, and exchange and electrostatic contributions, or

$$\Delta E = E_{AB,\,total} - E_{A,\,total} - E_{B,\,total}$$

$$= c_k \int \{\rho_{AB}t(\rho_{AB}) - \rho_A t(\rho_A) - \rho_B t(\rho_B)\}\, d\mathbf{r} - c_x \int \{\rho_{AB}e_{xc}(\rho_{AB})$$

$$- \rho_A e_{xc}(\rho_A) - \rho_B e_{xc}(\rho_B)\}\, d\mathbf{r} + \Delta E_{electrostatic} + E_{corr}(\rho_{AB}) - E_{corr}(\rho_A)$$

$$- E_{corr}(\rho_B) \qquad (9.47)$$

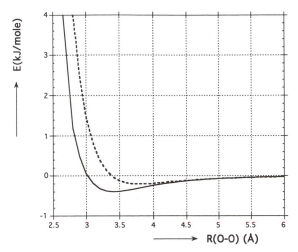

FIG. 9.2 Nonbonded potential for O—O. Broken line: according to the Gordon–Kim density functional model augmented with an attractive potential (Spackman 1986). Full line: empirical curve based on a fit to oxohydrocarbon crystal structures (Cox et al. 1981).

in which $\Delta E_{\text{electrostatic}}$ represents the Coulombic interactions, and the Thomas–Fermi–Dirac model [Eqs. (9.7) and (9.8)] has been used.[4]

This is the Gordon–Kim (1972) model. A number of applications have demonstrated that the Gordon–Kim model leads to a qualitatively valid description of potential energy surfaces between closed-shell subsystems. Spackman (1986a) has used the Gordon–Kim model, with an empirical scaling parameter for the exchange contribution, to derive a set of short-range repulsive potentials for homoatomic pairs up to the element krypton. The resulting potentials are complemented with an attractive r^{-6} term representing the dispersion forces.

The Gordon–Kim interaction functions may be compared with empirical potential functions derived by energy- or net-force minimization methods using known crystal structures. The O—O Gordon–Kim potentials are more repulsive, as illustrated in Fig. 9.2. Spackman points out that the empirical potentials likely contain a significant attractive component because of the inadequate allowance for electrostatic interactions in their derivation. This attractive component is included in the electrostatic interaction in the density functional model.

The application of the Gordon–Kim model to open-shell systems, which must include the interaction between unpaired spins, appears less successful (Kim and Gordon 1974).

Expressions of the forms in Eqs. (9.45) and (9.46) have been extremely useful in molecular modeling studies. An extensive literature exists on the choice of

---

[4] Note that in this equation an extra term is included to account for correlation energy. The term is derived from the topology of ρ.

coefficients and on the application of the expressions in structure calculations (Wampler 1994, Cornell et al. 1995).

9.4.2 Interactions Between Molecules in Crystals

Like the interaction energy between two molecules, the total lattice energy of a molecular crystal contains several contributions from the different types of interactions. We may write

$$U_{\text{molecular}} = E_{\text{repulsive}} + E_{\text{electrostatic}} + E_{\text{Debye}} + E_{\text{dispersion}} \tag{9.48}$$

In this expression, the dipole–dipole interactions are included in the electrostatic term rather than in the van der Waals interactions as in Eq. (9.43). Of the four contributions, the electrostatic energy can be derived directly from the charge distribution. As discussed in section 9.2, information on the nonelectrostatic terms can be deduced indirectly from the charge density. The polarizability α, which occurs in the expressions for the Debye and dispersion terms of Eqs. (9.41) and (9.42), can be expressed as a functional of the density (Matsuzawa and Dixon 1994), and also obtained from the quadrupole moments of the experimental charge density distribution (see section 12.3.2). However, most frequently, empirical atom–atom pair potential functions like Eqs. (9.45) and (9.46) are used in the calculation of the nonelectrostatic contributions to the intermolecular interactions.

9.4.2.1 Hydrogen-bonding

Hydrogen-bonding is one of the prime interactions determining the packing motif in molecular crystals. As discussed in section 12.3.3, topological analysis of the total charge distribution indicates normal hydrogen-bonding to be a closed-shell interaction, with a very low density at the critical point, and $\nabla^2\rho$ at the critical point being invariably positive. But for very short hydrogen-bonds, covalent contributions will be of increasing importance. Nevertheless, simple electrostatic models can to a large extent explain the energetics of hydrogen-bond formation. In the work of Spackman (1986b), hydrogen-bonding is accounted for by electrostatic forces, combined with the omission of the repulsive term in the hydrogen-acceptor interaction.

9.4.3 Expressions for the Evaluation of the Electrostatic Contribution to the Lattice Energy of Molecular Crystals

The electrostatic energy of a molecular crystal can be evaluated with summation over the structure factors in Eq. (9.15). But to obtain the cohesive energy of a molecular crystal with such a summation, we would have to subtract the molecular electrostatic energies, which are implicitly included in the result. An alternative is to perform the calculation in direct space.

With the molecular densities ρ_A and ρ_B obtained with one of the partitioning methods discussed in chapter 6, the electrostatic interaction energy is given by

$$E_{\text{lattice, es}} = \sum_B \int \frac{\rho_A(\mathbf{r})\rho_B(\mathbf{r}')}{(\mathbf{r}-\mathbf{r}')} \, d\mathbf{r} \, d\mathbf{r}' \tag{9.49}$$

where A stands for a unique central molecule, and B represents the molecules in the remainder of the crystal. If there is more than one molecule in the asymmetric unit, a term for the interaction of each additional molecule, A', with its environment, is to be added. The doubly counted A–A' interactions must be subtracted.

Expression (9.49) can be evaluated by substituting the expression for the potential in terms of the charge distribution [Eq. (8.2)], which gives

$$E_{\text{lattice, es}} = \sum_B \int \Phi_A(\mathbf{r})\rho_B(\mathbf{r}) \, d\mathbf{r} \tag{9.50}$$

For a slowly varying charge distribution, the potential can be expanded in a Taylor series with $\Phi(0)$, the potential at the origin of the distribution B, as leading term (Buckingham 1959, 1970, 1978; Jackson 1974):

$$\Phi(\mathbf{r}) = \Phi(0) + \mathbf{r}\cdot\nabla\Phi(0) + \frac{1}{2}\sum_i\sum_j r_i r_j \frac{\partial^2\Phi(0)}{\partial r_i \, \partial r_j} + \frac{1}{6}\sum_i\sum_j\sum_k r_i r_j r_k \frac{\partial^3\Phi(0)}{\partial r_i \, \partial r_j \, \partial r_k} + \cdots \tag{9.51}$$

Since, following Eq. (8.4), $-\nabla\Phi$ is the external field $\mathbf{E}$, Eq. (9.51) can also be written as

$$\Phi(\mathbf{r}) = \Phi(0) - \mathbf{r}\cdot\mathbf{E}(0) - \frac{1}{2}\sum_i\sum_j r_i r_j \frac{\partial E_j}{\partial r_i} - \frac{1}{6}\sum_i\sum_j\sum_k r_i r_j r_k \frac{\partial^2 E_j}{\partial r_i \, \partial r_k} - \cdots \tag{9.52}$$

As we are evaluating the potential at a point outside the distribution A, the Poisson equation gives $-\nabla^2\Phi = \nabla\cdot\mathbf{E} = \sum_i(\partial E_i/\partial r_i) = 0$. Expression (9.52) is therefore equivalent to

$$\Phi(0) - \mathbf{r}\cdot\mathbf{E}(0) - \frac{1}{6}\sum_i\sum_j(3r_i r_j - r^2\delta_{ij})\frac{\partial E_j}{\partial r_i} + \cdots$$

$$= \Phi(0) - \mathbf{r}\cdot\mathbf{E}(0) - \frac{1}{3}\sum_i\sum_j \hat{\Theta}_{ij}\frac{\partial E_j}{\partial r_i} - \frac{1}{15}\sum_i\sum_j\sum_k \hat{\Omega}_{ijk}\frac{\partial E_j}{\partial r_i \, \partial r_k} \cdots \tag{9.53}$$

using the definitions for Θ and Ω given in Eq. (7.2), and the caret above the letters indicating the corresponding operator.

To obtain the interaction energy, Eq. (9.53) is substituted into Eq. (9.50). When the charge density B is expressed in terms of one or more spherical harmonic expansions, only terms of like symmetry will integrate to nonzero values, and we obtain the expression for the interaction between two distributions as

$$E_{\text{es}} = q\Phi_0 - \mu_\alpha E_\alpha - \tfrac{1}{3}\Theta_{\alpha\beta}E'_{\alpha\beta} - \tfrac{1}{15}\Omega_{\alpha\beta\gamma}E''_{\alpha\beta\gamma} + \cdots \tag{9.54}$$

or, in terms of the atom-centered multipole expansions and explicit notation for the summations,

$$E_{es} = \sum_j q_j \Phi_0 - \mu_{\alpha j} E_{\alpha j} - \tfrac{1}{3}\Theta_{\alpha\beta} E'_{\alpha\beta j} - \tfrac{1}{15}\Omega_{\alpha\beta\gamma j} E''_{\alpha\beta\gamma j} + \cdots \qquad (9.55)$$

in which the sum j is over all atomic centers, and E' and E'' are the first and second derivatives of the components of $\mathbf{E}$.

To evaluate this expression for distributions expressed in terms of their multipolar density functions, the potential Φ and its derivatives must be expressed in terms of the multipole moments. The expression for Φ outside a charge distribution has been given in chapter 8 [Eq. (8.54)]. Since the potential and its derivatives are additive, a sum over the contributions of the atom-centered multipoles is again used. The resulting equation contains all pairwise interactions between the moments of the distributions A and B, and is listed in appendix J.

9.4.4 Calculated Lattice Energies of Molecular Crystals

It is evident that the electrostatic interactions constitute a major component of the lattice energy of ionic crystals. According the treatment for NaF described above, the ratio of absolute values of the electrostatic and repulsive forces to the lattice energy is $1:1/n$, where n is the Born coefficient. With $n^{eff} = 7.445$ (Table 9.2), the electrostatic contribution is $\approx 115\%$ of the total interaction energy. On the other hand, for small nonpolar molecules in the gas phase and dipole–dipole interactions averaged over all mutual orientations, the dispersion forces contribute 90–100% of the interaction (Israelachvili 1992, p. 95).

In molecular crystals, the relative importance of the electrostatic, repulsive, and van de Waals interactions is strongly dependent on the nature of the molecule. Nevertheless, in many studies the lattice energy of molecular crystals is simply evaluated with the exp-6 model of Eq. (9.45), which in principle accounts for the van der Waals and repulsive interaction only. As underlined by Desiraju (1989), this formalism may give an approximate description, but it ignores many structure-defining interactions which are electrostatic in nature. The electrostatic interactions have a much more complex angular dependence than the pairwise atom–atom potential functions, and are thus important in defining the structure that actually occurs.

Hirshfeld and Mirsky (1979) evaluated the relative contributions to the lattice energy for the crystal structures of acetylene, carbon dioxide, and cyanogen, using theoretical charge distributions. Local charge, dipole and quadrupole moments are used in the evaluation of the electrostatic interactions. When the unit cell dimensions are allowed to vary, inclusion of the electrostatic forces causes an appreciable contraction of the cell. In this study, the contributions of the electrostatic and van der Waals interactions to the lattice energy are found to be of comparable magnitude.

Spackman et al. (1988) have used experimental charge densities to sum

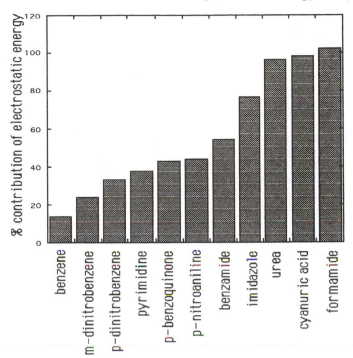

FIG. 9.3 Relative contribution of electrostatic interactions to the total lattice energy for a number of molecular crystals. *Source*: Coombes and Price (1995).

interactions between electrostatic atomic moments up to and including octupoles.[5] Repulsion and dispersion terms are calculated with the parameters based on the Gordon–Kim model as described above. Crystals analyzed include those of urea, imidazole, cytosine monohydrate, and 9-methyl adenine. For a pairwise molecule–molecule interaction in cytosine including two hydrogen bonds, the interaction energy is found to be -75 (27) kJ mol^{-1}, compared with -57.4 kJ mol^{-1} from a theoretical calculation of the dimer in the same configuration. The crystal lattice energy of urea is estimated as -66 (24) kJ mol^{-1}, in quite reasonable agreement with the value of -93 (6) kJ mol^{-1} derived from the experimental sublimation energy.

A systematic analysis of the electrostatic interactions in the crystals of 40 rigid organic molecules was undertaken by Price and coworkers (D. S. Coombes et al. 1996). In this work, distributed (i.e., local) multipoles up to hexadecapoles, obtained from SCF calculations with 6-31G** basis sets, scaled by a factor of 0.9 to allow for the omission of electron correlation, are used in the evaluation of the electrostatic interactions. The experimental lattice constants and structures are reproduced successfully, the former to within a few percent of the experimental

---

[5] Note that in Spackman (1986b), the energy is subdivided in contributions labeled as electrostatic, penetration, repulsion, and dispersive terms. The first two of these are due to electrostatic interactions.

values. In agreement with the statement by Desiraju (1989), very poor results are obtained when the multipolar electrostatic interactions are neglected, confirming that electrostatic forces dominate the anisotropy of the intermolecular interactions (Hurst et al. 1986). Though experimental lattice energies from heats of sublimation are generally not known with very good accuracy, many of the lattice energies are predicted within ≈ 15 kJ mol^{-1}; but lattice energies of hydrogen-bonded crystals tend to be underestimated.

Of interest is the relative contribution of the electrostatic interactions to the total calculated lattice energy. Some of the results are reproduced in Fig. 9.3 (Coombes and Price 1995). It is clear that the contribution increases rapidly for the more polar molecules, and can be pronounced. For formamide, the electrostatic contribution is more than 100% of the lattice energy, as the repulsive and the van der Waals r^{-6} forces are of approximately equal magnitude and sum to a small, opposite, contribution.

Several issues remain to be addressed. The effect of the mutual penetration of the electron distributions should be analyzed, while the use of theoretical densities on isolated molecules does not take into account the induced polarization of the molecular charge distribution in a crystal. In the calculations by Coombes et al. (1996), the effect of electron correlation on the isolated molecule density is approximately accounted for by a scaling of the electrostatic contributions by a factor of 0.9. Some of these effects are in opposite directions and may roughly cancel. As pointed out by Price and coworkers, lattice energy calculations based on the average static structure ignore the dynamical aspects of the molecular crystal. However, the necessity to include electrostatic interactions in lattice energy calculations of molecular crystals is evident and has been established unequivocally.

10

Charge Density Studies of Transition Metal Compounds

10.1 The Study of Transition Metal Complexes

The electron density in transition metal complexes is of unusual interest. The chemistry of transition metal compounds is of relevance for catalysis, for solid-state properties, and for a large number of key biological processes. The importance of transition-metal-based materials needs no further mention after the discovery of the high-Tc superconducting cuprates, the properties of which depend critically on the electronic structure in the CuO_2 planes.

The results of theoretical calculations of systems with a large number of electrons can be ambiguous because of the approximations involved and the frequent occurrence of low-lying excited states. The X-ray charge densities provide independent evidence from a technique with very different strengths and weaknesses, and thus can make significant contributions to our understanding of the properties of transition-metal-containing molecules and solids.

In inorganic and organometallic solids, the average electron concentration tends to be high. This means that absorption and extinction effects can be severe, and that the use of hard radiation and very small crystals is frequently essential. Needless to say that the advent of synchrotron radiation has been most helpful in this respect. The weaker contribution of valence electrons compared with the scattering of first-row-atom-only solids implies that great care must be taken during data collection in order to obtain reliable information on the valence electron distribution.

10.2 On the Electronic Structure of Transition Metal Atoms

10.2.1 Crystal Field Splitting of d Orbitals

When the field exerted by the atomic environment is not spherically symmetric, as is the case in any crystal, the degeneracy of the d-electron orbitals is lifted. In the electrostatic *crystal field* theory, originally developed by Bethe (1929) and Van Vleck (1932), all interactions between the transition metal atom and its ligands are treated electrostatically, and covalent bonding is neglected. Since the ligands are almost always negatively charged, electrons in orbitals pointing towards the ligands are repelled more strongly, and the corresponding orbitals will be higher in energy. The discussion is the simplest for the one d-electron case, in which d–d electron repulsions are absent.

For a *cubic field* exerted by ligands along the x, y, and z axes, the d_z^2 and $d_{x^2-y^2}$ orbitals are destabilized relative to the d_{xy}, d_{xz}, and d_{yz} orbitals. The d_{z^2} and $d_{x^2-y^2}$ orbitals form the basis for the e_g representation of the cubic point group, and are therefore referred to as e_g orbitals, while the d_{xy}, d_{xz}, and d_{yz} orbitals are t_{2g} orbitals, which transform like the t_{2g} representation. The magnitude of the splitting between the e_g and the t_{2g} orbitals depends on the strength of the field. For an array of point charges, it can be evaluated by a simple electrostatic calculation. For the detailed calculation, the reader is referred to texts on the subject (Sugano et al. 1970, Ballhausen 1962).

The splitting is expressed in units of Dq, where D depends on the magnitude and distance of the ligand charges, and q on the radial extent of the d-electron functions on the central atom. The total splitting is defined as $\Delta = 10$ Dq. Since the splitting does not affect the energy averaged over the levels, the two e_g orbitals are destabilized by $3/5\Delta$, $=6$ Dq, and the three t_{2g} orbitals are stabilized by $2/5\Delta = 4$ Dq, relative to the d-orbital energy in the average spherical field (Fig. 10.1).

In a *tetrahedral* field the splitting is reversed as the ligands are now located in directions away from the cubic axes, rather than along the axes. In a *square-planar* (D_{4h}) field, further splitting occurs (Fig. 10.1). The relative ordering of the levels is strongly dependent on the nature of the coordination. If the axial ligand is absent, or weak electrostatically, the d_{z^2} orbital in the D_{4h} complex will be more stabilized than shown in the figure.

In the trigonal point group 3, the axis of quantization is chosen along the three-fold axis, while the x and y axes may be selected anywhere in the plane perpendicular to the z axis. In the point group $3m$, which occurs for many distorted octahedral complexes, there are also vertical mirror planes. The relation to the cubic axes is described by the transformation

$$\begin{pmatrix} x_t \\ y_t \\ z_t \end{pmatrix} = \begin{pmatrix} \sqrt{\frac{1}{2}} & -\sqrt{\frac{1}{2}} & 0 \\ \sqrt{\frac{1}{6}} & \sqrt{\frac{1}{6}} & -\sqrt{\frac{2}{3}} \\ \sqrt{\frac{1}{3}} & \sqrt{\frac{1}{3}} & \sqrt{\frac{1}{3}} \end{pmatrix} \begin{pmatrix} x_c \\ y_c \\ z_c \end{pmatrix} \tag{10.1}$$

Since this transformation is unitary, it applies to both the axes and the coordinates x, y, and z. The new z axis along the body diagonal of the cube,

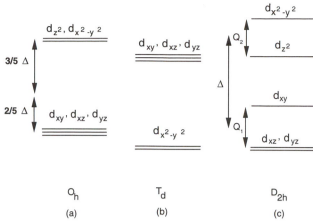

FIG. 10.1 Schematic drawing of the splitting of the d-electron levels in: (a) an octahedral field, (b) a tetrahedral field, and (c) a square-planar field. The size of the splittings and the relative values of Δ, Q_1, and Q_2 depend on the interatomic distances and the nature of the ligands.

which is the three-fold axis. In the case of the point group $3m$, x is perpendicular to one of the vertical mirror planes, while y lies within one of the vertical mirror planes.

In point group 3 (C_3 in spectroscopic notation), z^2 belongs to the representation a, and the pairs yz, xz, and xy, and $x^2 - y^2$ to the double degenerate representation e_g; in other words, the levels can be classified as a_g, e_g, and e_g. On the lowering of the symmetry from cubic to trigonal, the t_{2g} orbitals split into the single a_g and a double degenerate e_g level, while the e_g orbitals retain their symmetry characteristic (Fig. 10.2). Since the two sets of e_g orbitals belong to the same representation of the point group, they can be combined by a unitary transformation to give different sets of e_g orbitals. The symmetry-adapted set that correlates with the octahedral orbitals consists of combinations of $x^2 - y^2$ with xy, and of xz with yz. They are given by (Ballhausen 1962)

$$a_{1g} = d_{z^2}$$

$$e_{g+} = \sqrt{\tfrac{2}{3}} d_{x^2-y^2} - \sqrt{\tfrac{1}{3}} d_{xz}$$

$$e_{g-} = \sqrt{\tfrac{2}{3}} d_{xy} + \sqrt{\tfrac{1}{3}} d_{yz}$$

$$e'_{g+} = \sqrt{\tfrac{1}{3}} d_{x^2-y^2} + \sqrt{\tfrac{2}{3}} d_{xz}$$

$$e'_{g-} = \sqrt{\tfrac{1}{3}} d_{xy} - \sqrt{\tfrac{2}{3}} d_{yz}$$

(10.2)

The pseudo-octahedral e_g orbitals of Eq. (10.2) have z components, so they are lifted out of the xy plane.

For a transition metal element with more than one d electron, the atomic energy levels are more complex. As the electrons interact with each other, the

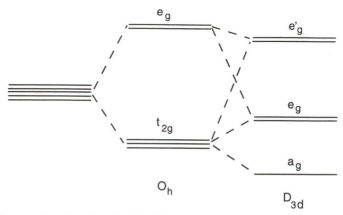

FIG. 10.2 Diagram showing the d orbitals in an octahedral field and the splitting of the t_{2g} orbitals that occurs upon trigonal distortion.

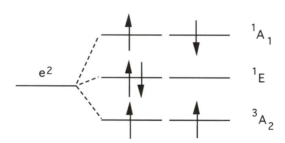

FIG. 10.3 Energy levels arising from the e^2 electron configuration.

energy levels are dependent on the filling of the levels. As an example, the e^2 electron configuration splits into three atomic levels, two of which are singlets and one is a triplet level (Fig. 10.3). In the singlet ^1E level, both electrons are in the same e orbital, while in the $^1A_1$ and $^3A_2$ levels, the electrons are distributed over the two different e orbitals, resulting in different energies. The electron distributions are, of course, different for the A and E levels, so a distinction can be made if the charge distribution is known with sufficient accuracy. The difference between the charge density distributions of the $^1A_1$ and $^3A_2$ levels is much smaller, but the spin density distribution, which can be measured with polarized neutron experiments on spin-oriented materials, is dramatically different.

10.2.2 Effect of Covalency on the Orbital Populations

The electrostatic theory of the preceding section is the starting point for a more complete treatment of the bonding in transition metal complexes, in which the covalency of the interactions is taken into account.

The metal valence orbitals combine with linear combinations of ligand orbitals of the same symmetry to give symmetry-adapted *molecular* orbitals. A schematic

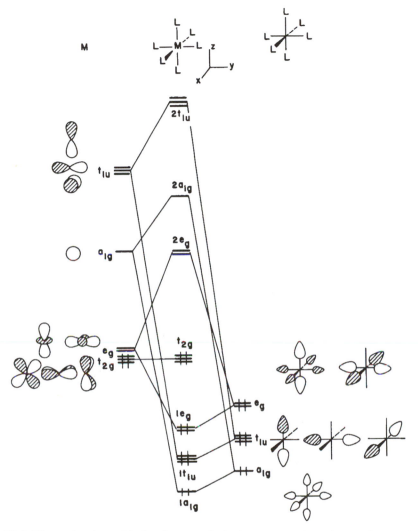

FIG. 10.4 The molecular orbitals of an octahedral ML_6 complex where L is an arbitrary σ-donor ligand. *Source*: Albright et al. (1985).

level diagram of an octahedral complex with σ-metal–ligand bonding is shown in Fig. 10.4. On the left are the metal orbitals, and on the right the symmetric combinations of the ligand orbitals. The t_{2g} metal orbitals are unaffected because no linear combinations of σ-ligand orbitals with the same symmetry exist. The crystal-field-destabilized e_g orbitals combine with the ligands to give a lower-lying bonding MO and a high-lying antibonding MO. Since the ligand orbitals are of lower energy, the occupied lower orbital will have mainly, but not solely, ligand character, that is, $c_2 \gg c_1$ in $\chi = c_1 \phi_{3d} + c_2 \phi_{\text{ligand}}$. The covalent bonding therefore results in a partial population of the crystal-field-destabilized transition metal orbitals. The size of this population is a measure of the importance of covalent bonding and of the relative energy of the ϕ_{3d} and ϕ_{ligand} orbitals in the LCAO expression. If the ligand orbital is much lower in energy, the bonding will be weaker, and the σ-donation into the metal atomic orbital will be less important.

If the transition metal atom has more than the six d electrons indicated on the diagram, the antibonding $2e_g$ molecular orbitals will also be populated and the metal–ligand bonding will be weakened. An example of this quite-prevalent effect is encountered in the series FeS_2, CoS_2, NiS_2, discussed in section 10.4.2.

The crystal-field-stabilized metal orbitals can combine with empty π orbitals on suitable ligands, leading to π-bonding, which is not included in Fig. 10.4. Since the ligand orbitals participating in this molecular orbital are generally empty, electrons are transferred to the ligands by so-called π-back-donation. This effect will lead to a lower population than predicted electrostatically for the crystal-field-stabilized metal orbitals, and, if the donation is into a ligand antibonding orbital, it leads to a lengthening of the bond on the ligand.

10.2.3 The Relation Between the Occupancies of Transition-Metal Valence Orbitals and the Multipole Population Parameters

In general, the atom-centered density model functions describe both the one-center valence density and the two-center density resulting from overlap of the valence orbitals. In the case of transition metal complexes, the overlap density in the metal–ligand bond is frequently quite small. An ab-initio theoretical calculation of the cobalt–porphine complex with a Co–N distance of 1.987 Å, for instance, gives, for the $^2A_{1g}$ ground-state, σ- and π-metal–ligand overlap populations of 0.02 and 0.04 e, respectively (Kashiwagi et al. 1978). This means that we can equate, to a good approximation, the multipolar density centered on the transition metal atom with the population of the outer valence shells of the atom.

If $\phi(d_i)$ is the atomic d-orbital basis set, the $3d$ density can be written as

$$\rho_d = \sum_i \sum_{j \geq i} P_{ij} \phi(d_i) \phi(d_j) \tag{10.3}$$

The cross terms $\phi(d_i)\phi(d_j)$, with $i \neq j$ in Eq. (10.3) do not appear in the case of the isolated atom for which the electron density is the sum of the square of the atomic orbitals. In the molecular case, the cross terms will only be nonzero for orbitals belonging to the same representation of the point group of the molecule, like the e_g orbitals in the case of trigonal site symmetry discussed above. In the square-planar point group $D_{4h}(4/m \, mm)$, the orbitals have a_{1g}, b_{1g}, b_{2g}, and e_g symmetry, and no such mixing occurs.

We recall that in the multipolar expansion, the $3d$ density is expressed in terms of the density-normalized spherical harmonic functions d_{lmp} as

$$\rho_d = \sum_{l=0}^{l_{\text{max}}} \kappa'^3 R_l(\kappa' r) \sum_{m=0}^{l} \sum_p P_{lmp} d_{lmp}(\mathbf{r}/r) \tag{10.4}$$

Since the d orbitals are invariant with respect to inversion through the nuclear position, only l even terms will occur in this summation.

Equating Eqs. (10.3) and (10.4) leads to an especially simple expression if the radial dependence of the density is equal in both descriptions. In this case, the radial parts cancel and we obtain

$$\mathbf{P}_{ij}\mathbf{Y}_{ij} = \mathbf{P}_{lmp}\mathbf{d}_{lmp} \tag{10.5}$$

Here, $\mathbf{Y}_{ij}$ is the 15-element column vector of the *angular* part of the $\phi(d_i)\phi(d_j)$ orbital products, $\mathbf{P}_{ij}$ is the row vector of the 15 unique elements of the symmetric 5×5 matrix of the coefficients in Eq. (10.3), and $\mathbf{P}_{lmp}$ is the row vector containing the coefficients of the 15 spherical harmonic density functions d_{lmp} with $l = 0, 2$, or 4. Density functions with other l values do not contribute to the d-orbital density.

The spherical harmonic functions constitute a complete set of functions in the spherical point group. A product of two spherical harmonics such as $y_i y_j$ must therefore be a linear combination of spherical harmonic functions. An example of such an expression is

$$y_{20}y_{20} = 0.241\,795\,54y_{40} + 0.180\,223\,75y_{20} + 0.282\,094\,79y_{00} \tag{10.6a}$$

For our purpose, it is preferable to express the right-hand side of this equation in terms of the spherical harmonic density functions. Use of the ratio of orbital- and density-function normalization factors gives the result

$$y_{20}y_{20} = 0.368\,48d_{40} + 0.274\,93d_{20} + 1.0d_{00} \tag{10.6b}$$

Expression (10.6) shows that the product of two identical d_{z^2} orbitals contains hexadecapolar, quadrupolar, and monopolar density functions. The product equations can, in general, be written as $\mathbf{Y}_{ij} = \mathbf{Ld}_{lmp}$. The elements of the matrix $\mathbf{L}$ are the coefficients in expressions like Eq. (10.6). A complete set of the equations for $l \leq 2$ is given in Table E.3 of appendix E.

The equivalence for the density in Eq. (10.5) can then be written as

$$\mathbf{P}_{ij}\mathbf{Y}_{ij} = \mathbf{P}_{ij}\mathbf{Ld}_{lmp} = \mathbf{P}_{lmp}\mathbf{d}_{lmp} \tag{10.7}$$

We obtain, for the relation between the coefficients P_{ij} and P_{lmp},

$$\mathbf{P}_{lmp} = \mathbf{P}_{ij}\mathbf{L} \tag{10.8a}$$

or

$$\mathbf{P}_{lmp}^T = \mathbf{L}^T\mathbf{P}_{ij}^T \equiv \mathbf{MP}_{ij}^T \tag{10.8b}$$

The d-orbital occupancies are derived from the experimental multipole populations by the inverse expression (Holladay et al. 1983)

$$\mathbf{P}_{ij}^T = \mathbf{M}^{-1}\mathbf{P}_{lmp}^T \tag{10.9}$$

The matrix $\mathbf{M}^{-1}$ is given in appendix I. In all but triclinic point groups, site-symmetry restrictions limit the allowed functions beyond the l even requirement. The symmetry-allowed multipolar density functions are given by the "index-picking" rules of appendix D, section D.3, and are listed in Table 10.1.

The $\mathbf{M}^{-1}$ matrices specific for higher point groups are obtained by omission of symmetry-forbidden columns in the full 15×15 matrix. This leads to rows with zero elements for the nonallowed cross products between d orbitals, which are subsequently omitted to recover a reduced matrix. The matrix for the point group $D_{4h} = 4/m\,mm$ is shown as an example in Table 10.2.

TABLE 10.1 Symmetry-Allowed Multipole Functions Describing d-Orbital Density

Point Group	Allowed Values of l, $m_\pm$ of M^{-1}[a]	Dimension of $\mathbf{M}$
1, $\bar{1}$	$l = 0, 2, 4$, all m	15×15
2, m, $2/m$	00, 20. 22+, 22−, 40, 42+, 42−, 44+, 44−	9×9
222, $m2m$, mmm	00, 20, 22+, 40, 42+, 44+	6×6
4, $4/m$, $\bar{4}$	00, 20, 40, 44+, 44−	5×5 (4×4)[b]
422, $\bar{4}2m$, $4mm$, $4/mmm$	00, 20, 40, 44+	4×4
3, $\bar{3}$	00, 20, 40, 43+, 43−	5×5 (4×4)[b]
32, $3m$, $\bar{3}m$	00, 20, 40, 43 +	4×4
6, $\bar{6}$, $6/m$, 622, $6mm$, $\bar{6}\,m2$, $6/mmm$	00, 20, 40	3×3
23, $m3$, 432, $\bar{4}3m$, $m3m$	00, 0.782 45 $(40+) + 0.579\,39\ (44+)$[c]	2×2

[a] Principle symmetry axis is z axis.
[b] Dimension can be reduced by rotation of coordinate system; see text.
[c] This function is usually described as the cubic harmonic K_4 (see appendix D).

10.2.3.1 The Relationships in Terms of Symmetry-Adapted Orbitals

For the point group D_{4h}, the atomic d-orbital functions belong to four different group-theoretical representations. When the same representation occurs more than once, as it does, for example, in trigonal point groups, $\mathbf{M}^{-1}$ will contain cross terms between orbitals of the same symmetry, as shown in Table 10.3(a). In this case, we are interested in the population of the symmetry-adapted orbitals y_k^s, such as defined for the trigonal case by expression (10.2). The symmetry-adapted orbitals are linear combinations of the original functions, that is,

$$y_k^s = \sum_i c_{ki} y_i \tag{10.10}$$

while the density products are given by

$$y_k^s y_l^s = \sum_i \sum_j c_{ki} c_{lj} y_i y_j \tag{10.11}$$

The populations P_{kl}^s of the symmetry-adapted orbital products follow, in analogy to Eq. (10.7), from

$$\mathbf{P}_{kl}^s \mathbf{Y}_{kl}^s = \mathbf{P}_{lmp} \mathbf{d}_{lmp}$$

TABLE 10.2 The Matrix $\mathbf{M}^{-1}$, Defining Orbital-Multipole Relations, for the Point Group D_{4h}

	Description		P_{00}	P_{20}	P_{40}	P_{44+}
P_{20}	z^2	a_{1g}	0.200	1.039	1.396	0.00
P_{21+}, P_{21-}	xz, yz	e_g^+, e_g^-	0.200	0.520	−0.931	0.00
P_{22+}	$x^2 - y^2$	b_{1g}	0.200	−1.039	0.233	1.570
P_{22-}	xy	b_{2g}	0.200	−1.039	0.233	−1.570

Source: Coppens and Becker (1992).

TABLE 10.3 The Matrix $\mathbf{M}^{-1}$ for Trigonal Point Groups

(a) In terms of d-orbital products

	P_{00}	P_{20}	P_{40}	P_{43+}	P_{43-}
P_{20}	0.200	1.039	1.396	0.00	0.00
P_{21+}	0.200	0.520	−0.931	0.00	0.00
P_{21-}	0.200	0.520	−0.931	0.00	0.00
P_{22+}	0.200	−1.039	0.233	0.00	0.00
P_{22-}	0.200	−1.039	0.233	0.00	0.00
$P_{21+/22+}$	0.00	0.00	0.00	2.094	0.00
$P_{21+/22-}$	0.00	0.00	0.00	0.00	2.094
$P_{21-/22+}$	0.00	0.00	0.00	0.00	2.094
$P_{21-/22-}$	0.00	0.00	0.00	−2.094	0.00

(b) In terms of products of symmetry-adapted orbitals[a,b]

	P_{00}	P_{20}	P_{40}	P_{43+}
$P(a_{1g})$	0.200	1.039	1.396	0.00
$P(e_g)$	0.400	−1.039	−0.310	−1.975
$P(e'_g)$	0.400	0.00	−1.087	1.975
$P(e_{g+}e'_{g+} + e_{g-}e'_{g-})$	0.00	−2.942	2.193	1.397

[a] The orbital expressions are given in Eq. (10.2).
[b] The signs given here imply a positive e'_g lobe in the positive xz quadrant; the coordinate system should be defined such that this lobe points towards a ligand atom.

or, following the derivation given above,

$$\mathbf{P}^s_{kl} = (\mathbf{L}^{s-1})^T \mathbf{P}^T_{lmp} = \mathbf{M}^{s-1} \mathbf{P}^T_{lmp} \tag{10.12}$$

The elements of $\mathbf{L}^s$ are obtained as linear combinations of expressions of the type of Eq. (10.6), using the product of coefficients from the expansion of Eq. (10.11). The results for the trigonal point groups $(3, \bar{3}, 32, 3m, \bar{3}m)$ are illustrated in Table 10.3(b). For exact cubic symmetry, the cross term $P(e_g, e'_g) = 0$, and $P(e'_g) = 2P(a_g)$.

In the symmetry-adapted formulation, the P_{43-} term no longer occurs because the d-orbital density contains a vertical mirror plane even if such a plane is absent in the point group. This is illustrated as follows. Point groups without vertical mirror planes differ from those with vertical mirror planes by the occurrence of both d_{lm+} and d_{lm-} functions, with m being restricted to n, the order of the rotation axis. But the coordinate system can be rotated around the main symmetry axis such that P_{4n-} becomes zero. As proof, we write the φ dependence as

$$\rho(\varphi') = P_{ln+} \cos n\varphi + P_{ln-} \sin n\varphi \tag{10.13}$$

which has the maxima and minima for

$$\partial\rho/\partial\varphi = -nP_{ln+} \sin n\varphi + nP_{ln-} \cos n\varphi = 0 \tag{10.14}$$

or, $\tan n\varphi = P_{ln-}/P_{ln+}$.

Thus, a rotation of the coordinate system by

$$\varphi_0 = \frac{1}{n} \tan^{-1}(P_{ln-}/P_{ln+}) \tag{10.15}$$

eliminates the antisymmetric component $\sin n\varphi$ represented by d_{lm-}. In the new coordinate system, the φ dependence is

$$\rho(\varphi') = P'_{ln+} \cos n\varphi \tag{10.16a}$$

with

$$P'_{ln+} = P_{ln+} \cos n\varphi_0 + P_{ln-} \sin n\varphi_0 = (P^2_{ln+} + P^2_{ln-})^{1/2} \tag{10.16b}$$

and

$$P'_{ln-} = 0 \tag{10.16c}$$

or, conversely,

$$P_{ln-} = P'_{ln+} \sin n\varphi_0 \qquad P_{ln+} = P'_{ln+} \cos n\varphi_0 \tag{10.17}$$

In the new coordinate system, the x axis coincides with one of the vertical mirror planes of the density, reducing the size of $\mathbf{M}^{-1}$ from 5×5 to 4×4.

Application of the transformation of Eq. (10.15) to tetragonal and trigonal point groups reduces the number of distinct sets of point groups in Table 10.1 from nine to seven.

10.3 The Electric Field Gradient at the Nucleus of a Transition Metal Atom

10.3.1 Electric Field Gradient Expressions for Transition Metal Elements

For transition metal atoms, the dominant contribution to the electric field gradient at a nucleus originates in the valence shell centered on that nucleus. The expressions for this central contribution are given in Eq. (8.43) for Slater-type, exponential, density functions, defined as in Eq. (3.34):

$$R_l(r) = \kappa^3 \frac{\zeta^{n_l+3}}{(n_l + 2)!} (\kappa r)^{n(l)} \exp(-\kappa\zeta r) \tag{10.18}$$

Expressions (8.43) for the elements of the traceless quadrupole tensor can be written in a slightly different form as

$$\nabla E_{xx} = +(3/5)(\pi P_{22+} - 3^{1/2}P_{20})Q_r$$

$$\nabla E_{yy} = -(3/5)(\pi P_{22+} + 3^{1/2}P_{20})Q_r$$

$$\nabla E_{zz} = +(6/5)(3^{1/2}P_{20})Q_r$$

$$\nabla E_{xy} = +(3/5)(\pi P_{22-})Q_r \tag{10.19}$$

$$\nabla E_{xz} = +(3/5)(\pi P_{21+})Q_r$$

$$\nabla E_{yz} = +(3/5)(\pi P_{21-})Q_r$$

where Q_r is the expectation value of r^{-3} defined as

$$Q_r = \langle r^{-3} \rangle_{3d} = \int_0^\infty (R(r)/r)\, dr \tag{10.20}$$

Substitution of Eq. (10.18) in Eq. (10.20), with the value of $n(2)$ in Eq. (10.18) equal to 4 for $3d$ valence-shell density functions of first-row transition metal atoms (chapter 3), gives[1]

$$Q_r = (\kappa\zeta)^3/[n_2(n_2 + 1)(n_2 + 2)] = (\kappa\zeta)^3/120 \tag{10.21}$$

The value of Q_r is sensitive to the nature of the radial function. The Q_r for the radial dependence of the Hartree–Fock isolated atom function can be evaluated analytically using the Clementi–Roetti Slater-type expansions, defined by expression (8.38) (Clementi and Roetti 1974). The result is a weighted sum over terms of the type $(\kappa\zeta)^3/[n_l(n_l + 1)(n_l + 2)]$, each with the appropriate expansion coefficient. For the isolated Fe atom, one obtains

$$Q_r(\text{Fe}) = 4.978 \text{ au}^{-3} \tag{10.22}$$

For comparison, the crude "best single ζ" value of Clementi and Raimondi (1963) for the Fe ($3d$) orbital is 3.7266 au^{-1} (appendix F), or, for the orbital exponent of the density function, $\zeta = 7.4532$ au^{-1}. This gives, with Eq. (10.21), a value of 3.4502 au^{-3}, illustrating the strong dependence of Q_r on the quality of the radial functions.

Marathe and Trautwein (1983) quote values of 5.09 and 5.73 au^{-3} from Hartree–Fock calculations on Fe^{2+} ($3d^6$) and Fe^{3+} ($3d^5$), respectively, showing that the radial contraction (i.e., the κ parameter in the diffraction formalism) has a pronounced effect on the $\langle r^{-3} \rangle_{3d}$ values.

For atoms in sites of low symmetry, the EFG tensor must be diagonalized to obtain its principal components ∇E_{ij}. Since the tensor as defined by Eq. (8.9) is traceless, two values will define all three principal elements. They are commonly chosen as ∇E_{33}, the principal component with the largest magnitude, and the asymmetry parameter $\eta = |\nabla E_{22} - \nabla E_{11}|/|\nabla E_{33}|$, which has the range $0 \leq \eta \leq 1$.

10.3.2 Comparison with Results from Mössbauer Spectroscopy

For a number of nuclei, the electric field gradient at the nucleus can be obtained very accurately from spectroscopic measurements using Mössbauer or magnetic resonance techniques as mentioned in chapter 8.

Mössbauer spectroscopy is based on transition between energy levels of nuclei with different values of the nuclear spin quantum number I. When a nucleus emits a γ-ray, the energy of the emitted radiation is lowered by the recoil of the nucleus. Conversely, the energy needed for absorption is higher than that needed for transition, because the absorbing nucleus absorbs energy in the recoil process. For nuclei tightly bound in solids, however, the effective mass of the emitter and

---

[1] Using the standard integral $\int_0^\infty x^n e^{-\mu x}\, dx = n!\mu^{-n-1}$.

absorber is that of the crystal, and recoil effects are absent for most of the emitting and absorbing nuclei. Under such conditions, the energy emitted by a nucleus can be absorbed by an identical nucleus exposed to the emitted radiation.

The ^{57}Fe nucleus has two levels with nuclear spin quantum numbers $I = 1/2$ and $I = 3/2$. The nuclear energy levels for a given nucleus are slightly affected by three types of interactions which are: (1) the interaction between the electron density *at the nucleus* and the positive nuclear charge, (2) quadrupole interactions between the electric field gradient of the charge distribution and the nuclear quadrupole moment, and (3) magnetic interactions. To record such changes it must be possible to vary the energy of the incoming radiation.

In Mössbauer spectroscopy, an energy range is obtained by *Doppler shifting*. By moving the source relative to the sample, the energy of the γ-rays can be varied over the range of energy differences arising from the interactions with the environment.

The relation between the velocity of the emitter and the energy shift of the emitted electromagnetic radiation is given by the Doppler expression

$$\Delta E_\gamma = \frac{\Delta v}{c} E_\gamma \tag{10.23}$$

where c is the speed of light. Thus, for a ^{57}Fe nuclear emitter with $E_\gamma = 14.4$ keV, a velocity of 1 mm s^{-1} corresponds to a shift of $4.80 \cdot 10^{-8}$ eV, or $7.69 \cdot 10^{-27}$ J. The difference of the nuclear transitions in the Fe^{3+} cation, and the hexacoordinated Fe atom in $Fe[Fe(CN)_6]$ is $2 \cdot 10^{-8}$ eV, and is thus of the same magnitude.

When the electric field gradient at the nucleus exerted by the electrons is nonzero, the nuclear levels will be split. The eigenvalues of the quadrupolar interaction Hamiltonian are given by

$$E_{QS} = -\frac{\nabla E_{33}\, eQ}{4I(2I - 1)} [3m_I^2 - I(I + 1)](1 + \tfrac{1}{3}\eta^2)^{1/2} \tag{10.24}$$

in which eQ is the nuclear quadrupole moment of the excited nucleus, and ∇E_{33} is the largest component of the diagonalized EFG tensor. Analogous to the quantum numbers for the electron, the magnetic spin quantum number m_I has the values $+I, +I - 1, \ldots, -I + 1, -I$. Substitution of $I = 0$ and $I = 1/2$ in Eq. (10.24) shows that *the energy level splitting occurs only for $I \geq 1$ nuclei*. Thus, Mössbauer spectroscopy is limited to a subset of the available isotopes.

For Fe, the $I = 3/2$ level splits under the influence of the field gradient into levels with quantum number $m_I = 3/2, 1/2, -1/2$, and $-3/2$. Thus, two different energy levels exist for the $I = 3/2$ nucleus in a nonzero electric field gradient, but the $I = 1/2$ level remains a single, though doubly degenerate $(2I + 1)$, level (Fig. 10.5). From Eq. (10.24), the splitting of the $I = 3/2$ level is equal to

$$\Delta E_{QS} = -1/2 \nabla E_{33}\, eQ(^{57}Fe^m)(1 + \eta^2/3)^{1/2} \tag{10.25}$$

where the superscript m indicates that Q is the quadrupole moment of the excited state of ^{57}Fe.

As noted by Tsirel'son et al. (1987), there is considerable scatter in the reported values for $Q(^{57}Fe^m)$. A theoretical value of Q has been reported as

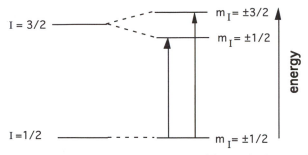

FIG. 10.5 Nuclear energy levels for the ground state and first excited state of ^{57}Fe. Left: zero field gradient, right: nonzero field gradient.

$(0.156 \pm 0.02) \cdot 10^{-28}$ m^2 (Litterst et al. 1978). Calculation of the EFG for αFe$_2$O$_3$, and solving Eq. (10.25) with the observed value of the quadrupole splitting has led to a number of values, including a recent one reported as $\approx 0.11 \cdot 10^{-28}$ m^2 (Nagel 1985). In a similar manner, Q can be derived from the diffraction value of the electric field gradient at the Fe nucleus. From the value for αFe$_2$O$_3$ and the published splitting, Tsirel'son et al (1987) obtain $Q(^{57}\text{Fe}^m) = 0.13 \cdot 10^{-28}$ m^2, while Su and Coppens (1996), by combining four sets of accurate diffraction results, obtain a best estimate of $Q(^{57}\text{Fe}^m) = 0.12(3) \cdot 10^{-24}$ cm^2. All these results are clustered in the same range. With additional accurate diffraction studies, it will be possible to reduce the standard deviation of the X-ray value of the ^{57}Fe nuclear quadrupole moment.

10.3.2.1 Conversion of Units Between Diffraction and Spectroscopic Results

The EFG is obtained from the diffraction experiment in units of eÅ^{-3} [or e au^{-3}, if ζ in Eq. (8.43) is in au^{-1}]. Conversion to the SI unit Cm^{-3} requires the use of $4\pi\varepsilon_0$, the permittivity of free space (chapter 8) ($4\pi\varepsilon_0 = 1.112\,626\,5 \cdot 10^{-10}$ C^2 N^{-1} m^{-2}, $1/(4\pi\varepsilon_0) = 8.9877 \cdot 10^9$ C^{-2} N^1 m^2). We thus have for ΔE_{QS}, indicating the dimensions in square brackets,

$$e[\text{C}] \cdot \nabla E_{33}[\text{Cm}^{-3}] \cdot Q[\text{m}^2]/4\pi\varepsilon_0[\text{C}^2 \text{ N}^{-1} \text{ m}^{-2}] = [\text{Nm}] = [\text{J}] \quad (10.26)$$

The energy difference in J can be converted to the Mössbauer unit of mm s^{-1} using the relation 1 mm s^{-1} = $7.69 \cdot 10^{-27}$ J, derived above. Numerically, starting with ∇E_{33} in eÅ^{-3}, we obtain for the conversion factor,

$$\Delta E_{QS} \text{ (J)} = 1/2 \cdot 1.602 \cdot 10^{-19} \cdot 8.9877 \cdot 10^9 \cdot Q[\text{m}^2] \cdot 1.602 \cdot 10^{11} \nabla E_{33}[\text{eÅ}^{-3}](1+\eta)^{1/2}$$

$$= 115.32 Q \text{ (m}^2) \nabla E_{33} \text{ (eÅ}^{-3})(1+\eta^{2/3})^{1/2} \quad (10.27)$$

where 1 e = $1.602 \cdot 10^{-19}$ C, and 1 eÅ^{-3} = $1.602 \cdot 10^{11}$ Cm^{-3} have been used. Assuming that $Q(^{57}\text{Fe}^m) = 0.15 \cdot 10^{-28}$ m^2, and substituting the conversion factor from mm s^{-1} to J, we get for Fe,

$$\Delta E_{QS} \text{ (mm s}^{-1}) = 0.225 \nabla E_{33} \text{ (eÅ}^{-3})(1+\eta^{2/3})^{1/2} \quad (10.28a)$$

or, equivalently

$$\Delta E_{QS} \text{ (mm s}^{-1}) = 1.520 \, \nabla E_{33} \text{ (e au}^{-3})(1 + \eta^{2/3})^{1/2} \tag{10.28b}$$

As in chapter 8, we will refer to contributions from the electron density centered on the nucleus as *central* contributions, and to the remainder as *peripheral* contributions. In the spectroscopic literature, the latter are commonly referred to as lattice contributions, a term we will avoid as it conflicts with the common definition of the lattice as a mathematical concept.

In the case that the axis of quantization is the z axis, and η is zero, the central contribution to ΔE_{QS} can be directly related to P_{20} by substitution of the expression (10.18) for ∇E_{zz} into Eq. (10.28b). With $Q_r(\text{Fe}, \kappa = 1) = 4.978 \text{ au}^{-3}$, we obtain

$$\Delta E_{QS} \text{ (mm s}^{-1}) = 15.72 \kappa^3 \, P_{20} \tag{10.29}$$

One of the earliest applications of this expression was to the mineral pyrite, FeS_2, in which the iron site has site symmetry $\bar{3}$, and the iron atoms are in a distorted octahedral environment (Fig. 10.6) (Stevens et al. 1980). The asphericity of the electron density is evident in the deformation density section through the Fe atom, shown in Fig. 10.7. Because of the local symmetry of the Fe site, d_{20} is the only allowed quadrupolar term, and the asymmetry parameter η is zero. The value of ∇E_{33} for the Fe atom was reported as $1.7 \cdot 10^{15}$ esu cm^{-3}, or 3.5 eÅ^{-3} (1 e = $0.4802 \cdot 10^{-9}$ esu), equal to $5.6 \cdot 10^{11}$ Cm^{-3}. With Eq. (10.28a), ΔE_{QS} is obtained as 0.8 mm s^{-1}, which is, within the experimental error of the diffraction experiment, equal to the Mössbauer values of 0.634 (6) mm s^{-1} (Finklea et al. 1976).

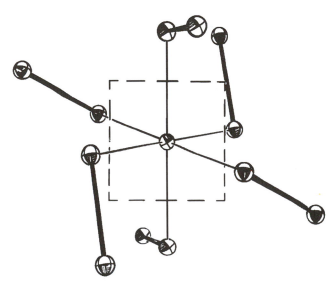

FIG. 10.6 Coordination of disulfide ions about iron in pyrite. Ellipsoids are plotted at 90% probability. The dashed lines represent the plane containing the axial Fe–S bonds and bisecting the equatorial bonds in which the electron density is plotted in Fig. 10.7. *Source:* Stevens et al. (1980).

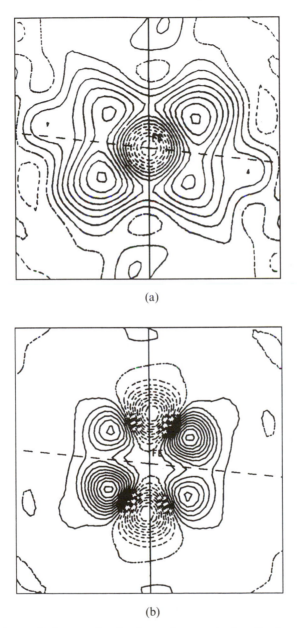

(a)

(b)

FIG. 10.7 Experimental electron density in a plane containing the iron atom in pyrite, defined in Fig. 8.4, showing the accumulation of deformation density in the diagonal directions. Contours are at $0.2 \, e\text{Å}^{-3}$. Zero and negative contours are broken lines. (a) The experimental deformation density, (b) the model deformation density. *Source*: Stevens et al. (1980).

A second, more recent, example is the analysis of the combined synchrotron/sealed-tube data on Cu_2O (Kirfel and Eichhorn 1990). The compound Cu_2O exhibits one of the largest known electric field gradients at the Cu nucleus. As the Cu atoms are located on a three-fold axis in the cubic crystals, the EFG tensor is diagonal, and, as in pyrite, is described by a single parameter. The most comprehensive refinement of the data gives a value of VE_{zz} of $+1.31\,(7)\cdot10^{22}\,Cm^{-3}$, in excellent agreement with an NMR value of $1.34\cdot10^{22}\,Cm^{-3}$. The X-ray study resolved a controversy about the sign of the EFG, which was predicted to be positive by the ionic point charge model, but not by a later more advanced calculation of a cluster model for the structure.

10.3.2.2 The Sternheimer Shielding and Antishielding Factors

Sternheimer has pointed out that the quadrupolar components of the charge distribution induce a polarization of the charge density which affects the effective field gradient at the nuclear positions (Sternheimer and Foley 1956). As the inner shells are close to the nucleus, the effect on the electric field gradient can be large. Both shielding and antishielding occur, the latter corresponding to an enhancement of the EFG. The Sternheimer *antishielding* factor $\gamma(r)$ is dependent on the distance to the nucleus for small r, but is quite constant beyond about 1–1.5 Å from the nucleus.

The effective peripheral contribution to the EFG, including antishielding, is given by

$$VE_{ij}^{\text{per, effective}} = (1 - \gamma_\infty)\,VE_{ij}^{\text{per}} \tag{10.30}$$

Quantum-mechanical perturbation theory calculations show γ_∞ to have large negative values for most atoms. Values for Fe, Fe^{2+}, and Fe^{3+}, given in Table 10.4, range between -8 and -11. It follows that the peripheral contribution is more than nine times enhanced by antishielding.

The calculations show that the central contribution to the EFG, due to the valence electrons, is shielded rather than antishielded, but the effect is less pronounced. The *shielding factor* R is the density-weighted average of $\gamma(r)$, $\langle\gamma(r)r^{-3}\rangle_{\text{core, valence}}/\langle r^{-3}\rangle_{\text{core, valence}}$, where the average is to be taken separately over the core and valence shells, depending on the shell in which polarization is induced.

For the neutral Fe atom, the value of $R = 0.07$ is often used in spectroscopic work (Ray and Das 1977). Recent values for both Fe and its free ions are listed in Table 10.4. Values of R_{core} for $\kappa_{\text{valence}} \neq 1$ can be obtained by applying the polynomial in the first footnote of the table.

The total effective EFG is obtained from the expression

$$VE_{ij}^{\text{eff}} = (1 - R)\,VE_{ij}^{\text{val}} + (1 - \gamma_\infty)\,VE_{ij}^{\text{per}} \tag{10.31a}$$

If the electron density were known at high resolution, the antishielding effects would be represented in the experimental distribution, and the correction in Eq. (10.31a) would be superfluous. However, the experimental resolution is limited, and the frozen-core approximation is used in the X-ray analysis. Thus, for consistency, the R_{core} shielding factor should be applied in the conversion of the

TABLE 10.4 Sternheimer Nuclear Quadrupole
Factors R and γ_∞ for Fe, Fe^{2+}, and Fe^{3+} ($\kappa = 1.0^a$)

	R	γ_∞
Fe, $\langle r^{-3} \rangle_{3d} = 4.979$ au		
Core	0.0730	-8.933
Valence[b]	0.0521	-1.294
Total core + valence	0.1251	-10.227
Fe^{2+}, $\langle r^{-3} \rangle_{3d} = 5.086$ au		
Core	0.0704	-8.681
Valence	0.0442	-2.354
Total core + valence	0.1146	-11.035
Fe^{3+}, $\langle r^{-3} \rangle_{3d} = 5.728$ au		
Core	c	7.974
Valence		-1.453
Total core + valence		-9.427

[a] For Fe^{2+}, the dependence of R_{core} on $\kappa_{valence}$ is well fitted by the polynomial $R_{core}(Fe^{2+}, \kappa) = 1.0686 - 2.4955\kappa + 2.061\,00\kappa^2 - 0.56369\kappa^3$. The corresponding expression for neutral Fe is $R_{core}(Fe, \kappa) = 1.1061 - 2.5832\kappa + 2.1352\kappa^2 - 0.585\,19\kappa^3$.
[b] Including the two $4s$ electrons.
[c] Spherical atom, $R = 0$.
Source: Su and Coppens (1996).

X-ray EFG values to spectroscopic splittings. But the polarization of the valence shell is at least in part accounted for in the aspherical multipole description. To the extent that the model is sufficiently flexible, the shielding factor $R_{valence}$ is not needed, and the correction equation becomes

$$\nabla E_{ij}^{eff} = (1 - R_{core})\,\nabla E_{ij}^{val} + (1 - \gamma_{\infty,core})\,\nabla E_{ij}^{per} \qquad (10.31b)$$

Though the peripheral contribution is considerably enhanced by the anti-shielding, it is nevertheless small relative to the central contribution, except for very short metal–ligand distances (Coppens 1990).

Applications of the core shielding factors in X-ray studies of pyrite, Fe(II) phthalocyanine, and bis(pyridine)(*meso*-tetraphenylporphinato)iron(II) generally improve agreement with spectroscopic values (Su and Coppens 1996).

10.4 Electron Density Studies of Octahedral and Distorted Octahedral Complexes

10.4.1 Complexes with CO, CN, and NH_3 Ligands

Octahedral and distorted octahedral complexes of first-row transition metal atoms were subjected to X-ray charge density analysis in the pioneering studies of Iwata and Saito in the early 1970s (Iwata and Saito 1973).

TABLE 10.5 Comparison of Orbital Populations in a Number of Octahedral and Distorted Octahedral Complexes (From Treatment with Zero $4s$ Population)

	$Co(NH_3)_6^{3+}$	$Co(NH_3)_6^{3+}$	$Co(CN)_6^{3-}$	$Cr(CN)_6^{3-}$	$Cr(CO)_6$	HPNO—Co
Reference	[a]	[b]	[b]	[a]	[c]	[d]
$a_g(\%)$	26.4 ⎫ 76.0	24.3 ⎫ 76.5	25.0 ⎫ 74.2	19.4 ⎫ 69.3	70.8	16.4 ⎫ 66.3
$e_g(\%)$	49.6 ⎭	52.2 ⎭	49.2 ⎭	49.9 ⎭		49.9 ⎭
$e_g'(\%)$	24.1	23.5	25.9	30.7	29.2	33.6
e_g'(electrons)	1.89	1.75	1.88	1.61	1.37	2.41
Total $3d$ population	7.84	7.44	7.26	5.26	4.69	7.27

[a] Data from Iwata and Saito (1973), room temperature study, analysis from Holladay et al. (1983).
[b] Data from Iwata (1977), liquid nitrogen temperature study, analysis from Holladay et al. (1983).
[c] Rees and Mitschler (1976), liquid nitrogen temperature study.
[d] Wood (1995), hexapyridine N-oxide Co(II) perchlorate, liquid nitrogen temperature study.

Many additional studies have since been made. A summary of the results for a number of octahedral complexes is given in Table 10.5. The predictions of ligand field theory are clearly borne out by the results, which show pronounced depopulation of the field-destabilized e_g' orbitals and increased population of the stabilized $t_{2g}(e_g, a_g)$ orbitals relative to the distribution in the high-spin spherical atom.

The population of the destabilized e_g' orbitals is larger for the Co complexes than for the Cr compounds listed, a trend with increasing number of electrons reproduced in the sulfides discussed in the following section. A population of more than two electrons of the e_g' orbitals implies population of the antibonding metal–ligand orbitals, a state only reached in the Co(II) complex listed in the last column. The total number of d electrons, however, seems to correlate more with the element than with the specific valence state of the element, as there is no systematic difference between the Co(II) amd Co(III) complexes. But the number of available studies is still too small to allow more general conclusions.

10.4.2 First-Row Transition Metal Sulfides

Iron pyrite, FeS_2, was among the first transition metal complexes of which the charge density was studied by X-ray methods. It has a simple rock-salt-type structure with alternating Fe^{2+} and Se_2^{2-} ions being located at sites of symmetry $\bar{3}$ in the space group $Pa\bar{3}$. The experimental charge density shows a pronounced preferential occupation of the stabilized a_g and e_g orbitals (Fig. 10.7), which point into the voids between the ligands. The features are sharpened in the static density map, shown in part (b) of the figure. The populations from a refinement including higher cumulants for the thermal motion treatment, but keeping the total Fe population equal to six, are given in the first column of Table 10.6.

The results of a rerefinement of the FeS_2 data and of more recent accurate data on CoS_2 and NiS_2 (Nowack et al. 1991) are listed in Table 10.6. Anharmonic

TABLE 10.6 Electron Population in First-Row Transition Metal Sulfides

	FeS_2[a]	FeS_2[b]	CoS_2[b]	NiS_2[b]
$P(a_g)$	1.59 (6)	1.68 (7)	1.73 (8)	1.94 (6)
$P(e_g)$	2.81 (7)	2.98 (9)	3.52 (12)	3.84 (10)
$P(e_g')$	1.60 (7)	1.30 (9)	2.19 (12)	2.81 (10)
$P(e_g, e_g')$	−0.09 (14)	−0.02 (15)	−0.01 (14)	−0.23 (12)
Total electrons	6.0	5.97 (15)	7.44 (23)	8.58 (19)
$R(F)$	0.018	0.011	0.012	0.014
$\hat{\sigma}^2$[c]	1.42	1.13	1.21	1.15

[a] From population parameters in Stevens et al (1980); total Fe population fixed at 6 electrons; with anharmonic cumulant temperature factors.
[b] Su and Coppens (to be published); with Gram–Charlier temperature parameters, and anomalous scattering corrections as given by Kissel and Pratt (1990).
[c] Goodness of fit [Eq. (4.27)].

Gram–Charlier temperature factors (chapter 2) were applied in this analysis, which made use of the XD programming package (Su and Coppens, to be published). In accordance with the results on H_3PO_4, discussed in chapter 3, the values of n_l were chosen as 6, 6, 7, and 7 for $l = 1, 2, 3,$ and 4, respectively. However, the differences between the results from this choice and those of two alternative selections (4, 4, 4, and 4; and 6, 6, 6, and 6) are within one standard deviation.

Except in the Ni compound, the a_g and e_g orbitals are less than fully populated in agreement with the π-metal-to-ligand back-donation concept discussed in section 10.2.2. The trend of increased population of the destabilized e_g' orbitals in the Fe → Co → Ni sequence is evident. The destabilized orbitals, which point towards the ligands, participate in σ-metal–ligand bonding, corresponding to σ-ligand-to-metal electron donation. The population of these orbitals beyond two electrons for the Co and Ni chalcogenides indicates that antibonding is of increasing importance towards the right of the periodic table. This in accordance with the increase of the M–S distances in the same sequence from 2.2633 (2) to 2.3252 (1) to 2.3987 (3) Å, and provides an explanation for the generally observed increase in low-spin radii of the transition metals towards the right of the periodic table. The e_g, e_g' cross term is insignificant, which indicates that the vertical mirror planes containing the three-fold axis are well preserved on the lowering of the symmetry of the coordination sphere from octahedral to trigonal.

The model deformation densities for all three compounds show a peak between the metal atom and the bonded sulfur, located closer to the sulfur atoms, in a region not covered by Fig. 10.7. A local density functional calculation of pyrite by Zeng and Holzwart (1994) gives a theoretical deformation density which closely reproduces the features of the experimental densities, including the Fe—S bond peak. No orbital populations are as yet available from this calculation.

10.5 Electron Density Studies of Iron(II) Porphyrins

10.5.1 The Electronic Structure of Iron(II) Tetraphenyl Porphyrins

Unlike the metal atoms in the chalcogenides discussed in the previous section, the iron atoms in iron(II) tetraphenyl porphyrins and the related iron(II) phthalocyanine generally occupy crystallographic sites with symmetry $\bar{1}$ or 1 only. The exception is the iron site in iron(II) tetraphenyl porphyrin (FeTPP), space group $I\bar{4}2d$, which has $\bar{4}$ symmetry.

The electronic structure varies widely in the iron porphyrin family, and depends on the presence and the nature of axial ligands. When FeTPP is axially substituted by two pyridine ligands, the complex has a singlet ground state; but on tetrahydrofurane (THF) substitution, a quintet state is obtained. Unsubstituted FeTPP and iron(II) phthalocyanine have triplet ground states, reported variously as $^3B_{2g}$, ^3E, and ^3A. The reason for the ambiguity is that the electronic levels are closely spaced, which means that configuration interaction must be taken into account to obtain reliable theoretical results. On the other hand, the charge densities vary widely among the alternative configurations, so the X-ray method provides a suitable probe for resolving any controversy.

10.5.2 Experimental Results for Iron Porphyrins and Comparison with Theory

The deformation density in the plane of the porphyrin ring of *bis(pyridine)*(meso-*tetraphenylporphinato)iron(II)* (bPyFeTPP), shown in Fig. 10.8, shows density accumulation near the iron atom in the diagonal directions bisecting the iron–pyrrole nitrogen-bonds. This indicates preferential occupancy of the d_{xy} orbitals (Li et al. 1988, Mallinson et al. 1988),[2] as confirmed by the d-orbital population analysis (Table 10.7). This feature is quite generally observed for low- and intermediate-spin transition metal porphyrins, and reproduced in theoretical calculations, including those using density functional methods (Berkovitch-Yellin and Ellis 1981). But it does not occur in the *high-spin* bis-tetrahydrofurane complex discussed below.

Density accumulation is also evident between the nitrogen atoms and the transition metal atom, quite close to the former. This accumulation represents the ligand participation in the metal–ligand σ-bonding orbitals, and is as observed in the charge density maps of the metal sulfides.

The populations of the crystal-field-stabilized d_{xz}, d_{yz}, and d_{xy} orbitals are in very good agreement with results of an Extended Hückel (EH) calculation (third column of Table 10.7), which is perhaps unexpected, given the approximate nature of the calculation. Reasonable agreement is also obtained for d_{z^2}, but not for the destabilized $d_{x^2-y^2}$ orbital, for which the EH method overestimates the population

---

[2] The d-orbital population analysis was performed with both the harmonic and a more complete anharmonic thermal motion treatment, as discussed in section 10.7.3. The harmonic map is shown here.

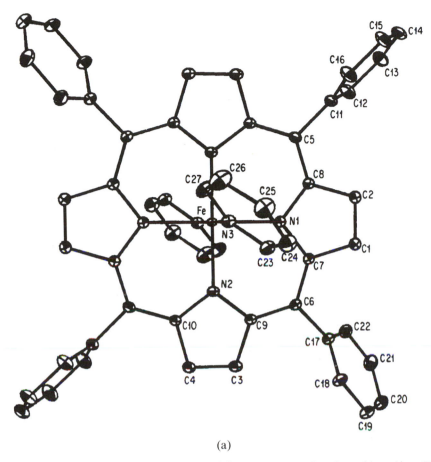

(a)

FIG. 10.8(a) Perspective drawing of bis(pyridine)(*meso*-tetraphenylporphinato)iron(II).
Source: Mallinson et al. (1988).

of the bonding orbital. The overestimate of the σ-donation from the ligand to the metal orbitals is a recurring result, characteristic for approximate calculations, which tend to underestimate the stability of the ligand orbitals. Since the ligand orbitals lie below the metal orbitals (Fig. 10.4), this reduces the energy difference between the two sets, and thus increases the σ-donation.

Bis(tetrahydrofurane)(meso-*tetraphenylporphinato*)*iron(II)* (bTHF FeTPP) is the only known six-coordinate high-spin Fe(II) complex. Its THF ligands are rather loosely bound. Crystals slowly lose THF when exposed to the atmosphere, while the iron is five-coordinate in a solution in benzene (Reed et al. 1980). The magnetic susceptibilty of bTHF FeTPP is temperature dependent (Lecomte et al. 1986) and the axial Fe—O bonds shrink on cooling from 1.35 Å at ambient temperature to 1.29 Å at nitrogen temperature.

The deformation density in the plane of the macrocyclic ligand, measured at 100 K (Fig. 10.9), differs radically from the corresponding map for bPy FeTPP.

CONTOUR INTERVAL = .10   E/A3

0 |1   2A

(b)

FIG. 10.8(b) Deformation electron density in the porphyrin plane in bis(pyridine)(*meso*-tetraphenylporphinato)iron(II). Contours are at 0.10 eÅ^{-3}. Negative contours are broken lines. First positive contour is 0.05 eÅ. *Source*: Li et al. (1988).

TABLE 10.7 *d*-Electron Orbital Populations in bis(Pyridine)(*meso*-phenylporphinato) iron(II) (the *z* axis is perpendicular to the molecular plane, the *x* and *y* axes point towards the ligands)

Orbital	Experimental	Extended Hückel Calculation[a]
$d_{x^2-y^2}$	0.35 (4.8%)	0.81 (11.0%)
d_{z^2}	1.05 (14.4%)	0.724 (9.8%)
d_{xz}, d_{yz}	1.93 (26.5%)	1.93 (26.1%)
d_{xy}	2.02 (27.7%)	1.99 (27.0%)
Total	7.29	7.40

[a] W. R. Scheidt, private communication.

FIG. 10.9 Deformation density in the plane through the iron atom and the pyrrole ring in bis(tetrahydrofurane)(*meso*-tetraphenylporphinato)iron(II) after averaging over the molecular *mmm* symmetry. Contours are at 0.5 eÅ^{-3}. *Source*: Lecomte et al. (1986).

The *d*-orbital peaks near the iron atom now occur in the direction of the pyrrole–nitrogen ligand atoms, and thus lie along the bonds, indicating the effect of covalence (σ-donation) superimposed on the cylindrical distribution of the d^6 ion.

Comparison of the orbital populations with the idealized ionic states (Table 10.8) shows reasonable agreement with the ${}^5E_g$ state. The depopulation of the $d_{xz, yz}$ orbitals and an excess population of $d_{x^2-y^2}$ relative to the ion is as expected from the σ-donation, π-back-donation concept applied to a high spin complex.

As noted above, the nature of the electron ground states of *iron(II) porphyrin* (FeP) and the related complex *iron(II) phthalocyanine* (FePc) have been controversial, with assignments ranging from ${}^3B_{2g}$ based on magnetic data (Barraclough et al. 1970) to ${}^3E_g$ and ${}^3A_{2g}$ from theoretical calculations and from techniques such as NMR and circular dichroism. Some of the theoretical results are summarized in Table 10.9. The calculations show that the spacing of the ${}^3A_{2g}$ and ${}^3E_g$ levels is only a few tenths of an electron volt (eV).

Experimental orbital populations for FePc, obtained at 110 K (Coppens and Li 1984), are given in Table 10.10, together with values for the ionic configurations. The main difference between the ${}^3E_g A$ and ${}^3A_{2g}$ states is a shift of one electron from the $d_{xz, yz}$ orbitals to the d_{z^2} orbital. The experimental populations are close to the almost 3:1 ratio of the $d_{xz, yz}/d_{z^2}$ populations predicted for ${}^3E_g$. Compared

TABLE 10.8 Iron Atom d-Orbital Populations in bis(Tetrahydrofurane) (*meso*-tetraphenylporphinato)iron(II). Axes as defined in Table 10.7

Symbol	Term				
	$^5B_{2g}$	$^5A_{1g}$	$^5B_{1g}$	^5E_g	Exp.
$d_{x^2-y^2}$	1 (16.7%)	1 (16.7%)	2 (33.3%)	1 (16.7%)	1.42 (24%)
d_{z^2}	1 (16.7%)	2 (33.3%)	1 (16.7%)	1 (16.7%)	1.04 (17.5%)
$d_{xz, yz}$	2 (33.3%)	2 (33.3%)	2 (33.3%)	3 (50%)	2.52 (42.6%)
d_{xy}	2 (33.3%)	1 (16.7%)	1 (16.7%)	1 (16.7%)	0.93 (15.7%)
Total	6	6	6	6	5.92

TABLE 10.9 Theoretical Results for the $(^3E_g - {}^3A_{2g})$ Energy Difference for Iron(II) Porphyrin

Reference	Type[a]	$\Delta E(^3E_g - {}^3A_{2g})$ (eV)
Zerner and Gouterman (1966), Zerner et al. (1966)	EH	Negative
Obara and Kashiwagi (1982)	SCF	0.32
Obara and Kashiwagi (1982)	SCF after configuration mixing	0.08
Sontum et al. (1983)	SCF–Xα	0.2
Rohmer (1985)	SCF–CI	0.27
Edwards et al. (1986)	INDO	0.27
Edwards et al. (1986)	INDO–CI	0.03
Rawlings et al. (1985a, 1985b)	SCF	0.29
Rawlings et al. (1985a, 1985b)	SCF–CI	−0.47

[a] EH: Extended Hückel, SCF: Self-Consistent Hartree Fock, INDO: Intermediate Neglect of Differential Overlap, CI: Configuration Interaction.

TABLE 10.10 d-Electron Orbital Population in Iron(II) Phthalocyanine. Axes as defined in Table 10.7

Symbol	Term				
	$^3E_g A$	$^3A_{2g}$	$^3B_{2g}$	$^3E_g B$	Exp. X-ray
$d_{x^2-y^2}$					0.70 (7) (12.9%)
d_{z^2}	1 (17%)	2 (33%)	1 (17%)	2 (33%)	0.93 (6) (17.1%)
$d_{xz, yz}$	3 (49%)	2 (33%)	4 (67%)	3 (49%)	2.12 (7) (39.1%)
d_{xy}	2 (33%)	2 (33%)	1 (17%)	1 (17%)	1.68 (10) (30.9%)

with ^3E_g there is a depopulation of the $d_{xz,yz}$ and an excess population of $d_{x^2-y^2}$, as observed for bTHF FeTPP, and, again as expected, from covalent σ-donation, π-back-donation. Theoretical results on high-spin six-coordinate bis-NH_3FeP (Rawlings et al 1985a, b) show a similar effect.

The distinction between the two states is readily illustrated in a deformation density section perpendicular to the porphyrin plane through the iron atom. Because of the transfer of an electron into the d_{z^2} orbital in going from ^3E_gA, and to $^3A_{2g}$, the former configuration shows a deficiency and the latter an excess of density above and below the nitrogen atom. This is confirmed by theoretical deformation density maps (Fig. 10.10) (Rohmer 1985). The experimental map for FePc shows the deficiency along the z axis, as expected from the d-orbital populations listed in Table 10.10.

On the other hand, the experimental results on (*meso-tetraphenylporphinato*)-*iron(II)* (FeTPP) (Li et al. 1990) are in reasonable accord with the $^3A_{2g}$ assignment, particularly with regard to the crucial d_{z^2} occupancy (Table 10.11). The population of the $d_{x^2-y^2}$ orbital is small, in agreement with the SCF–CI results, indicating a smaller metal contribution to the σ-bonding orbital compared with FePc. This is compatible with the lengthening of the Fe—N (pyrrole) bond length to 1.967 Å from 1.927 Å in FePc. We note that the d_{xy} population is significantly lower than predicted by the calculations, and that the Extended Hückel results again overemphasize σ-covalence relative to the ab-initio calculations and the experimental results.

It is evident that FeTPP and FePc have different ground states in the crystalline modifications studied. Apart from the different Fe—N bond lengths, there is evidence that intermolecular interactions play a role. Theoretical studies of substituted porphyrins provide evidence for the strong sensitivity of the ground state to axial ligation (Mispelter et al. 1980). In monoclinic FePc, *meso*-nitrogen atoms (i.e., nitrogen atoms in the bridging bonds) of neighboring molecules are located at 3.42 Å above and below the iron atoms. They constitute an axial "pseudoligand." In tetragonal FeTPP, the molecular planes are perpendicular to the $\bar{4}$ axis of the space group $I\bar{4}2d$, and no such intermolecular approach exists.

TABLE 10.11 *d*-Electron Orbital Populations in (*meso*-Tetraphenylporphinato)iron (II). Axes as defined in Table 10.7

| | | Theoretical (Nowack et al. 1991) | | |
	Experimental	$^3A_{2g}$–SCF	^3E_gA–SCF	EH
$d_{x^2-y^2}$	0.24 (15) (3.9%)	0.18 (2.9%)	0.18 (2.9%)	0.90 (12.8%)
d_{z^2}	2.10 (14) (33.7%)	1.94 (31.7%)	0.99 (16.1%)	1.07 (15.3%)
$d_{xz,yz}$	2.28 (18) (36.5%)	1.98 (32.3%)	2.96 (48.1%)	3.05 (43.5%)
d_{xy}	1.52 (15) (25.9%)	2.02 (33.0%)	2.02 (32.8%)	1.99 (28.4%)
Total	6.24 (31)	6.12	6.15	7.02

FIG. 10.10 Computed static deformation density map of FeP. Contours at $0.10\,\mathrm{e\mathring{A}^{-3}}$. (a) $^3A_{2g}$ in plane, (b) ^3E_g in plane.

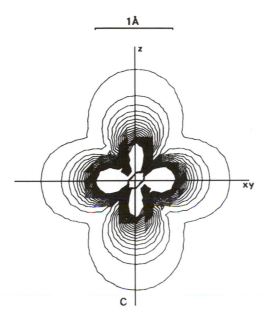

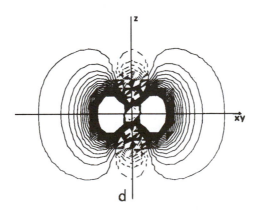

FIG. 10.10 *continued* (c) $^3A_{2g}$ bisecting plane, (d) ^3E_g bisecting plane. Reprinted from M. M. Rohmer, *Chem. Phys Lett.* **116** 44 (1985) with permission from Elsevier Science-NL, Sara Burgerhartstraat 25, 1055 KV, Amsterdam, The Netherlands.

10.6 The Electron Density in Metal–Metal Bonds of Transition Metal Complexes

10.6.1 Metal–Metal Bonding

Metal–metal bonding in transition metal complexes of low nuclearity (i.e., with only a few metal atoms) tends to be more directed and therefore stronger than the bonding in metals discussed in chapter 11. Accordingly, the metal–metal bonds in transition metal complexes are often localized and considerably shorter than those in most extended solids. Charge accumulations are frequently observed in metal–metal bonding regions of deformation density maps.

Many of the currently available studies of metal–metal bonding were completed before the multipole model and the topological analysis of the total density were fully developed. For this reason, the discussions reported below focus on the deformation density distributions, and their comparison with theoretical results, though a more quantitative analysis is now possible and would be of considerable interest.

10.6.2 Bonding Between Chromium Atoms

Quadruply bonded Cr—Cr compounds like the dichromium tetracarboxylates have metal–metal bond lengths which vary by as much as 0.7 Å between compounds (Cotton and Stanley 1977, Cotton 1978). A "pure" quadruple bond is a $\sigma^2\pi^2\delta^2$ bond, with one σ ($d_{z^2} - d_{z^2}$), two π ($d_{xz} - d_{xz}$ and $d_{yz} - d_{yz}$), and a δ ($d_{xy} - d_{xy}$) component, z being selected along the Cr–Cr axis. Theoretical calculations show that the formal bond order of four can be drastically reduced by strong correlation effects involving excited-state configurations with antibonding character for the Cr—Cr interaction (Bénard 1978a). For dichromium tetraformate, calculated with a double ζ basis set for the transition metal valence shells and a minimal basis set for the other orbitals, the noncorrelated SCF wave function is found to have very small weight (18%) in the multideterminant expansion of the CI (configuration interaction) wave function. Large contributions of antibonding Cr—Cr orbitals from the additional Slater determinants, added to the wave function, tend to reduce the strength of the Cr—Cr bond. The Cr—Cr pairwise potential is found to be very shallow when plotted as a function of the Cr—Cr distance, thus explaining the large variation of observed Cr—Cr distances.

The CI deformation density maps have a maximum of about 0.1 eÅ^{-3} in the Cr—Cr bond region, compared with 0.2 eÅ^{-3} according to the SCF calculation. The overall theoretical charge distribution is in good qualitative agreement with the experimental results on dichromium tetraacetate dihydrate, $[Cr(CH_3COO)_2] \cdot 2H_2O$, averaged over equivalent regions (Fig. 10.11) (Bénard et al. 1980), except on the bond axis, where differences may be due to the absence of an axial ligand in the complex on which the calculation was performed. The Cr—Cr region does not show a peaked accumulation of the electron density, but there is a broad area of excess electron density off the bond axis, compatible with overlap density between diffuse d orbitals. The accumulations are statistically significant in view of the

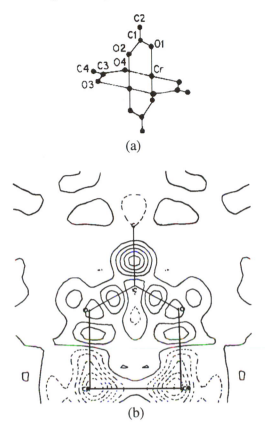

(a)

(b)

FIG. 10.11 Electron density in the metal–ligand plane of dichromium tetraacetate. (a) Molecular diagram, (b) deformation density through Cr—Cr and the acetyl group averaged over equivalent regions. Contours are at $0.10\ e\text{Å}^{-3}$. Negative contours are broken lines. *Source*: Bénard et al. (1980).

estimated SD of $0.02\ e\text{Å}^{-3}$ in the averaged maps, and they support the presence of π- and δ-bonding between the chromium atoms.

The Cr—Cr bond length in dichromium tetraacetate dihydrate is 2.362 (1) Å (Cotton et al. 1971), compared with 2.498 Å in Cr metal. A much shorter Cr—Cr bond of 1.879 Å length exists in tetrakis (μ-2-hydroxy-6-methylpyridine)dichromium, $Cr_2(mhp)_4$; this bond has been labeled "supershort." The bulky bridging ligands in this complex are coordinated through the pyridine nitrogen to one chromium atom, and through the hydroxy oxygen atom to the second chromium atom (Fig. 10.12). The existence of methyl groups in the 6 position of the hydroxypyridine ring limits access to the chromium atoms, and thereby prevents axial ligation, which appears to strengthen the metal–metal bond. The deformation density maps show a striking accumulation of density between the Cr atoms, with a maximum at bond midpoint, and extension over the π and δ regions. The peak heights (experimental, $0.4\ e\text{Å}^{-3}$; theoretical for $[H_2P(CH_2)_2]_4Cr_2$, $0.3\ e\text{Å}^{-3}$) are large and comparable to bonds between first-row atoms.

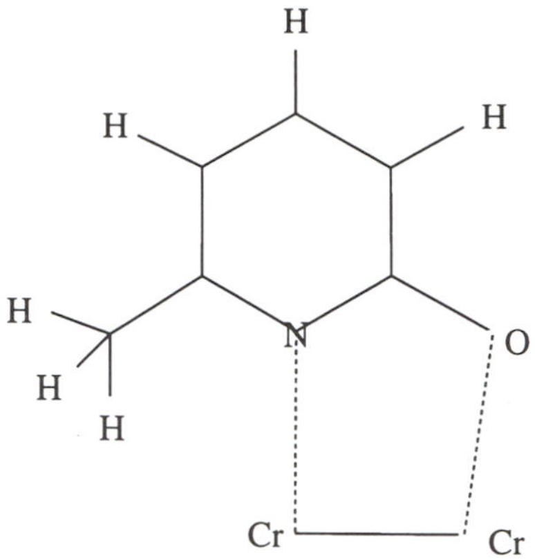

FIG. 10.12 The 2-hydroxy-6-methylpyridine ligand its bonding to the Cr atoms in tetrakis (μ-2-hydroxy-6-methylpyridine)dichromium, $Cr_2(mhp)_4$.

10.6.3 Mn—Mn and Fe—Fe Bonding

Dimanganese decacarbonyl, $Mn_2(CO)_{10}$, is a simple dimeric compound with a metal–metal bond unsupported by bridging ligands. The existence of the metal–metal bond satisfies the 18-electron rule and accounts for the diamagnetism of the complex. The X—X deformation density, based on both AgKα and MoKα data collected at 78 K, shows only a very diffuse maximum of 0.05 eÅ^{-3} around the bond midpoint, not significantly different from zero, given the experimental standard deviation (Martin et al. 1982). While accumulation in the standard deformation density is not a good criterion to judge bond strength, especially for atoms with more than half-filled valence shells (see chapter 5), the weakness of the bond is supported by HFS—Xα theoretical calculations, which show the populations in $Mn_2(CO)_{10}$ to be very similar to those in the $Mn(CO)_5$ fragment (Heijser et al. 1980). Density accumulation in the metal–metal bond, with a peak height ≈ 0.1 eÅ^{-3}, becomes visible in the theoretical *fragment deformation map*, in which the density of two $Mn(CO)_5$ fragments is subtracted from the total density. The weak bonding is found to be entirely due to interaction between the $3d_{z^2}$ and $4p_z$ metal orbitals.

A similar lack of density accumulation in the metal–metal bonding region in the X-ray standard deformation density map has been observed for (μ-methylene) bis[dicarbonyl(η^5-cyclopentadienyl)manganese], (μ-CH$_2$) [CpMn(CO)$_2$]$_2$ (Clemente et al. 1982). In the homologous complexes $Fe_2(CO)_9$ and $Co_2(CO)_8$, the theoretical results show the bonding to arise entirely from interactions through the bridging ligands.

A complex with an Fe—Fe bond for which the electron density is available is bis(dicarbonyl-π-cyclopentadienyl iron), [C$_5$H$_5$Fe(CO)$_2$]$_2$ (Fig. 10.13) (Mitschler et al. 1978). The 18-electron rule again requires only a single bond to explain the observed diamagnetism. The combined X-ray and neutron study indicates a complete absence of density accumulation in the metal–metal bonding region of the standard deformation density, in agreement with an SCF theoretical density,

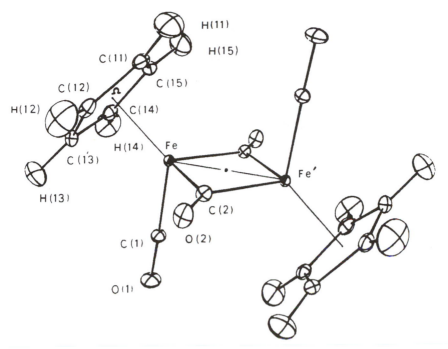

FIG. 10.13 Diagram of the bis(dicarbonyl-π-cyclopentadienyl iron) molecule at 74 K. The 50% probability ellipsoids are shown. *Source*: Mitschler et al. (1978).

and similar to the results for dimanganese decacarbonyl. The Mulliken population analysis of the SCF wave functions gives a small negative value for the overlap population between the iron atoms, and thus supports a multicenter metal–metal interaction involving the two bridging carbonyl groups, rather than a direct linkage between the metal atoms (Bénard 1978b).

10.7 Anharmonic Thermal Motion and the Bonding Anisotropy of Transition Metal Complexes

10.7.1 Atomic Asphericity and Anharmonic Thermal Motion

Though the anharmonic components of the thermal motion decrease rapidly with temperature, as described in chapter 2, they will be present to some extent even if the motion is reduced to zero-point vibrations.

The vibrational displacements corresponding to the anharmonic terms in the potential are most pronounced in the directions away from the stronger bonding interactions, in which restoring forces are weaker. Thus, for the tetrahedral site symmetry of the diamond structure, the anharmonicity causes a larger mean-square displacement in directions opposite to the covalent bonds. At lower temperatures, at which the bonding anisotropy dominates the asphericity, this will reduce the asphericity of the thermally averaged charge density distribution, as discussed further in chapter 11.

In low-spin transition metal complexes, the preferential occupancy of the d orbitals in the crystal field tends to create excess density in the voids between the bonds, which means that anharmonicity tends to *reinforce* the electron density asphericity. We will discuss, in the following sections, to what extent the two effects can be separated by combined use of aspherical atom and anharmonic thermal motion formalisms.

10.7.2 A Model Study on Theoretical Structure Factors of $Fe(H_2O)_6$

Mallinson et al. (1988) have performed an analysis of a set of static theoretical structure factors based on a wave function of the octahedral, high-spin hexa-aquairon(II) ion by Newton and coworkers (Jafri et al. 1980, Logan et al. 1984). To simulate the crystal field, the occupancy of the orbitals was modified to represent a low-spin complex with preferential occupancy of the t_{2g} orbitals, rather than the more even distribution found in the high-spin complex. The complex ion (Fig. 10.14) was centered at the corners of a cubic unit cell with $a = 10.000$ Å and space group $Pm3$. Refinement of the 1375 static structure factors ($\sin \theta/\lambda \leq 1.2$ Å^{-1}) gave an agreement factor of $R = 4.35\%$ for the spherical-atom model with variable positional parameters (Table 10.12). Addition of three anharmonic thermal

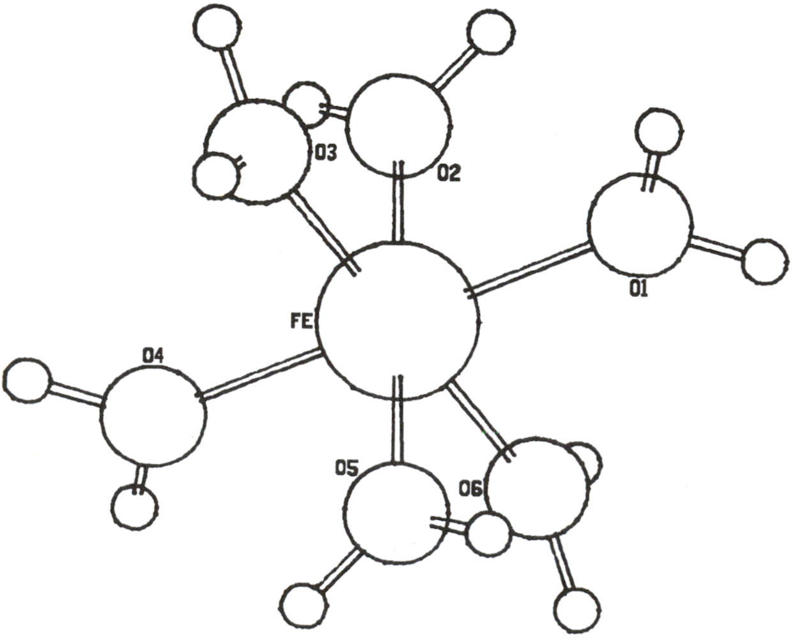

FIG. 10.14 Structure of the hexaaquairon(II) ion. *Source*: Mallinson et al. (1988).

TABLE 10.12 Refinements of Static Theoretical Structure Factors of the Hexaaquairon(II) Ion

	Spherical, $U_{ij} = 0$	Spherical, Anharmonic Fe	Aspherical Fe, $U_{ij} = 0$	Aspherical Fe, Anharmonic Fe
N_0	1375	1374	1375	1375
N_v	3	6	6	9
$wR(F)$	0.0435	0.0417	0.0401	0.0398

parameters for the iron atom (U_{11}, c^{1111}, and c^{1122}; other parameters being related by the symmetry of the cubic site) reduced the R factor to 4.17%, compared with $R = 4.01\%$ for the multipole model with the three variables κ'', P_{00}, and the population of the cubic harmonic K_{44+} [$= 0.78245d_0 + 0.57939d_{44}$, see appendix D, section D.3(c)], and no thermal parameters. Though the multipole model gives

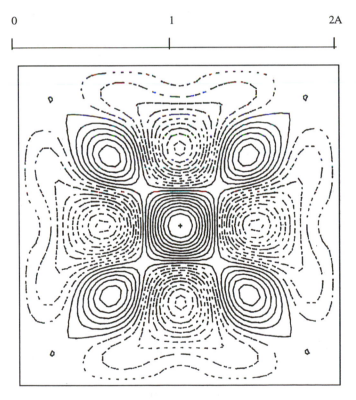

FIG. 10.15(a) Model maps in the FeO_4 plane in low-spin $[Fe(H_2O)_6]^{2+}$, obtained by subtraction of the calculated structure factors from a conventional refinement of theoretical structure factors from those of a refinement in which the Fe atom is treated with the multipole formalism, and subsequent Fourier transformation. Contours are at $0.20\ e\text{Å}^{-3}$. Based on theoretical data. First positive contour is at $0.1\ e\text{Å}^{-3}$. Negative contours are broken lines. Oxygen atoms are at 2.131 Å from the iron atom in horizontal and vertical directions. *Source:* Mallinson et al. (1988).

FIG. 10.15(b) Model maps in the FeO_4 plane in low-spin $[Fe(H_2O)_6]^{2+}$, obtained by subtraction of the calculated structure factors from a conventional refinement of theoretical structure factors from those of a refinement in which the Fe atom is treated anharmonically, and with subsequent Fourier transformation. Based on theoretical data. Contours are at 0.20 eÅ^{-3}. First positive contour is at 0.1 eÅ^{-3}. Negative contours are broken lines. *Source*: Mallinson et al. (1988).

the better fit, the model deformation maps for the two refinements are remarkably similar (Fig. 10.15), showing that the gross features of the electron asphericity can be mimicked well by the anharmonic thermal vibrational model.

A final refinement of the static data with both sets of parameters gave a further small decrease in the R factor to 3.98%, and, reassuringly, temperature parameters smaller than three times their estimated standard deviations. At least for static data, the refinement properly attributes the asphericity to the multipole functions.

10.7.3 An Experimental Example: Anharmonic Refinement of bis(Pyridine) (*meso*-Tetraphenylporphinato)Iron(II)

Bis(pyridine)(*meso*-tetraphenylporphinato)iron(II), discussed in section 10.5.2, was reanalyzed to evaluate the importance of anharmonic motion in nitrogen-temperature transition metal studies. A number of different refinements are summarized in Table 10.13. It is striking that treating the Fe atom as spherical

TABLE 10.13 Summary of Refinements of
bis(Pyridine)(*meso*-tetraphenylporphinato)iron(II)

	I Spherical, Harmonic	II Spherical Anharmonic[a]	III Aspherical Harmonic[b]	IV All atoms Aspherical, Anharmonic[c]
N_0	8497	8497	8497	8497
N_v	353	349	368	489
$wR(F)$	0.0416	0.0365	0.0398	0.0277

[a] Anharmonic in this table refers to the Fe atom only; harmonic temperature parameters of the other atoms are fixed at values from refinement I.

[b] Only the Fe atom is treated with an aspherical valence charge distribution.

[c] Starting with anharmonic parameters from the high-order refinement, as described in the text.

with anharmonic thermal motion (refinement II) gives a lower agreement factor than the converse treatment in which the Fe asphericity is taken into account, but thermal motion is restricted to be harmonic (refinement III), even though the number of parameters is larger in the latter case.

In a subsequent refinement, neutral spherical-atom parameters, with the symmetry-allowed Gram–Charlier anharmonic parameters for Fe, were fitted to 3105 high-angle reflections with $\sin \theta / \lambda \geq 0.8$ Å^{-1}, even though some effects of atomic asphericity will persist beyond this cut-off. A final refinement of all multipole and Gram–Charlier parameters (IV), starting with the Fe anharmonicity of the high-order refinement, converged satisfactorily. As expected, the introduction of the anharmonicity reduces the occupancy of the crystal-field stabilized orbitals. In this example, the population of d_{xy} decreases from 2.02 to 1.82 and that of $d_{xz,yz}$ from 3.86 to 3.38 electrons (Table 10.14). Qualitatively, the conclusions of the harmonic treatment are not affected, particularly when the percentage occupancies are considered. Nevertheless, for quantitative analysis of data collected at liquid-nitrogen temperatures, the deconvolution of anharmonic thermal motion and

TABLE 10.14 Iron *d*-Orbital Populations
(Electrons) and Percentages of the Total
Population of bis(Pyridine)(*meso*-tetra-
phenylporphinato)iron(II) Without and With
Anharmonic Treatment of the Fe Atom. Axes
as Defined in Table 10.7

	Harmonic	Anharmonic
$d_{x^2-y^2}$	0.35 (4.8%)	0.39 (6.1%)
d_{z^2}	1.04 (14.4%)	0.75 (11.8%)
d_{xz}, d_{yz}	3.86 (53.0%)	3.38 (53.3%)
d_{xy}	2.02 (27.7%)	1.82 (28.8%)

transition metal anharmonicity requires attention. It is important to note that the effect of anharmonicity on the charge distribution will be more serious when the core electrons constitute a larger fraction of the total distribution, as is the case for second- and third-row transition metal atoms.

Collection of diffraction data at liquid-helium temperatures is important to reduce thermal motion and its anharmonicity. Similarly, the use of shorter wavelengths at such low temperatures makes data at higher values of $\sin \theta/\lambda$ accessible, which facilitates deconvolution of thermal motion and bonding effects. Both very low temperatures and hard radiation are becoming more readily available, and are expected to play a crucial role in future studies.

11

The Charge Density in Extended Solids

Extended solids encompass all solids in which no well-defined molecular entities can be distinguished. This is the case for metals and alloys, covalently bonded solids like diamond and silicon, and ionic crystals of which the alkali halides are prototypes. Intermediate cases are common, such as crystals consisting of a charged covalent network with counterbalancing cations or anions. Silicates and their analogues are a prime example of often charged networks with partially covalent bonding. An increasing number of solids are known in which both an extended framework and molecular entities exist, with the molecules being embedded in the extended framework. Graphite intercalation compounds and a variety of host/guest complexes are examples of this class.

The bonding features in the charge density are pronounced in crystals with extended covalent networks. The availability of perfect silicon crystals has allowed the measurement of uncommonly accurate structure factors, of millielectron accuracy. The data have served as a test of experimental formalisms for charge density analysis, and at the same time have provided a stringent criterion for quantum-mechanical methods.

We will start the discussion in this chapter with silicon and its analogues, diamond and germanium, and proceed with the treatment of silicates, and metallic and ionic crystals.

11.1 Covalently Bonded Extended Solids

11.1.1 Silicon

11.1.1.1 The Structure Factor Formalism for the Diamond-Type Structures

In the face-centered cubic structure of silicon, atoms are located at 1/8 1/8 1/8 and at the center-of-symmetry related position of $-1/8$ $-1/8$ $-1/8$. The static structure factor can therefore be expressed simply as

$$F(hkl) = 4f_{Si}^{A} \exp 2\pi i\left(\frac{h+k+l}{8}\right) - 4f_{Si}^{B} \exp -2\pi i\left(\frac{h+k+l}{8}\right) \quad (11.1)$$

where f_{Si}^{A} and f_{Si}^{B} are the scattering factors of the two center-of-symmetry-related Si atoms. In the spherical atom approximation, $f_{Si}^{A} = f_{Si}^{B}$, and Eq. (11.1) reduces to

$$F(hkl) = 8f_{Si} \cos 2\pi\left(\frac{h+k+l}{8}\right) \quad (11.2)$$

It is clear that in this approximation $F(hkl)$ equals zero for all reflections for which $h + k + l = 4n + 2$. This is the reason that the observation of the (222) reflection of diamond led Bragg to conclude that bonding effects are detectable by X-ray diffraction (see chapter 3). If the Si atoms are not spherical, and their density contains antisymmetric components, such as dipolar or octupolar valence density functions, f_{Si}^{A} will be the complex conjugate of f_{Si}^{B} [1] and Eq. (1.12) is no longer valid. We can write $f_{Si}^{A} = f_c + if_a$ and $f_{Si}^{B} = f_c - if_a$, where c stands for the symmetric and a for the antisymmetric component of the atomic rest density. This gives

$$F(hkl) = 8f_c \cos 2\pi\left(\frac{h+k+l}{8}\right) - 8f_a \sin 2\pi\left(\frac{h+k+l}{8}\right) \quad (11.3)$$

which leads to nonzero intensity for the $h + k + l = 4n + 2$ reflections.

As first shown by Dawson (1967), Eq. (11.3) can be generalized by inclusion of anharmonicity of the thermal motion, which becomes pronounced at higher temperatures. We express the anharmonic temperature factor of the diamond-type structure [Chapter 2, Eq. (2.45)] as $T(\mathbf{H}) = T_c(\mathbf{H}) + iT_a(\mathbf{H})$, in analogy with the description of the atomic scattering factors. Incorporation of the temperature

---

[1] This is no longer strictly true when resonance scattering is taken into account, which adds the same imaginary component to both atomic scattering factors. In the case of silicon, it is straightforward to correct the structure factors for anomalous scattering, using reasonably accurate temperature parameters.

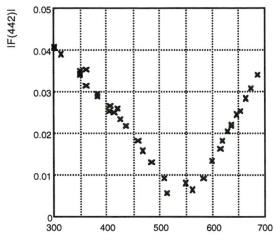

FIG. 11.1 Temperature dependence of the magnitude of the (442) structure factor of silicon. *Source*: Data from Tischler (1983).

parameters into Eq. (11.1), and writing out all terms, gives

$$F(hkl) = 8f_c\left[T_c \cos 2\pi\left(\frac{h+k+l}{8}\right) - T_a \sin 2\pi\left(\frac{h+k+l}{8}\right)\right]$$

$$- 8f_a\left[T_a \cos 2\pi\left(\frac{h+k+l}{8}\right) + T_c \sin 2\pi\left(\frac{h+k+l}{8}\right)\right] \quad (11.4)$$

With $T_a = 0$ and $T_c = 1$, Eq. (11.3) is retrieved as required.

For the $h + k + l = 4n + 2$ reflections, this becomes

$$F(h + k + l = 4n + 2) = (-1)^{n+1}(f_c T_a + f_a T_c) \quad (11.5)$$

from which it is clear that the intensity of reflections such as (222) and (442) is entirely due to bonding effects and anharmonic thermal motion. The chemical bonding leads to an accumulation of overlap density into the tetrahedrally arranged covalent bonds, but the thermal displacements are larger in the directions opposite the covalent bonds, as the stretching of the bonds is easier than their compression. The quantities f_a and T_a therefore have opposite signs. Because thermal motion increases with temperature, the two contributions to Eq. (11.5) cancel at a "crossover" temperature. This is elegantly illustrated by a series of measurements of the temperature dependence of the $(4n + 2)$ reflections of silicon by Batterman and coworkers (Roberto and Batterman 1970, Trucano and Batterman 1972, Tischler 1983, Tischler and Batterman 1984). Tischler and Batterman (1984) measured the temperature dependence of the (442) and (622) reflections of Si and Ge, using synchrotron radiation. Their results for Si-(442) are plotted in Fig. 11.1. The balance between the two effects that causes the intensity minimum at about 525 K is evident.

For higher-order "forbidden" reflections, the bonding effects will be less and the thermal effects larger, so the crossover will shift to lower temperatures, or completely disappear. The latter is the case for Si-(622) (sin $\theta/\lambda \approx 0.61$ Å^{-1}), the integrated intensity of which is found to increase monotonically with temperature. Analysis of the temperature dependence leads to a bond charge contribution $f_a(622)$ of only $+0.0088 \pm 0.0011$ electrons, an order of magnitude smaller than the value for (442).

The appearance of reflections in the diffraction pattern due to anharmonicity of thermal motion is not limited to the diamond-type structures, and is observed, for example, for the A15-type structure of the low-temperature superconductor V_3Si (Borie 1981), and for zinc (Merisalo et al. 1978). It has been described as *thermal excitation of reflections*, though no excitation in the spectroscopic sense of the word is involved.

11.1.1.2 Experimental Structure Factors

The exceedingly accurate structure factors on silicon are obtained with perfect crystals. According to dynamical theory of diffraction, the interference between the incident and Bragg reflected waves in a perfect crystal creates a standing wave, which is visible as alternating dark and light *fringes*, with a spacing inversely proportional to the structure factor of the Bragg reflection. The first structure factor measurements based on *Pendellösung fringes* were made with wedge-shaped perfect crystals (which must be differently cut for each reflection!) by Hattori *et al.* in 1965, followed by more accurate measurements at a level described as $\approx 0.1\%$ by Tanemura and Kato (1972) and Aldred and Hart (1973). Somewhat later work by Teworte and Bonse (1984) used intensity fluctuations in high-precision double-crystal rocking curves. Saka and Kato (1986, 1987) rotated a flat crystal in the symmetric Laue geometry around an axis in the plane of the crystal, to vary the effective crystal thickness and therefore the diffracted intensity.

The five room-temperature and two liquid-nitrogen temperature data sets obtained in these experiments were reanalyzed by Cummings and Hart (1988). Where necessary, they introduced improved corrections for residual strain, resonance, and nuclear scattering to arrive at a set of mean structure-factor values with typical errors of 3–5 millielectrons. This is extremely good, but as noted, for some reflections the errors are significantly larger than the 0.1% claimed earlier. Additional accurate values for individual reflections are available for γ-ray (Alkire et al. 1982), X-ray (Roberto and Batterman 1970, Trucano and Batterman 1972), and synchrotron X-ray (Tischler 1983, Tischler and Batterman 1984), measurements. Combined, these measurements provide a superbly accurate set for analysis of the charge density in silicon.

The structure factors on diamond and germanium are less accurate, but are adequate for comparison with the silicon results. For diamond, Pendellösung (Takama et al. 1990) and powder data (Göttlicher and Wölfel 1954) are available, while for germanium, Pendellösung observations (Tamaka and Sato 1981, Deutsch et al. 1990) and reflection-profile based measurements of the structure factors (Matshushita and Kohra 1974) have been made. The relative merits of the different sets have been discussed by Lu, Zunger, and Deutsch (Lu and Zunger 1992, Lu et al. 1993).

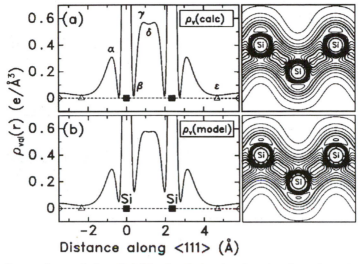

FIG. 11.2 Comparison of the ab-initio local density functional static valence density for Si (top) with the model valence density based on the structure factor compilation by Cummings and Hart (1988) (bottom half of the figure). Contour interval is $0.05 \, \text{e} \, \text{Å}^{-3}$. *Source*: Lu and Zunger (1992), Lu et al. (1993).

11.1.1.3 Analysis of the Density and Comparison with Theory

Two detailed analyses of the Si structure factors are available. The first, by Spackman (1986), was made before the consolidated list of reflections of Cummings and Hart (1988) became available. The second, by Deutsch (1992) and by Lu et al. (1993) is a comprehensive analysis of the available data, not only on silicon, but also on diamond and germanium, and includes a comparison with a large number of theoretical calculations. Notwithstanding the general agreement between the two analyses, there are differences in detail. While Spackman corrects for the effect of anharmonicity on the structure factors before calculating valence and deformation density maps, Lu et al. found no evidence for anharmonicity in their refinement of the Si (and C) room-temperature data, which in both cases excluded the anharmonicity-sensitive $(4n + 2)$ reflections, except (222). As a low-order reflection, (222) is less sensitive to anharmonic effects. Notwithstanding the difference in the treatments, the features in the density maps are generally the same in the two studies. The bond peak in the valence density is elongated along the bond direction and shows a double-maximum "camel back" feature with a slight dip between the maxima (Fig. 11.2). This feature is also observed in diamond and germanium. On the other hand, as a result of the subtraction of the spherical atomic valence densities, the bond peak in the deformation density is elongated perpendicular to the bond axis (Fig. 11.3).

There is almost quantitative agreement between the experimental model valence density (lower half of Fig. 11.2) and the result of an ab-initio local density functional calculation (upper part of Fig. 11.2). This agreement is also evident in

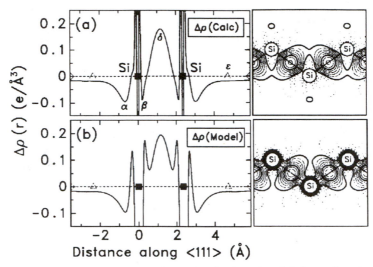

FIG. 11.3 Comparison of the ab-initio local density functional deformation density for Si with the experimental static model deformation density. Contour interval is $0.025 \, \text{e} \, \text{Å}^{-3}$. Negative contours are dashed lines. *Source*: Lu and Zunger (1992), Lu et al. (1993).

reciprocal space. The *R*-factor between theory and experiment (after correction for isotropic thermal motion) for the 18 reflections is only 0.21%, with an rms deviation of $0.012 \, \text{e} \, \text{atom}^{-1}$. For C and Ge, the agreement is poorer, with rms deviations of 0.017 and $0.170 \, \text{e} \, \text{atom}^{-1}$, respectively, reflecting the lower quality of the data sets. It is noteworthy that, for Ge, use of *relativistic* scattering factors produces a superior fit to the theoretical structure factors.

The valence *M*-shell of silicon is found to expand by about 6% ($\kappa_M = 0.9382$), in agreement with the results of a much earlier aspherical atom refinement, which gave $\kappa = 0.956$ (9) (Hansen and Coppens 1978). Lu, Zunger and Deutsch (1993) (LZD) also conclude that the core *L*-shell is expanded by 0.5% ($\kappa_L = 0.9949$), the first such observation of the effect of bonding on the inner electrons. However, these results are not independent of the treatment of the temperature parameters. In the LZD analysis, the $K + L$ and M shells are assigned separate isotropic thermal parameters; the latter are found to be significantly smaller with $B_{\text{core}} = 0.4585 \, \text{Å}^2$ (no standard deviations are given), and $B_{\text{valence}} \leq 0.11 \, \text{Å}^2$, leading the authors to the conclusion that the atoms do not vibrate rigidly.

That the vibrational displacements of the valence shell electrons may be smaller than those of the core electrons can be qualitatively understood by considering the vibrations of two identical, strongly bonded atoms. When the atoms vibrate in phase, they behave as a rigid body, so all shells will vibrate equally. But when they vibrate out of phase, the density near the center of the bond will be stationary, assuming the average static overlap density to be independent of the vibrations. This apparently invariant component of the valence density would contribute to a lowering of the outer-shell temperature parameter. Adequate separation of this effect from the valence shell expansion due to chemical bonding will likely require accurate studies at a number of temperatures.

11.1.2 Charge Density Studies of Silicates

11.1.2.1 The Structure of Silicates

Silicates comprise more than 95% by weight of the earth's crust and mantle, and are widely used in glasses, ceramics, sieves, catalysts, and electronic devices. Crystals of silicates are often hard, and may show considerable extinction in their diffraction pattern, which means not only that small samples must be used, but also that ambient temperatures may be adequate for charge density studies.[2]

Silicates form a large family of compounds and therefore provide a fertile ground for comparative studies. The understanding of the structure of silicates was one of the early triumphs of X-ray crystallography (Bragg et al. 1965). Orthosilicates like Mg_2SiO_4 contain negatively charged isolated SiO_4 tetrahedra, while pyroxenes like $Mg_2Si_2O_6$ and $LiAlSi_2O_6$ contain chains of edge-sharing SiO_4 tetrahedra, represented by the formula $n(SiO_3)^{2-}$. The chains are linked sideways into ribbons in the amphiboles, and into sheets with a hexagonal-type network of composition $n(Si_2O_5)^{2-}$, in "flaky" minerals such as talc, $Mg_3(Si_4O_{10})(OH)_2$. The many known modifications of SiO_2, including quartz, coesite, cristobalite, and tridymite, are framework silicates, in which every oxygen atom links two four-coordinated silicon atoms. Substitution of Si^{4+} by Al^{3+} in the tetrahedra is widespread, and fully described in the literature. In the sheet structures, it leads to minerals such as phlogopite $KMg_3(AlSi_3O_{10})(OH)_2$, and mica $KAl_2(AlSi_3O_{10})(OH)_2$, in the framework structures to feldspars, like sanidine $KAlSi_3O_8$, and zeolites, of which leucite $KAlSi_2O_6$ and natrolite $Na_2(Al_2Si_3O_{10})\cdot 2H_2O$ are examples. Substitution by B^{3+} similarly leads to the borosilicates.

The "Resource Book of Crystal Structures" by Mak and Zhou (1992) includes a description of the rules governing silicate structure, as well as a number of detailed examples, to which the reader is referred for necessary detail.

11.1.2.2 Charge Density Studies

As early as 1939, Pauling, on the basis of electronegativity differences between Si and O, came to the conclusion that each Si—O bond, rather than being purely ionic, has a $\approx 50\%$ covalent character. Notwithstanding this early insight, the detailed nature of the Si—O bond remains a subject of discussion. A recent comprehensive review (Gibbs et al. 1994) summarizing theoretical and experimental geometry and charge density studies on silicates bears the title "The Elusive SiO Bond".

---

[2] As noted below, the temperature parameters for zeolites are often considerably larger than those for simple silicates. Disorder may contribute to this difference.

The study of the nature of the Si—O bond through analysis of its charge density is based on net ionic charges, heights of peaks in atom deformation density maps, and, more recently, topological analysis of the total charge density.

There is considerable spread in the reported net charges on Si and O, in part because of variations in bonding, but also because basis functions may vary between analyses. However, the κ-refinement of experimental data provides a standard for comparison of Si and O atoms in different bonding environments. Net charges on the oxygen atoms are similar in orthosilicates and the *chain-structure* pyroxenes: κ-refinement values for five orthosilicates and five pyroxenes, listed in the survey by Tsirelson et al. (1990), average to -1.32 e and -1.28 e for the two classes. The corresponding numbers for Si are $+1.68$ e and $+2.34$ e. This corresponds to a net charge of -3.60 for the SiO_4 ion and of -3.00 for the Si_2O_6 group, a possibly significant difference. Both are counterbalanced by two divalent cations. In natrolite (Ghermani et al. 1996), $Na_2(Al_2Si_3O_{10})\cdot 2H_2O$, the oxygen charges vary from -0.90 (5) e to -1.21 (5) e, the largest of which occurs for an O atom within the Na^+ coordination sphere. The charges on the Si atoms are $\approx +1.75$ e.

The *framework silicates* lack counterions, therefore two oxygen charges must exactly counterbalance the charge on silicon, leading to lower net oxygen charges. The reanalysis of the data on coesite (Geisinger et al. 1987), a high-pressure polymorph of SiO_2, by Downs (1995), gives an average oxygen net charge of -0.74 (6) e for the five nonequivalent oxygen atoms. In the very-high-pressure polymorph of SiO_2, stishovite, stable above 10 GPa, Si is octahedrally coordinated, while oxygen is bonded to three Si atoms in a rutile-type structure. The oxygen charge from a κ-refinement of stishovite is -0.86 (15) e (Hill et al. 1983), in quite good agreement with the coesite results, even though the octahedral six-coordinate structure of stishovite is acknowledged to be much more ionic than the low-pressure tetrahedral structure (Cohen 1994). All the net charges are considerably less than the ionic values, in qualitative support of Pauling's (1939) concepts.

Coesite is an especially interesting case for the analysis of the Si—O bond. It is centrosymmetric, and contains eight nonequivalent Si—O bonds, participating in five different SiOSi groups, with angles at the oxygen varying from 137.22 (2)° to 180°. Both the dynamic and static deformation densities show peaks of 0.4–0.5 eÅ^{-3} in all Si—O bonds; the smallness of the differences between the dynamic and static values can be attributed to the low vibrational amplitudes in the bonded network. The peak heights exceed those in stishovite, which are ≈ 0.3 eÅ^{-3} according to experiment, and 0.35 eÅ^{-3} according to a linearized augmented plane-wave (LAPW) calculation (Cohen 1994). This difference supports the stronger ionic character of the high-pressure polymorph.

The deformation density peaks in the Si—O bonds of coesite are located away from the midpoint of the bond towards the oxygen atom, and also tend to be displaced towards the interior of the SiOSi angle. Downs reports a bending of the Si—O bonds of 9° and 14° relative to the 137.22° SiOSi angle, based on the peaks in the deformation density. This would indicate a bending of the Si—O bonds due to O$\cdots$O nonbonded repulsion, like the bending of the bonds in small ring compounds described in the following chapter. However, according to the topological analysis of the total density, the bond paths and the internuclear vectors coincide.

The experimental deformation density features, including the displacement of the bond peaks from the bond axes, are well reproduced in theoretical maps on the molecule $H_6Si_2O_7$, which contains two linked SiO_4 tetrahedra in a doubly eclipsed configuration. The results for different values of the central SiOSi angle are shown in Fig. 11.4.

The topological analysis of the total density has the advantage of being independent of a reference model. In coesite, the bond critical points are found at about 0.67 Å from Si and ≈ 0.94 Å from O, in contrast to the deformation density peaks which are closer to the oxygen atoms. The values of $\nabla^2\rho$ at the bond critical point are positive, as the contraction of the density perpendicular to the bond direction is overbalanced by the sharp decrease in density along the bond path. The $\nabla^2\rho$ values in coesite are reported as $+20.3$ eÅ^{-5} and $+12$ eÅ^{-5}, depending on the details of the radial density functions used in the refinements. Positive values are not typical for pure covalent bonds, and indicate a significant ionic contribution to the Si—O bonding in silicates. But the negative values of $\nabla^2\rho$, which occur for covalent bonds between first-row atoms, may not be typical for covalent bonds involving heavier atoms like Si.

Covalent contribution to the bonding is confirmed by analysis of the components of $\nabla^2\rho$. The average values of λ_1 and λ_2, representing the contraction of the density in the directions perpendicular to the bond (chapter 6), are found to be -9 eÅ^{-5}, compared with -4 eÅ^{-5} for the independent-atom model (IAM) density and -6 eÅ^{-5} for the ionic model. In addition, the density at the bond critical point ρ_b is ≈ 1.1 eÅ^{-3}, higher than 0.85 eÅ^{-3} and 0.68 eÅ^{-3} calculated for the IAM and ionic models, respectively. Results for the borosilicate danburite $CaB_2Si_2O_8$, are similar, with $\rho_b(SiO) \approx 0.95$ eÅ^{-3}, and large positive values of $\nabla^2\rho$ for the Si—O bonds (Downs and Swope 1992).

The electrostatic potential in the silicate minerals is of importance for understanding of the nature of electrophilic reactions, and of the position of ions and host molecules located in cavities of microporous silicates such as the zeolites. The zeolites have the general formula $M_r(I)M_s(II)[Al_pSiO_{2(p+q)}] \cdot mH_2O$, with $r + 2s = p$. The water molecules are loosely bound and can easily be removed by heating, or replaced by other small molecules. Ghermani et al. (1996) find in their analysis of natrolite $Na_2(Al_2Si_3O_{10}) \cdot 2H_2O$, that the sodium ion is exactly located at the minimum of the potential calculated from the framework charge density without the inclusion of the cations. The depth of the minimum is -1.5 eÅ^{-1}, corresponding to an electrostatic binding energy of 21.6 eV mol^{-1}. By comparison, the potential minimum in dehydrated sodium zeolite A, studied by Spackman and Weber (1988) using earlier low-resolution data, is -0.48 eÅ^{-1}. Though the significance of such differences remains to be assessed, and is affected by variation in $V(0)$ between solids (chapter 8), the understanding of the framework–guest interactions in zeolites is of obvious practical importance. Spackman and Weber (1988) note that the pictures of the potential based on point charges at the atomic sites are very different from the more detailed information revealed by the diffraction experiment.

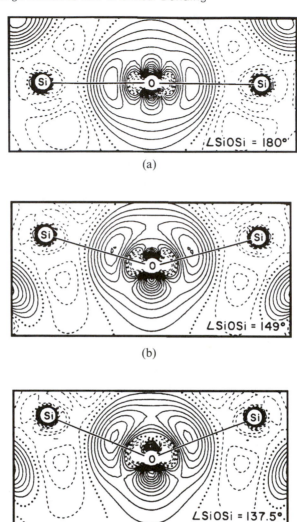

FIG. 11.4 Theoretical deformation electron density for the molecule $H_6Si_2O_7$ (basis set: 6-31G* on Si, 6-31G on O, and 3-1G on H). (a) $R(SiO) = 1.595$ Å, SiOSi $= 180°$; (b) $R(SiO) = 1.610$ Å, SiOSi $= 149°$; (c) $R(SiO) = 1.621$ Å, SiOSi $= 137.5°$. Contour interval is 0.05 eÅ^{-3}. Solid-line contours are positive; zero and negative contours are dotted and dashed lines, respectively. *Source*: Geisinger et al. (1987).

It is noteworthy that the zeolites have significantly higher thermal motion than the simpler silicates. The room-temperature equivalent isotropic temperature factors for Si and O in sodium zeolite A are reported as 1.85 and 3.0 Å^2, respectively, compared with 0.49 and 0.99 Å^2 in quartz.

11.2 Metallic Solids

11.2.1 Specific Aspects of Metals

Metals, and to a lesser extent metallic alloys, tend to crystallize in highly symmetric space groups. The high symmetry and the relatively simple chemical composition makes metals and alloys especially suited for Compton scattering studies, in which the electron momentum distribution is derived from the measurement of the Compton incoherent X-ray scattering (Williams 1977, Cooper 1985). Compton scattering occurs as a result of a collision between an X-ray photon and an electron, and causes a change of momentum of the photon that is dependent on the momentum of the scattering electron. The dependence of the Compton scattering on the direction of the scattering vector of the photons can be analyzed to yield the three-dimensional momentum distribution against which theoretical calculations can be tested. Other techniques of great importance in the study of the electronic structure of metals, beyond the scope of this treatment, include Fermi surface mapping through positron annihilation measurements.

The outer electrons in metals such as Li and Na have a very low ionization energy, and are largely delocalized. Such electrons are described as constituting a "nearly free electron gas." It may be noted, though, that this description is somewhat misleading as the behavior of the electrons is dominated by the exclusion principle, while the molecules in normal gases can be described by classical statistical mechanics.

The Fermi surface plays an important role in the theory of metals. It is defined by the reciprocal-space wavevectors of the electrons with largest kinetic energy, and is the highest occupied molecular orbital (HOMO) in molecular orbital theory. For a free electron gas, the Fermi surface is spherical, that is, the kinetic energy of the electrons is only dependent on the magnitude, not on the direction of the wavevector. In a free electron gas the electrons are completely delocalized and will not contribute to the intensity of the Bragg reflections. As a result, an accurate scale factor may not be obtainable from a least-squares refinement with neutral atom scattering factors.

The bonding in transition metals involves *d* orbitals, which are considerably more directional. In transition metals, Fermi surfaces are markedly nonspherical, and cohesive energies show regularities which can be understood in terms of bonding involving hybridized orbitals. The cohesive energies of transition metals (defined as the energy relative to the free atoms) are found to increase sharply in a row of the periodic table, up to about the middle of the row, near Cr, Mo, and W, for the first, second, and third rows, respectively, and then decrease towards the right of the periodic table. The increase in the first half of the rows can be explained in terms of half-filled or less-than-half-filled *d* bands, and the decrease on the right-hand side by the occupancy of antibonding orbitals. This is completely analogous to the reduced bond order of the O_2 molecule compared with N_2, and the decrease in the metal–sulfur bond strength toward the right of the periodic table in transition metal sulfides, discussed in the previous chapter. The exact nature of the hybrid orbitals forming the bands in metals differs with crystal

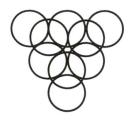

FIG. 11.5 The closed-packed arrangement of atoms, showing an octahedral interstitial hole in the center of the drawing, surrounded by three tetrahedral holes.

structure, but the use of even g-type (d and s) orbitals is favored energetically (Adams 1974).

Because the effect of bonding on the electron density of metals is relatively small, charge density studies on metals require very careful collection of intensity data. The number of available studies is limited, but important conclusions have been reached.

11.2.2 The Charge Density in Beryllium Metal

11.2.2.1 The Structure of Beryllium Metal

Beryllium is a hexagonally closed packed (hcp) metal in which each Be atom is surrounded by 12 neighbors, six of which are located in a plane perpendicular to the hexagonal c axis, and three each in the planes above and below the central plane. In this packing arrangement there are tetrahedral holes directly above and below each Be atom at 0.625 times the interplanar spacing, and octahedral holes at a height of half the interplanar spacing directly above and below the triangles formed by three in-plane Be atoms (Fig. 11.5).

Two of the four electrons of the beryllium atom are valence electrons. The bonding in the metallic solid must be accomplished by a combination of $2s$ and $2p$ orbitals; if not, the s band would be completely filled and Be would be an insulator, or perhaps not a room-temperature solid at all. In the orbital description, the two valence electrons of each Be atom participate in two $2s2p$ hybrid electron pair bonds, spread over the 12 nearest neighbors of the hcp structure.

11.2.2.2 Experimental Data

Beryllium metal has been the subject of careful charge density studies. The first set of X-ray data containing 27 reflections was collected with $AgK\alpha$ radiation by Brown (1972), using two single-crystal plates. In subsequent work, Larsen and Hansen (LH) (1984) used a small crystal of $0.1 \times 0.25 \times 0.30$ mm and both $MoK\alpha$ and $AgK\alpha$ radiation. Short-wavelength neutron structure factors, of importance for separation of thermal motion and charge density effects, were measured (Larsen et al. 1980), while the absolute scale was established with eight low-order 0.03 Å γ-ray reflections from a ^{198}Au source (Hansen et al. 1984), and with a larger $\lambda = 0.12$ Å set collected with a ^{157}Sm source (Hansen et al. 1987).

In a metal like Be, extinction can be a cumbersome effect, especially if unrecognized, as appears to have been the case of the early 1972 data. It was very small in the LH data collected on a small sample, but significant in the 0.12 Å

γ-ray data, which required a large crystal ($1.70 \times 1.90 \times 2.16$ mm) because of the weakness of the available sources.

11.2.2.3 Charge Density Analysis

The reliable experimental information on the absolute scale and thermal vibrations of beryllium metal made it possible to analyze the effect of the model on the least-squares scale factor, and test for a possible expansion of the 1s core electron shell. The 0.03 Å γ-ray structure factors were found to be 0.7% lower than the LH data, when the scale factor from a high-order refinement ($\sin \theta/\lambda > 0.65$ Å^{-1}) is applied. Larsen and Hansen (1984) conclude that because of the delocalization of the valence electrons, "it is doubtful that diffraction data from a metallic substance can be determined reliably by high-order refinement, even with very high $\sin \theta/\lambda$ cut-off values." This conclusion, while valid for the lighter main-group metals, may not fully apply to metals of the transition elements, which have much heavier cores and show more directional bonding.

A κ-refinement of the 0.12 Å γ-ray data reproduces the absolute scale poorly when the neutron U_{ij} thermal parameter values are used (Hansen et al. 1987). The discrepancy can be removed by introduction of a core-κ-parameter, which refines to $\kappa_{core} = 0.988$ (2), corresponding to a 1.2% linear expansion. This is supported by a similar result obtained with the LH X-ray data, and related to the scale factor discrepancy noted above. Hansen, Schneider, Yellon, and Pearson (1987), conclude that without independent knowledge of either the scale or the thermal parameters, good agreement with experiment can be achieved, but the resulting scale factor may be in error by as much as 2.5%.

Though the core expansion leads to the appropriate fit, it may not be the proper explanation for the scale factor discrepancy. Hansen et al. (1987) note that the expansion of the core would lead to a decrease of ≈ 7.5 eV in the kinetic energy of the core electrons, at variance with the HF band structure calculations of Dovesi et al. (1982), which show the decrease to be only about 1.5 eV. An alternative interpretation by von Barth and Pedroza (1985) is based on the condition of orthogonality of the core and valence wave functions. The orthogonality requirement introduces a core-like cusp in the s-like valence states, but not in the p-states. Because of the promotion of electrons from $s \to p$ in Be metal, the high-order form factor for the crystal must be lower than that for the free atom. It is this effect that can be mimicked by the apparent core expansion.

The most conspicuous feature in the deformation density maps is accumulation of charge in the tetrahedral holes of the hcp structure (Figs. 11.6 and 11.7). This feature is reproduced in all analyses, including that of the earlier 1972 data (Yang and Coppens 1978), and has a maximum with height of 0.046 eÅ^{-3} according to the LH X—$X_{high\,order}$ and 0.043 eÅ^{-3} according to the X—γ—N analysis. The peak in the X—γ—N density integrates to 0.013 (2) e for the AgKα, and 0.012 e for the MoKα data.

Larsen and Harsen describe the bonding as resulting from sp^3 hybridization with one lobe $\frac{1}{2}(s + 3p_z)$ parallel to the c axis, and the other three lobes pointing into the tetrahedral holes arranged trigonally below the atom. The latter combine with the orbitals of adjacent atoms, similarly pointing into the tetrahedral holes. A second resonance structure is obtained by reflecting the hybrids in the horizontal

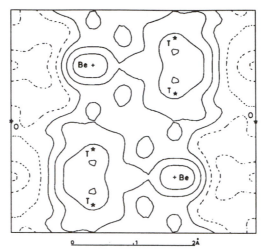

FIG. 11.6 An X—$X_{\text{high-order}}$ deformation map in the (110) plane of Be metal; z direction is vertical, showing the electron deficiency along the channels formed by adjoining octahedral holes, marked O, and the surplus of charge in the bipyramidal space formed by two tetrahedral holes, marked T. Contours are at 0.015 eÅ^{-3} intervals. Negative contours are broken lines. The first solid line is the zero contour. *Source*: Larsen and Hansen (1984).

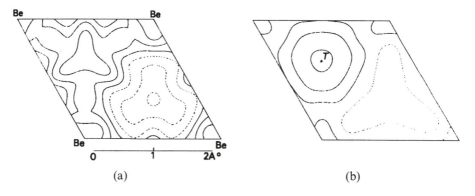

(a)   (b)

FIG. 11.7 An X—$X_{\text{high-order}}$ deformation map in sections parallel to (001). Contours are as in Fig. 11.6. (a) Section at $z = 0.75$ containing the atomic position. (b) Section at $z = 0.625$ through the tetrahedral position (marked T). *Source*: Larsen and Hansen (1984).

plane, leading to a $\frac{1}{2}(s - 3p_z)$ hybrid, and three hybrids pointing up into the tetrahedral holes at 0.625 the interplanar spacing above the central atom. This attractive scheme is in agreement with the density observed in the tetrahedral holes and the complete absence of density in the octahedral holes of the structure. The bonding through the tetrahedral holes is reminiscent of the two-electron, three-center bonds encountered in electron-poor molecules such as B_2H_6. As each tetrahedral hole has four neighbors and the two Be valence electrons are distributed over eight equivalent hybrids in the proposed scheme, it can be described as *one-electron, four-center bonding.*

A number of theoretical calculations are available for comparison with the experimental results on Be metal. The increase of the valence density in the tetrahedral holes is well reproduced by both the early augmented plane wave (APW) calculation of Inoue and Yamashita (1973), and the all-electron HF-LCAO calculation of Dovesi et al. (1982), but the latter gives somewhat better agreement with the experimental results.

The comparison between theory and experiment may alternatively be based on the (static) deformation atomic form factors for each of the reflections (i.e., $\Delta f = f_{\text{true}} - f_{\text{spherical}}$), as was done by Hansen et al. (1987). Unlike the direct comparison of the valence density distribution, the form-factor analysis is affected by the theoretical treatment of the core electrons. They are not specifically included in pseudopotential calculations, though the potential due to the core electrons is, of course, accounted for. The core electron scattering must therefore be added into the theoretical form factors. The linearized augmented plane-wave (LAPW) calculation of Pindor et al. (1986) includes all electrons, but uses the local density approximation, which is less successful when the density is rapidly varying, as it is in the core region. In general, this calculation, the local pseudopotential calculation of Chou et al. (1983), and the nonlocal density functional calculation of von Barth and Pedroza (1985) agree well with each other and with experiment in the low-order region. The von Barth and Pedroza values agree better with experiment at high angles beyond $\sin \theta/\lambda \approx 0.60$ eÅ^{-1}, due to the correction for nonorthogonality of the s and p states mentioned above. The high-angle form factors are typically 0.02 e below the free atom values.

11.2.3 The Charge Density in Vanadium and Chromium

Vanadium and chromium have body-centered cubic (bcc) structures, in which each atom is surrounded by eight nearest neighbors along the cube diagonals and six next-nearest neighbors along the cube axial directions. In the bcc structure, the difference between the nearest and next-nearest contacts is only 17% of the nearest-neighbor distance. Hybrids compatible with the bcc structure are sd^3 and d^4 for the nearest neighbors, and d^3 for the next-nearest neighbors (Altmann et al. 1957).

Careful measurements of the structure factors of vanadium (Ohba et al. 1981) and chromium (Ohba et al. 1982) up to $\sin \theta/\lambda = 1.72 \text{ Å}^{-1}$, using AgK$\alpha$ radiation and small spherical crystals (≈ 0.2 mm diameter), have been reported. The bcc structure of these metals leads to pairs of reflections such as (330/441), (431/510), at identical values of $\sin \theta/\lambda$, which have the same intensity for a structure with one spherical atom per lattice point. This is no longer true when the t_{2g} and e_g orbitals of the cubic site are no longer equally occupied. This is easiest seen as follows.

The angular functions for the e_g orbitals are given by (appendix D)

$$y_{20} = (5/16\pi)^{1/2}(3z^2 - 1)$$
$$y_{22+} = (15/4\pi)^{1/2}(x^2 - y^2)/2 \tag{11.6}$$

where x, y, and z are direction cosines. Summing of the squares of these distributions gives an angular density function which peaks in the six directions of the cubic axis, and is described by

$$d(e_g) = (5/8\pi)[3z^4 + 3y^4 + 3z^4 - 1] \tag{11.7}$$

The corresponding expressions for the t_{2g} orbitals are

$$y_{21+} = (15/4\pi)^{1/2}xz$$

$$y_{21-} = (15/4\pi)^{1/2}yz \tag{11.8}$$

$$y_{22-} = (15/4\pi)^{1/2}xy$$

which gives, for the total t_{2g} density,

$$d(t_{2g}) = (15/4\pi)[(xz)^2 + (yz)^2 + (xy)^2] \tag{11.9}$$

The t_{2g} density is zero where two of the direction cosines x, y, and z are zero, that is, in the planes of the coordinate axes, but peaks along the eight cube diagonals. The density functions are Fourier-transform invariant, as discussed in chapter 3, and expressed by the equation

$$f_{lmp}(\mathbf{S}) = f[d_{lmp}(\theta, \phi)] = 4\pi i^l \langle j_l \rangle d_{lmp}(\beta, \gamma) \tag{3.43}$$

For reflections at the same $\sin \theta/\lambda$, the radial scattering factors $\langle j_l \rangle$ are equal, but the angular factors d_{lmp} differ for different occupation of the t_{2g} orbitals. According to Eqs. (11.7) and (11.9), the scattering factor $f(e_g) \sim 3(h^4 + k^4 + l^4) - 1$, and $f(t_{2g}) \approx (hl)^2 + (kl)^2 + (hk)^2$. As a result occupancy of the e_g orbitals in excess of the spherical population increases the intensity of reflections with larger values of $h^4 + k^4 + l^4$, the scattering vectors of which are closer to the directions of the e_g lobes, while occupancy of the t_{2g} orbitals beyond 3 electrons increases the reflections with larger values of $(hl)^2 + (kl)^2 + (hk)^2$ for which $h^4 + k^4 + l^4$ is smaller.

Experimental and theoretical ratios for paired reflections in Cr are listed in Table 11.1. The ratios are listed such that the reflection with the smallest value of $h^4 + k^4 + l^4$ is in the numerator. It is evident that the reflections with smaller values of $h^4 + k^4 + l^4$ are more intense by 1–2.5%, indicating a preferential occupancy of the t_{2g} orbitals, which in the bcc structure are directed towards the nearest neighbors.

This conclusion is confirmed by both experimental and theoretical deformation density maps. The experimental maps show a positive peak of height 1.4 (1) eÅ^{-3} at 0.25 Å from the Cr nucleus along the cube diagonals (Fig. 11.8), in qualitative agreement with a tight binding scheme involving d, or s and d orbitals.

For vanadium, the ratios are smaller, and the dynamic density maps do not show a distinct maximum in the cube direction. The difference is attributed to anharmonicity of the thermal motion. Thermal displacement amplitudes are larger in V than in Cr, as indicated by the values of the isotropic temperature factors, which are 0.007 58 and 0.004 07 Å^2 respectively. As in silicon, the anharmonic displacements are larger in the directions away from the nearest neighbors, and therefore tend to cancel the asphericity of the electron density due to bonding effects.

TABLE 11.1 Experimental and Theoretical Ratios of the Integrated Intensities of the Reflection Pairs for Chromium Metal

	Experimental		Theoretical	
$F^2/F^{2\,a}$	Ohba et al. (1982)	Diana and Mazzone (1972)	Ohba et al. (1982)	Rath and Callaway (1973)
330/411	1.018 (14)[b]	1.026 (14)	1.022	1.016
413/510	1.017 (10)		1.030	
433/530	1.013 (12)		1.016	
442/600	1.025 (16)	1.028 (14)	1.038	
532/611	1.013 (11)		1.025	

[a] The ratio of the paired reflections $h_1k_1l_1$ and $h_2k_2l_2$ is defined as $F^2(h_1k_1l_1)/F^2(h_2k_2l_2)$, where the value of $h_1^4 + k_1^4 + l_1^4$ is always selected to be less than that of $h_2^4 + k_2^4 + l_2^4$.
[b] Standard deviations of the experimental values are shown in parentheses.

Source: Ohba et al. (1982).

More quantitatively, the effect of the thermal motion follows from the anharmonic thermal motion formalisms discussed in chapter 2. In the bcc structure, the relevant nonzero anharmonic term in the one-particle potential is the anisotropic, cubic site-symmetry allowed, part of $u^j u^k u^l u^m$ in expression (2.39). The modified potential for the cubic sites is given by (Willis 1969, Willis and Pryor 1975)

$$V(u_1 u_2 u_3) = V_0 + \delta[(u_1^4 + u_2^4 + u_3^4) - \tfrac{3}{5}(u_1^2 + u_2^2 + u_3^2)^2] \qquad (11.10)$$

in which δ is the coefficient of the fourth-order term in the expansion (2.39).

Since the anharmonic term is small relative to the leading harmonic term, the corresponding temperature factor can be written as

$$T_{\text{anharmonic}}(\mathbf{H}) = T_{\text{harmonic}}(\mathbf{H})\{1 - C\delta T^3[(h^4 + k^4 + l^4) - \tfrac{3}{5}(h^2 + k^2 + l^2)^2]\} \qquad (11.11)$$

in which C is a constant, dependent on the harmonic potential and the lattice constant of the material, and the T^3 dependence follows from the temperature factor expansion, as in Eq. (2.43).

The second term in square brackets is the same for two members of a reflection pair. Thus, since the correction term is small,

$$T_1/T_2 = 1 + C\delta T^3[(h_2^4 + k_2^4 + l_2^4) - (h_1^4 + k_1^4 + l_1^4)] \qquad (11.12)$$

The coefficient δ will be negative, because the potential is softened in the direction of the more remote next-nearest neighbors. Thus, the larger displacement corresponds to a relative increase in the scattering for the second member of the reflection pair with the larger value of $(h^4 + k^4 + l^4)$, and therefore to a reduction of the ratios listed in Table 11.1.

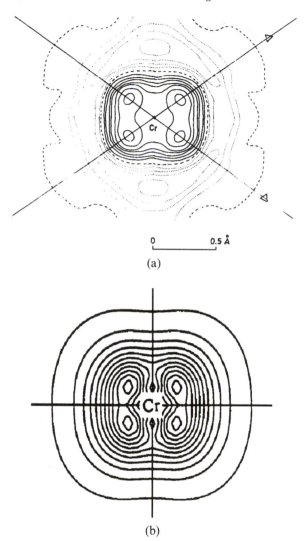

FIG. 11.8 (a) A section of the difference synthesis through the Cr nucleus, parallel to the (110) plane. Contours are drawn at intervals of 0.2 eÅ^{-3}. (b) Theoretical contour map of valence electron distribution on the (110) plane for chromium metal. Contours are drawn at intervals of 0.5 eÅ^{-3}. The lobes point towards the nearest neighbors in the body-centered cubic structure. *Source*: Ohba et al. (1982).

11.2.4 The Charge Density in Copper

Copper is a face-centered cubic (fcc) metal. Band structure calculations show the valence bands to be copper d bands and hybrid bands of sd, pd, and sp character. The hybridization is essential for the conductivity of copper, as some of the bands cross the Fermi surface and are thus only partially occupied (K. Schwarz, private communication).

Schneider, Hansen, and Kretschmer (SHK) (1981) have measured the 19 reflections with $\sin \theta / \lambda < 0.7$ eÅ^{-3} with 0.03 Å γ-radiation. Deformation densities based on these reflections show a small accumulation of charge of height 0.19 eÅ^{-3} at 1/4 1/4 0 and equivalent positions, which is between nearest neighbors located along the [110] directions, as well as an accumulation of similar height, but somewhat more extended, in the voids between the atoms at 1/4 1/4 1/2. This seems fully compatible with a hybrid bonding model.

The SHK low-order structure factors are systematically lower than those from a series of theoretical calculations on copper. Mackenzie and Mathieson (MM) (1992) have reanalyzed both the theoretical and the experimental results, and conclude that the absolute scale of the SHK data may have been affected by the use of the Darwin extinction theory for an intentionally deformed (to intercept the whole divergence of the γ-ray beam) single crystal. They compare an "educated" mean over all available X-ray and γ-ray structure factors, and structure factors derived from electron diffraction,[3] with the results of six theoretical calculations. Like for beryllium, the core electrons are not included in the available band structure calculations. They must be added a posteriori to allow comparison with the experimental structure factors. The most detailed calculation, by Ekardt et al. (1984), gives an excellent match to the photoemission spectra, which depend on the valence-state energy levels, but fits the X-ray data not as well as the less sophisticated calculations. Mackenzie and Mathieson show that the difference between the Ekardt et al. theoretical structure factors and the average over three other sets behaves very much like a fraction of the Cu $(K + L + M)$ core electron scattering factor (Fig. 11.9). They argue convincingly that a minor modification of the core contribution, or equivalently, a change in the pseudopotential which represents the effect of the core electrons, would improve the fit to the diffraction data but would not affect the relative energy levels which explain the photoelectron results. In other words, theoretical structure factors from band structure calculations will not be reliable in the high-order region, which is dominated by the core scattering. It is also pointed out by MM that, rather than compare theoretical and experimental results, it is more appropriate to use high-quality, absolute-scale experimental structure factors in a complementary fashion, to test the appropriateness of the pseudopotential used in the band structure calculations.

11.2.5 Intermetallic Compounds and the Critical-Voltage Electron Diffraction Method

11.2.5.1 Electron Structure Factors

Much of the available experimental information on intermetallic compounds comes from high-energy electron diffraction (HEED) measurements. In electron diffraction, the electron beams interact with the electrostatic potential in the crystal. The electron structure factor is therefore directly dependent on this

---

[3] See section 11.2.5.1 for the relevant expressions.

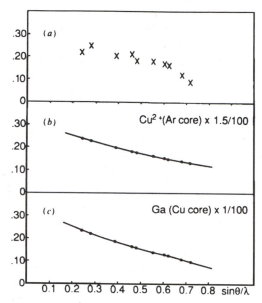

FIG. 11.9 Plots against $(\sin\theta)/\lambda$ (Å^{-1}) of the reflections for Cu up to (333/511). (a) The differences between the value of Ekardt et al. (1984) and the average of the four earlier theoretical calculations; (b) 1.5% of the argon core of Cu^{2+} from *International Tables for X-ray Crystallography* (1974); (c) 1% of the Cu core of Ga from *International Tables for X-ray Crystallography* (1974). *Source*: Mackenzie (1992).

potential including a nuclear contribution, as defined by Eq. (8.3). In SI units, and using the spherical-atom and isotropic-thermal-motion approximations, the electron structure factor is given by the Mott–Bethe expression

$$V(\mathbf{H}) = \frac{|e|}{4\pi^2\varepsilon_0 V_{\text{unit cell}}} \sum_i [Z_i - f_i^X(\mathbf{H})]/H^2 \exp(-B_i \sin^2\theta/\lambda^2) \exp(2\pi i\mathbf{H}\cdot\mathbf{r}_i)$$

$$(11.13)$$

in which e is the electron charge, ε_0 is the permittivity of free space (see chapter 8), Z is the atomic number, and f_i^X is the X-ray scattering factor of the ith atom. Conversion of an electron structure factor, measured at temperature T, to the corresponding X-ray structure factor involves subtraction of the nuclear terms in Eq. (11.13). It is clear that this requires knowledge of the structure and of the atomic temperature factors B_i.

In particular the critical-voltage technique (Spence 1993) can provide low-angle structure factors with an estimated accuracy as good as 0.1%. It requires an electron microscope whose accelerating voltage can be continuously varied over a large range, typically 100 kV to 1 MeV. Because of the interaction between the incident and diffracted beams in the crystal, the intensity of a second-order reflection will show a minimum as a function of the accelerating voltage, provided that $V(\mathbf{H})^2 > V(2\mathbf{H})^2$, where $V(\mathbf{H})$ is the electron structure factor defined by Eq. (11.13). The position of the intensity minimum depends sensitively on the ratio of the first- and second-order electron structure factors. It is commonly assumed that the second-order structure factor can be described in the independent-atom approximation, and that the ratio can be corrected properly for the difference in the temperature factors of the two reflections. With this assumption, the bonding effects in the first-order reflection can be analyzed.

11.2.5.2 The Charge Density in β' NiAl

The alloy β' NiAl is a solid with a CsCl-type structure in which one atom is located at the corners, and the second atom at the center of the unit cell. The valence-electron to atom ratio is often quoted as 1.5, using a counting scheme in which the transition metal has zero valence and the Al is considered as trivalent.

In such models, the bonding is considered to be partially ionic with a charge transfer from Al to the Ni $3d$ valence band. To explain the properties of β' NiAl at a more sophisticated level, Fox and Tabernor (1991) measured four low-angle structure factors by the HEED critical-voltage technique. The deformation density based on these four reflections shows a depletion of density around both the Ni and Al atoms, and a buildup of about 0.13 eÅ^{-3} along the [111] direction halfway between Ni and Al nearest neighbors.

The electron diffraction study was complemented by an all-electron theoretical calculation of Lu, Wei, and Zunger (LWZ) (1992), using the local density approximation for the exchange and correlation terms in the Hamiltonian. They find agreement within $\approx 0.6\%$ between the calculated and dynamic structure factor values for the lowest three reflections, (100), (110), and (111). But for (200), with $\sin\theta/\lambda = 0.3464$ Å^{-1}, the discrepancy is as large as 1.7%. The discrepancy is attributed to insufficiently accurate knowledge of the temperature factors in this diatomic crystal, which affect the derivation of the X-ray structure factor from the electron diffraction measurement, as well as the calculation of the dynamic theoretical structure factors needed for the comparison with experiment. For the monoatomic Si crystal for which the B values are well known, the agreement is 4–20 times better than for β' NiAl and within 0.2 e atom^{-1}.

As is not surprising, given the experience gained with X-ray charge densities, the deformation electron density of β' NiAl is significantly modified when the higher-order Fourier terms, not measured by HEED, are added. In the theoretical maps, there is still a slightly electron-positive region halfway between the Ni and Al, but the dominant features are lobes on the Ni atom along [111], pointing towards the Al atoms, much like the Cr metal deformation density of Fig. 11.8. Furthermore, there is electron buildup at the Ni and depletion at the Al positions, in support of an ionic contribution to the bonding. It was concluded by LWZ that NiAl exhibits both ionic and covalent bonding components, with the former dominating in the deformation density. The orbital populations from the calculation show a loss of Al sp charge of 0.27 e and a gain of Ni pd charge of 0.23 e. The strong directionality of the bonding explains why it is extremely difficult to disorder this alloy.

11.2.5.3 The Charge Density in γ-TiAl

The alloy γ-TiAl is a tetragonal solid in which face-centered Ti + Al (001) planes alternate, with a relative displacement of 1/2 along the a (or, equivalently, b) axis (Fig. 11.10). The TiAl alloy is considered a covalent intermetallic compound, compared with the more ionic character of NiAl. Lu et al. (1994) compared the

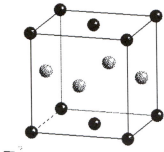

FIG. 11.10 Structure of γ-TiAl. Dark spheres: Ti; light spheres: Al.

electron diffraction values of seven low-order X-ray structure factors with theoretical results, using the same methods as applied to NiAl. The agreement for the low-order structure factors, derived with a single thermal parameter for both atoms, is within 0.7%, with a largest discrepancy of 0.022 e atom^{-1}. The deviation of the true structure factors from the spherical atom values is 1.1–1.5% for the lowest-order reflections, which, while referred to as a substantial difference by the authors, illustrates the high accuracy required for the charge density analysis of intermetallic compounds. An analysis of the electron density based on the low-order reflections only, leads to incorrect conclusions, but the fact that the experimental and theoretical structure factors agree gives confidence in the analysis of the high-resolution theoretical density.

The overall bonding pattern in γ-TiAl is described by nonspherical depletion of electrons from both the Al and Ti sites, with a considerable buildup in the d_{xy} lobes on the Ti atoms pointing towards nearest-neighbor Ti atoms along the [110] directions. This evidently suggests directional $3d$-bonding between the Ti atoms, rather than between the metal and Al atoms as in β'NiAl. The bonding between the nearest-neighbor Al atoms in the all-Al (001) planes, on the other hand, appears metallic, without pronounced accumulations of negative charge density. There is, however, a slight (<0.02 eÅ^{-3}) density buildup above and below the Al atoms, halfway between the Al and Ti (001) planes.

The picture that emerges from the available studies of transition metal bonding in metals and alloys is that of bonding lobes directed towards nearest neighbors, indicating Cr—Cr, V—V, Ti—Ti, and Ni—Al, but not Ti—Al interactions of at least partially covalent nature.

11.3 Ionic Solids

11.3.1 Do Completely Ionic Solids Exist?

The possibility of assessing the ionicity of atoms in crystals from the diffraction pattern has been among one of the most controversial topics of X-ray analysis[4]

---

[4] For some of the discussions, see Coppens and Feil (1991).

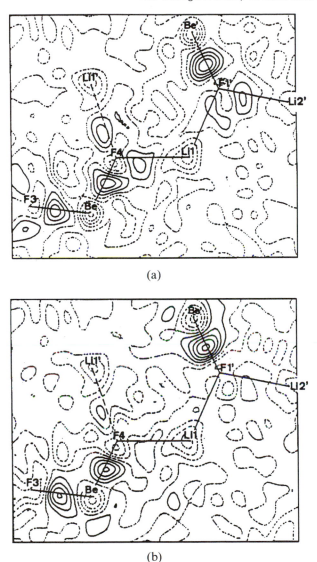

(a)

(b)

FIG. 11.11 Electron-density difference maps on Li_2BeF_4 calculated with all reflections $<$ $\sin \theta / \lambda = 0.9$ Å^{-1} (81 K). (a) Based on the neutral atom procrystal model, (b) based on the ionic model. Contour levels are drawn at intervals of 0.045 eÅ^{-3}.[5] Full lines for positive density, dashed lines for negative and zero density. The standard deviation, estimated from $[2\Sigma\sigma^2(F_0)]^{1/2}N$, is 0.015 eÅ^{-3}. *Source:* Seiler and Dunitz (1986).

[5] In the original publication, the contour interval was erroneously specified as 0.015 eÅ^{-3}, P. Seiler and J. D. Dunitz, private communication.

James 1948, in his famous treatise on X-ray diffraction, stated that

> any attempt to determine the state of ionization of the atoms in a crystal by means of the measurement of the atomic scattering factors is likely to fail, since the curves will differ appreciably only at angles for which no spectra exist.

This overly pessimistic conclusion has been refuted by many studies done in the past decades, including those described in this volume. Many crystals have much larger unit cells, and therefore produce a considerable number of reflections at small $\sin \theta / \lambda$ values below $0.25 \, \text{Å}^{-1}$. Examples are the silicates discussed above, for which atomic charges are now routinely obtained with high-quality data. It is worth noting that a change in net charge on an atom is accompanied by a radial expansion or contraction of the valence shell, the effect of which persists to higher scattering angles.

The question of the measurement of ionicity was specifically addressed in a study on lithium tetrafluoroberyllate Li_2BeF_4 by Seiler and Dunitz (1986). The primitive rhombohedral unit cell of this solid has a size 30 times larger than that of the primitive cell of LiF, resulting in 40 reflections with $\sin \theta / \lambda < 0.25 \, \text{Å}^{-1}$. Alternative refinements with neutral and ionic scattering factors for Li, Be, and F led to essentially identical agreement factors when all reflections were considered. However, for 11 weak, low-order reflections with $\sin \theta / \lambda < 0.25 \, \text{Å}^{-1}$, the R-factors were 0.043 for the neutral atom and 0.125 for the ionic model, a rather striking difference supporting the neutral atom model, and confirmed by further analysis of a larger low-order data set. The difference densities for both models (Fig. 11.11) show significant bond peaks in the Be—F bonds, indicating a covalent contribution, and some density between Li and F.

Seiler and Dunitz point out that the main reason for the widespread acceptance of the simple ionic model in chemistry and solid-state physics is its ease of application and its remarkable success in calculating cohesive energies of many types of crystals (see chapter 9). They conclude that the fact that it is easier to calculate many properties of solids with integral charges than with atomic charge distributions makes the ionic model more convenient, but it does not necessarily make it correct.

With the topological analysis of the total charge density, the distinction between a covalent and a closed-shell ionic interaction can be based on the value of the Laplacian and its components at the bond critical point. Such an analysis will be most conclusive when done on a series of related compounds, analyzed with identical basis sets, as the topological values of the model density from experimental data have been found to be quite dependent on the choice of basis functions.

12

Electron Density Studies of Molecular Crystals

12.1 Why Molecular Crystals?

12.1.1 The Importance of Molecular Crystals

Small molecules consisting of light-, few-electron atoms were the first species beyond atoms to yield to quantum-mechanical methods. Similarly, crystals of small light-atom molecules have served as most useful test cases of charge density mapping. The small number of core electrons in first-row atoms enhances the relative contribution of valence electron scattering to the diffraction pattern. Early studies, done just after automated diffractometers became widely available, were concerned with molecular crystals such as uracil (Stewart and Jensen 1967), s-triazine (Coppens 1967), oxalic acid dihydrate (Coppens et al. 1969), decaborane (Dietrich and Scheringer 1978), fumaramic acid (Hirshfeld 1971), glycine (Almlof et al. 1973), and tetraphenylbutatriene (Berkovitch-Yellin and Leiserowitz 1976). While thermal motion is often pronounced in molecular crystals, advances in low-temperature data collection have done much to alleviate this disadvantage. In recent years, subliquid-nitrogen cooling techniques have been increasingly applied.

Among the most interesting aspects of molecular crystals are the influence of intermolecular interactions on the electronic structure. Physically meaningful Coulombic parameters pertinent to a molecule in a condensed environment can be obtained from the diffraction analysis, and can be used in the modeling of macromolecules. The enhancement of the electrostatic moments relative to those of the isolated species has been noted in chapter 7. But, beyond these considerations, molecular crystals are important in their own right. For example, crystals of aromatic molecules substituted with π-electron donor and acceptor groups are among the most strongly nonlinear optical solids known, considerably exceeding the nonlinearity of inorganic crystals such as potassium titanyl phosphate (KTP); while mixed-valence organic components of low-dimensional solids can become superconducting at low temperatures. The relation between such properties of molecular crystals and their charge distribution provides a continuing impetus for further study.

12.1.2 The Suitability Factor

The suitability of light-atom crystals for charge density analysis can be understood in terms of the relative importance of core electron scattering. As the perturbation of the core electrons by the chemical environment is beyond the reach of practically all experimental studies, the frozen-core approximation is routinely used. It assumes the intensity of the core electron scattering to be invariable, while the valence scattering is affected by the chemical environment, as discussed in chapter 3. As an order-of-magnitude approximation, we write

$$\langle F^2_{core} \rangle = \sum_i^{\text{unit cell}} f^2_{core,i} \cong \sum_i^{\text{unit cell}} n^2_{core,i} \tag{12.1}$$

where $n_{core,i}$ is the number of core electrons of atom i. The relevant measure for the importance of the core scattering is the core-scattering intensity per unit volume. Its inverse may be selected as a gauge of the suitability of a crystal for X-ray charge density analysis (Stevens and Coppens 1976):

$$S = \frac{V}{\sum_i^{\text{unit cell}} n^2_{core,i}} \tag{12.2}$$

The value of the *suitability factor* S varies from typically 3–5 for first-row atom organic crystals to 0.1–0.3 for metals and alloys of first row transition metal elements. (Table 12.1). The implication is that much better accuracy will be

TABLE 12.1 Suitability Factors for Various Crystals

	V	V/Z	$\sum n^2_{core}$	S
Formamide	223.6	55.9	12	4.7
α-Glycylglycine	579.0	144.8	36	4.0
α-Oxalic acid dihydrate	255.4	127.7	32	4.0
Tetracyanoethylene	897.8	149.6	40	3.7
$Cr(CO)_3C_6H_6$	419.5	209.8	372	0.56
$Cr_2(O_2C_2H_3)_4 \cdot 2H_2O$	1392.2	348.1	720	0.48
$Ni_2(C_2H_2)(C_5H_5)_2$	1058.6	264.7	696	0.38
Cl_2	225.2	56.3	200	0.28
S_8 (orthorhombic)	3292.9	205.8	800	0.26
Si	157.5	19.7	100	0.19
Al	66.4	16.6	100	0.17
V_3Si	105.7	52.85	1072	0.05

Source: Stevens and Coppens (1976).

required to obtain meaningful results on crystals with low S values. In very low-S crystals containing second or third-row transition metal elements, for instance, electron density effects of anharmonic motion may dominate the asphericity in the density maps.

12.1.3 The Oxalic Acid Project

To examine the reliability of X-ray charge densities at a time of rapid development of new methods, the Commission on Charge, Spin and Momentum Densities of the IUCr organized a project under which a single substance, α-oxalic acid dihydrate, was studied in a number of laboratories using X-ray, neutron, and theoretical methods. The report by Coppens on the study, published in 1984, established unequivocally the qualitative reproducibility of chemically significant features in deformation density maps, which had not been generally accepted.

Four X-ray and five neutron data sets were collected and three sets of theoretical calculations were performed. The main discrepancies between theory and experiment, and among experiments, were found in the heights of the oxygen lone-pair peaks. Among the theoretical maps, significant discrepancies occurred also in the bond peak heights. Among the best experiments, peak heights in the deformation maps were reproducible within $0.15 \, e\text{Å}^{-1}$. Large differences in vibrational parameters were observed, indicating deficient temperature calibration and systematic bias of the parameters in some of the experiments. Positional parameters were found to be reproducible within $0.001 \, \text{Å}$, with average discrepancies between some experiments being as low as $0.0005 \, \text{Å}$.

With the more widespread use of subnitrogen cryogenic temperatures, use of smaller samples made possible by brighter sources, and rapid developments in detector technology and computational methods, the conclusions of the oxalic acid project are now of mainly historical importance. However, the project remains an example of the value of collaborative efforts in establishing the validity of a scientific method.

12.2 Transferability of Charge Density Parameters Among Related Atoms

The systematic bias introduced in the positional and thermal parameters by the spherical-atom approximation may be reduced, or at best eliminated, by the use of improved scattering formalisms. The development of aspherical atom scattering factors, typical for an element in a specific bonding environment, would be a highly desirable achievement. To what extent are atomic or fragment charge densities transferable? Can we obtain the charge density of macromolecules by a buildup of densities of atoms or small fragments?

Transferability of atomic densities was tested by Brock et al. (1991), who applied atomic charge density parameters from an accurate low-temperature study of perylene (I) to data collected at five and six different temperatures on naphthalene (II) and anthracene (III), respectively. The molecules are all aromatic hydrocarbons. To reduce the number of variables, all H atoms were assigned identical deformation densities, and the number of independent carbon atoms was constrained as indicated in the schematic. Only three of the four C-atom types of perylene, and the H atom, occur in the smaller naphthalene and anthracene molecules.

Perylene (I)   Naphthalene (II)   Anthracene (III)

A difficulty in transferring charge density parameters in this manner is that the reconstructed molecules may no longer be neutral. In the hydrocarbon study, this difficulty was circumvented by repeating the perylene refinement for each case with constraints designed to maintain electroneutrality in the smaller hydrocarbon molecule to which the aspherical scattering factors are transferred. The aspherical atom refinements with the perylene-derived aspherical scattering factors lead to a systematic increase of the in-plane molecular translation amplitudes, and a decrease in those normal in the plane. This is because the charge density is more diffuse in the plane, undoubtedly due to the bond overlap density, but more concentrated in directions perpendicular to the plane. In the spherical-atom treatment, such bonding features introduce systematic bias in the vibrational parameters, in accordance with earlier studies discussed in chapter 3. This is an example of the type of bias that could be reduced by the introduction of standard aspherical-atom scattering factors.

The difficulty of transferring charges between molecules was avoided in a later study by Pichon-Pesme et al. (1995a), who transferred the atomic asphericity but not the deviations from atomic neutrality and atomic expansion/contraction represented by the κ parameter. The experimental charge densities in the peptide, phenyl, and methyl groups in the polypeptides N-acetyl-L-tryptophan methyl-amide, N-acetyl-α,β-dehydrophenyl alanine, and Leu-enkephalin were used as a starting point. From the least-squares results, Pichon-Pesme et al. constructed a *data bank* of transferable population parameters for three different types of carbon atoms, the hydrogen atom, the peptide nitrogen, and C=O oxygen atoms. Remarkably, a small number of populated aspherical density functions is adequate to describe the density of each of these atoms (see Table 12.2). For the phenyl sp^2 carbon atoms, for example, the only significant nonzero deformation parameters are the P_{20} and P_{33+} populations. The negative population of the P_{20} spherical harmonic function describes the contraction of the density in the direction perpendicular to the plane of the three bonds, as in the hydrocarbons described above, while the positive value of the P_{33+} parameter represents the accumulation

TABLE 12.2 Atomic Multipole Parameters from Averaging over Three Polypeptides

	C'	O	N	C(sp^3), α	C(sp^3), β	C(sp^2)	H
P_{11+}	0.12	−0.10					0.15
P_{20}	−0.32	−0.06				−0.23	
P_{22+}	0.13	−0.06					
P_{31+}				−0.16	−0.19		
P_{31-}				−0.21	−0.24		
P_{33+}	0.43		0.27	0.25	0.21	−0.32	

Parameters not listed are equal to zero. C': carbon atom in C=O group; O: oxygen atom in C=O group; N: nitrogen atom in peptide bond; H; generalized hydrogen atom; C(sp^3)α: carbon atoms bonded to two C, one H, and one N atom, C(sp^3)β: bonded to two C and two H atoms; C(sp^2): phenyl carbon atom. Coordinate systems are defined in Fig. 12.1.

of density in the three bonds. The negative population of the d_{20} spherical harmonic is also evident for the C' atom of the C=O group, which similarly is linked to its neighbors by three coplanar bonds.

The deformation of the peptide nitrogen atom is described by a single d_{33+} ($=x^3 - 3xy^2$) function, the z axis being perpendicular to the C—N—C plane. The deformations of three types of atoms are illustrated in Fig. 12.2.

As a test, the standard aspherical form factors in their proper orientations were applied in a refinement of both room temperature and low temperature (125 K), accurate, but low-resolution, data (≈ 0.7 Å^{-1}) on the tripeptide pGlu-Phe-D-Pro-ϕ[CN$_4$]-Me (PPP). The improvements on introduction of the aspherical scattering factors are dramatic, the R-factors being reduced by 30–40% without the introduction of additional variables. Significant adjustments in the anisotropic

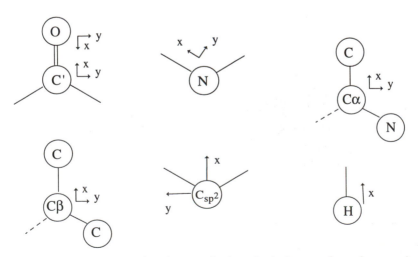

FIG. 12.1 Coordinate systems for the standard aspherical atom form factors given in Table 12.2.

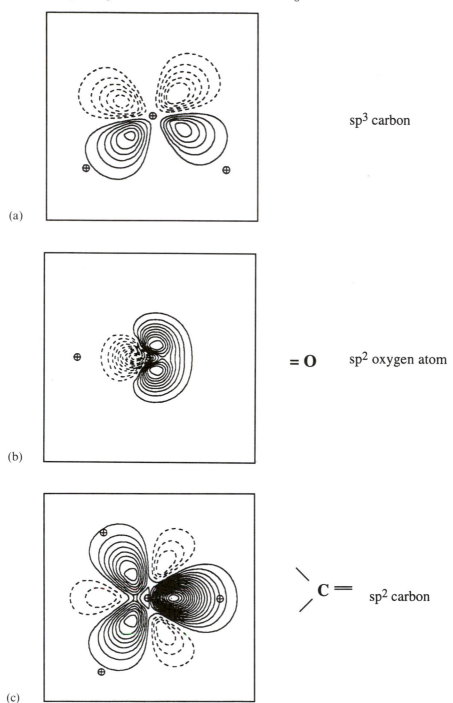

FIG. 12.2 Static deformation densities for (a) sp^3 carbon atom, (b) oxygen atom, and (c) sp^2 carbon atom in C=O double bond. Contour interval is 0.05 eÅ^{-3}. Negative contours are dashed lines; zero contours are omitted. *Source*: Pichon-Pesme et al. (1995)

thermal parameters occurred, while changes in the positional parameters were relatively small, but not insignificant (≤ 0.009 Å). In PPP, the U_{ij} thermal parameters are reported to decrease on the average by 10–20% upon introduction of the standard aspherical atom form factors, though for the sp^2 hybridized atoms an increase in the direction perpendicular to the bonds must occur, because of the contraction of the static atom density represented by the negative d_{z^2} population.

In subsequent work by the same authors, non-neutrality of the standard atoms was allowed by addition of a transferable P_v valence-shell monopole population, with neutrality being maintained by a slight adjustment of the hydrogen charges, and κ parameters refined after the transfer (Pichon-Pesme et al. 1995b).

The introduction of standard aspherical-atom scattering factors leads to a very significant improvement in Hirshfeld's rigid bond test. The results are a beautiful confirmation of Hirshfeld's (1992) statement that "an accurate set of nuclear coordinates" (and thermal parameters!) "and a detailed map of the electron density can be obtained, via X-ray diffraction, only jointly and simultaneously, never separately or independently".

A related theoretical approach to charge density transferability has been developed by Mezey and collaborators (Walker and Mezey 1993, 1994). But rather than composing a molecule of standard pseudoatoms, the density of large molecules, including proteins, is constructed from the density of a number of standard theoretical fragments. The fragment densities are defined by the distribution

$$\rho(\mathbf{r}) = \sum_\mu \sum_\nu P_{\mu\nu} \phi_\mu(\mathbf{r}) \phi_\nu(\mathbf{r}) \tag{3.7}$$

but with $P_{\mu\nu}(\text{fragment}) = P_{\mu\nu}(\text{molecule})/2$ if either ϕ_μ or ϕ_ν is centered on an atom not belonging to the fragment. The method is referred to as MEDLA, the molecular electron density Lego approach. It differs from the Lego construction in that the building blocks are diffuse like pseudoatoms, rather than having sharp boundaries like Lego blocks.

The main purpose of the method is to define molecular shapes through isodensity surfaces. Tests on a number of small molecules show that this aim is achieved with a great efficiency in computer time. Discrepancies between MEDLA densities and theoretical distributions, averaged over the grid points, are typically below 10% of the total density. While this does not correspond to an adequate accuracy for an X-ray scattering model, the results do provide important information on the shapes of macromolecules.

12.3 Selected Studies of Molecular Crystals

12.3.1 Bent Bonds

The concept of bond bending was introduced by Coulson and Moffit (1949) in discussions of angle strain in small-membered ring compounds such as cyclopropane. Since bonding hybrids made up of s and p contributions cannot be at angles of less than 90°, the hybrids cannot point along the edges of an equilateral triangle

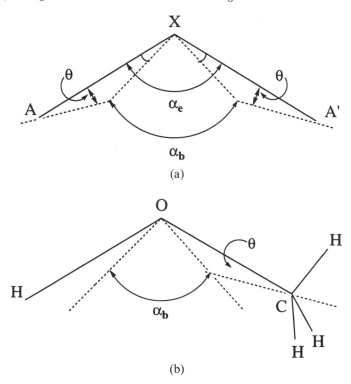

FIG. 12.3 (a) Illustration of bond bending due to nonbonded repulsions in an acyclic molecule. *Source*: Hirshfeld (1964). (b) Noncolinearity of the three-fold axis of the methyl group and the O—C internuclear vector as a result of bond bending. α_e is the angle defined by the nuclear positions; α_b is the angle defined by the bonding hybrids; θ and θ' are the bond-bending angles at the substituent atoms. Orbital axes are indicated by broken lines. *Source*: Eisenstein and Hirshfeld (1979).

as formed by the carbon atoms in cyclopropane. This *bending* of the bonds leads to a lowering of the bond energy, which is the cause of the strain in small-membered ring systems.

Hirshfeld (1964) pointed out that bond bending not only occurs in ring systems, but also results from steric repulsions between two atoms two bonds apart, referred to as 1–3 interactions. The effect is illustrated in Fig. 12.3. The atoms labeled A and A' are displaced from the orbital axes, indicated by the broken lines, because of 1–3 repulsion. As a result, the bonds defined by the orbital axes are bent inwards relative to the internuclear vectors. When one of the substituents is a methyl group, as in methanol [Fig. 12.3(b)], the methyl–carbon-atom hybrid reorients such as to maximize overlap in the X—C bond. This results in noncolinearity of the X—C internuclear vector and the three-fold symmetry axis of the methyl group. Structural evidence for such bond bending in acyclic molecules is abundant. Similarly, in phenols such as *p*-nitrophenol (Hirshfeld

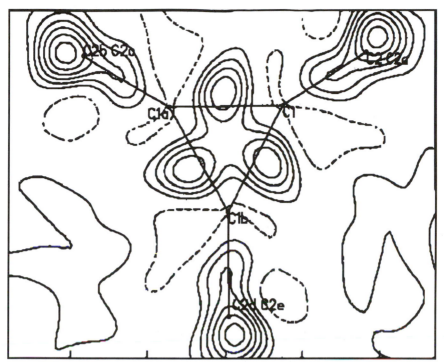

FIG. 12.4 Deformation density in the central cyclopropane ring of 3 [rotane]. Each carbon atom in the molecule is part of an additional cyclopropane ring oriented perpendicular to the central ring. Contour interval 0.05 eÅ^{-3}, negative contours broken. *Source*: Boese et al. (1991).

1964), the exocyclic C—O bond is bent by C$\cdots$H repulsion, leading to unequal C—C—O bond angles.

The first charge density observation of bond bending in cyclopropane was from the experimental charge densities of *cis*-1,2,3-tricyanocyclopropane (Hartman and Hirshfeld 1966) and 2,5-dimethyl-7,7-dicyanonorcaradiene (Fritchie 1966). It has been confirmed by a considerable number of other studies, including one on [3] rotane (Boese et al. 1991) (Fig. 12.4) and those on the three-membered heterocyclic ethyleneimine (Ito and Sakurai 1973) and ethyleneoxide (Matthews and Stucky 1971, Matthews et al. 1971), azirinidyl (CH_2CH_2N) (Cameron et al. 1994) and diazirine (N≡NC) rings (Kwiatkowski et al. 1994). Density-based evidence for the bending in acyclic molecules caused by intramolecular nonbonded repulsions is available from experimental studies on 2-cyanoguanidine (Hirshfeld and Hope 1980) and theoretical analysis of 2-cyanoguanidine, hydrazoic acid, cyanogen azide, formic acid, and diimide (Eisenstein and Hirshfeld 1979). The results have abundantly confirmed Hirshfeld's earlier conclusions based on molecular geometry.

That bond-bending strain is not confined to three-membered rings is evident from charge density studies on cyclobutane (Stein et al. 1992), cyclobutadiene

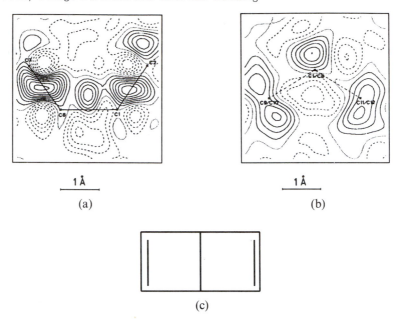

(a)

(b)

(c)

FIG. 12.5 Top: experimental difference density in the Dewar benzene derivative 3,4:5,6-dibenzo[6.2.2]propella-3,5,9,11-tetraene: (a) in the plane through the central bridgehead bond and two exocyclic bonds to the bridgehead carbon atoms; (b) in a section perpendicular to the bridgehead bond passing through its center and through the midpoints of the cyclobutene double bonds. Contours are at 0.05 eÅ^{-3}. Zero contours are dotted; negative contours are dashed. *Source*: Irngartinger and Deuter (1990), Irngartinger et al. (1990). (c) Schematic diagram of Dewar benzene.

(Irngartinger et al. 1977), and the hydro-bis(squarate) anion (Lin et al. 1994). Even in the deformation density maps of the five-membered pyrrole rings in transition metal tetraphenyl porphyrins (chapter 10), bond bending is visible.

Of particular interest is the electron density in polycyclic molecules containing several small condensed rings (see Table 12.3 for molecular diagrams). The central bond *bridgehead* bond in bicyclobutane was shown to be bent outwards in both theoretical and experimental deformation density maps (Eisenstein and Hirshfeld 1981, 1983). The same feature has been observed in bicyclo[2.2.0]hexadiene ("Dewar benzene") derivatives (Fig. 12.5). On carboxylate substitution at the bridgehead carbons, the substituents are found to be oriented such as to give optimal interaction of their π orbitals with the orbitals of the bent bridgehead bond (Irngartinger and Deuter 1990, Irngartinger et al. 1990).

The propellanes are a class of compounds with three condensed rings, either three- or four-membered, sharing a bridgehead bond. In two [1.1.1] propellane derivatives studied by Seiler et al. (1988), no peaks were observed in the deformation density of the central bridgehead bonds of lengths ≈ 1.58 Å, but peaks at the apex of the inverted (i.e., pyramidal) bridgehead carbon atoms are in agreement wih electrophilic attack at these positions.

TABLE 12.3 Topological Analysis of Theoretical Densities on Strained Ring Molecules

Molecule	Structure	Angle	Geometric Angle α_e (deg)	Bond Path Angle α_b (deg)	$\Delta\alpha$ $\alpha_b - \alpha_e$ (deg)	Strain Energy (kJ mol^{-1})
Cyclopropane		1	60.0	78.84	18.84	115
Cyclobutane		1	89.01	95.73	6.72	111
Bicyclo[1.1.0]butane		1	58.99	72.78	13.79	267
		2	60.50	76.62	16.12	
		3	97.91	105.07	7.16	
Bicyclo[1.1.1]pentane		1	74.44	84.72	10.27	285
		2	87.20	95.85	8.65	
[1.1.1]propellane		1	61.81	59.37	−2.44	410
		2	95.98	107.99	12.01	
		3	59.09	69.09	9.99	

Source: Bader (1990).

The large body of information on the electron density of strained cyclic systems now available can be analyzed quantitatively. An attractive tool for this purpose is provided by the topological analysis of the density, as discussed in chapter 6. For a number of theoretical densities, values of $\Delta\alpha$, the differences between the "geometric" bond angle α_e, defined by the nuclear positions, and the angle α_b, defined by the bond paths, are reproduced in Table 12.3, together with the molecular strain energies. In the [1.1.1] propellanes, analysis of the theoretical density shows a critical bond point between the the bridgehead atoms, notwith-standing the absence of density in the deformation densities. Substitition of hydrogen at the bridgehead carbon atom breaks the central bond and replaces the (3, −1) critical point by a (3, +1) cage critical point, typical for ring structures. The ellipticity of the bonds in three-membered rings is much larger than that in single bonds. According to the theoretical densities, the ellipticity in the ring bonds of cyclopropane is actually larger than that of the central bond in ethylene, indicating a pronounced concentration of density into the ring plane (Bader 1990). This observation may be compared with the contraction of the density into the plane of sp^2 hybridized carbon atoms, noted above in the discussion of atomic transferability.

12.3.2 Molecular Crystals with Nonlinear Optical Properties

Molecular crystals are among the most efficient second- and higher-harmonic generating materials. An external electric field **E**, upon interacting with a molecule, will induce a dipole moment μ. If the field is strong, the response may not be linear, in which case the components of μ can be developed in increasing powers of E as described by the expansion ($i = 1, 2, 3$)

$$\mu_i = \mu_{0i} + \alpha_{ij}E_j + \beta_{ijk}E_jE_k + \gamma_{ijkl}E_jE_kE_l + \cdots \tag{12.3}$$

In this expression, the Einstein convention of summation over repeated indices has been followed: μ_0 is the permanent dipole moment, while α_{ij}, β_{ijk}, and γ_{ijkl} are the tensorial elements of the linear polarizability, and the second- and third-order *hyperpolarizabilities* of the molecule, respectively.

The nonlinear optical properties are due to the higher-order terms in the expansion. For a molecular crystal, they are both a function of the molecule's properties and of the molecular packing. For the odd polarizabilities, represented by β_{ijk} and higher-order odd terms, to be nonzero, the crystal must be noncentro-symmetric. For strong nonlinearity, the relative orientation of the molecules must favor enhancement of the collective property. A strongly nonlinear molecular crystal, such as 2-methyl-4-nitroaniline, contains parallel chains of aligned molecules in the space group Cc.

Robinson (1967) has used the Unsöld approximation for the energy levels to express the polarizabilities in terms of the electrostatic moments of the ground-state electron distribution. The expressions have been applied to X-ray charge densities by Zyss, Baert, and coworkers (Fkyerat et al. 1995; F. Hamzaoui, F. Baert and J. Zyss, private communication). A detailed description of the derivation and the approximations involved is beyond the scope of this treatise. However, it should be mentioned that the severe approximations are made that all excited-state energy levels are equal, and that the exciting light frequency is equal to zero.

The resulting equations show that the first-order polarizabilities α_{ij} depend on the second moments of the distribution, while the second-order polarizabilities β_{ijk} are functions of both the second and third moments of the polarizable body, as in

$$\alpha_{ii} = \frac{4m}{\hbar^2}\,\mu_{ii}^2 \tag{12.4}$$

$$\alpha_{ij} = \frac{2m}{\hbar^2}\,\mu_{ij}\cdot(\mu_{ii} + \mu_{jj} + 2\mu_{ij}) \tag{12.5}$$

and

$$\beta_{iii} = \frac{12m^2}{\hbar^2} \mu_{ii}^2 \mu_{iii} \tag{12.6}$$

$$\beta_{iij} = \frac{4m^2}{\hbar^2} (2\mu_{ii}\mu_{jj} + \mu_{ii}^2)\mu_{iij} \tag{12.7}$$

where m is the electron mass, and μ_{ij} and μ_{ijk} are the elements of the second- and third-moment tensors of the charge distribution, respectively, as defined in chapter 7.

The expressions have been applied to the experimental electron distribution on N-(4-nitrophenyl)-L-prolinyl (NPP) (Fkyerat et al. 1995), and to 3-methyl-4-nitropyridine-N-oxide (Hamzaoui 1995; F. Hamzaoui, F. Baert and J. Zyss, private communication). Given the severity of the approximations involved, the agreement between the charge density results and theoretical isolated-molecule values for the largest component of β of NPP is quite reasonable. According to both theory and the charge-density derived result, β_{xxx}, with x being the long axis of the molecule, is by far the largest element. Theoretical and experimental values for β_{xxx} agree within 10%. The theoretical off-diagonal elements are small because of the approximate symmetry of the isolated molecule, but they are larger for the molecule in the crystal matrix, a result that may be interpreted as evidence for the influence of the molecular packing on the optical properties of a molecule.

In general, the experimental charge distribution has the advantage that it incorporates the effects of intermolecular interaction, which can be pronounced for suitably aligned molecules, as further discussed below.

12.3.3   The Effect of Intermolecular Interactions on the Charge Density

12.3.3.1   Hydrogen-Bonding

Oxalic acid dihydrate, studied by several laboratories as part of the IUCr oxalic acid project, contains a short hydrogen-bond of 2.481 Å O$\cdots$O distance, linking the oxalic acid and water molecules. All experiments are in agreement that the lone-pair peak of the water-molecule oxygen atom is polarized into the short hydrogen bond. The deformation density in the plane perpendicular to the water-molecule plane, bisecting the H—O—H angle, for one of the experiments is shown in Fig. 12.6.

The calculational results on oxalic acid dihydrate are not unequivocal Breitenstein et al. (1983) in an extended basis set HF calculation arrive at an asymmetry of the H_2O oxygen lone-pair electron density *away* from the donor H atom in the short hydrogen bond. This is at variance with the experimental results, and with density functional studies by Krijn, Graafsma, and Feil, (KGF) (Krijn and Feil 1988, Krijn et al. 1988) which are reported to "clearly indicate the polarization of the density towards the hydrogen atom of the bond donor" in the static density maps. In the study by KGF, overall agreement between theory and experiment is much improved by taking into account the effect of the crystal field,

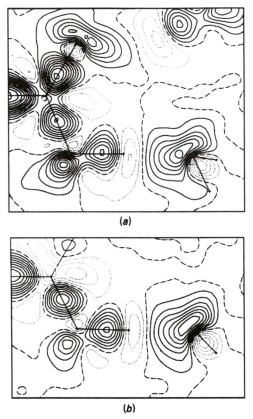

(a)

(b)

FIG. 12.6 Experimental dynamic model deformation density in oxalic acid dihydrate, using a scale factor derived from a comparison with theoretical results. Contour intervals are at $0.05\ \mathrm{e\AA^{-3}}$. (a) In the plane of the oxalic acid molecule. (b) In the plane perpendicular to the water molecule and bisecting the H(2)—O(3)—H(3) angle. *Source*: Dam et al. (1983).

which is simulated by the electrostatic potential exerted by the surrounding molecules, the multipole expansion of the difference density being used in the potential calculation. The improved agreement between theory and experiment is evident both in reciprocal and direct space. The two charge densities agree within a surprisingly low margin of $0.015\ \mathrm{e\ au^{-3}}$.

Ab-initio SCF calculations on the water molecule in various model complexes such as $(H_2O)_2$, $Li^+\cdot H_2O$, and $(Li^+)_2\cdot H_2O$ show a *depletion* of the density of the lone pair in the internuclear region for long bonds, while for short bonds, such as $Li^+\cdots O < \approx 1.6\ \text{Å}$ and $O(H)\cdots O < 2.6\text{–}2.7\ \text{Å}$, the effect is found to be reversed (Hermansson 1985), in accordance with the observations on oxalic acid dihydrate. The induced polarization of the acceptor density towards the Li^+ or H atom is apparently still present for the longer distances, but very diffuse and below the lowest contour of most maps (D. Feil, private communication). Exchange repulsion opposing the attractive effect becomes important for larger ions such as K^+, for which the oxygen lone pair penetrates the ion's electron cloud. For the hydrogen atom, the exchange repulsion and the polarization act in tandem; for all distances, a depletion of electron density at the donor H atom is observed.

Though the more recent results present a consistent pattern, systematic comparisons of theory and experiment over a larger range of interatomic distances would be most useful.

It is encouraging that the correction for thermal diffuse scattering (TDS), applied to the 123 K data set of Dam et al. (1983), has very little influence on the deformation density, though it significantly affects the thermal parameters, as may be expected.

The net density redistribution upon hydrogen bonding cannot be deduced from the experiment alone, as the density of the free molecules cannot be measured by X-ray diffraction. The effect of the coordination on the water molecule density is nevertheless evident because of the deviation from the idealized symmetry of the water molecule.

Hydrates containing several water molecules in different environments are a fruitful subject for the analysis of hydrogen bonding. Magnesium thiosulfate hexahydrate $MgS_2O_3 \cdot 6H_2O$ contains three independent water molecules in different environments (Bats and Fuess 1986). Two of the water molecules coordinate only to a Mg ion, with the oxygen atom in an approximately trigonal arrangement, while the oxygen atom of the third water molecule accepts two interactions, $O \cdots Mg$ and $O \cdots H - O$. The static model deformation densities show the third water molecule to have two pronounced and separated maxima in the oxygen lone-pair region, while the first two show one half-moon-shaped broad maximum. Very comparable results have been obtained in a study of magnesium sulfite hexahydrate (Bats et al. 1986).

Olovsson and coworkers have pointed out that the superposition of the electron density of adjacent molecules in the experimental deformation density may lead to modification of the contours in the lone-pair region of the water molecules (Fernandes et al. 1990, McIntyre et al. 1990). To avoid this complication, it is preferable to partition the crystal density through the multipole analysis, after which comparisons can be based on individual molecule or fragment densities.

The topological analysis of the total density in hydrogen bonds of normal length shows features typical of closed-shell interactions. Experimental results for $O \cdots H$ from the study on L-dopa are listed in Table 12.4 in order of increasing hydrogen bond strength (Howard et al. 1995). The density at the critical point ρ_b is very low, and $\nabla^2\rho$ at the critical point is invariably positive, unlike that for covalent bonds between first-row atoms. As discussed in chapter 6, these features correspond to an absence of the potential-energy lowering typical for covalent bonds, which results from shared interaction of the electrons with the two proximal nuclei. Indeed, the values of the topological parameters differ very little from those for an assembly of noninteracting molecules, listed in the second row for each entry in Table 12.4. The charge density evidence points to a largely electrostatic interaction for the hydrogen bonds listed.

Two recent studies show that this conclusion does not apply to very short hydrogen bonds. In methylammonium hydrogen succinate monohydrate, three "normal" hydrogen bonds, with O—H between 1.72 and 1.86 Å exist, while a very

TABLE 12.4 Hydrogen-Bond Critical Point Parameters in L-Dopa. Top Row: Experimental Values. Bottom row: Values from Multipole Refinement of Theoretical Structure Factors on a Crystal of Noninteracting Molecules

Bond	O···H Length (Å)	Hessian Eigenvalues			ρ (eÅ^{-3})	$\nabla^2\rho$ (eÅ^{-5})	Ellipticity
O···H(N)	1.96	−0.715	−0.641	4.007	0.145	2.65	0.11
		−0.616	−0.521	3.932	0.126	2.80	0.18
O···H(N)	1.94	−0.919	−0.886	4.716	0.192	2.91	0.04
		−0.865	−0.667	4.162	0.161	2.63	0.30
O···H(O)	1.88	−1.450	−1.046	5.996	0.200	3.50	0.39
		−1.063	−1.009	5.113	0.190	3.03	0.05
O···H(N)	1.83	−1.259	−1.255	6.149	0.243	3.64	0.00
		−1.436	−1.312	6.068	0.239	3.32	0.10
O···H(O)	1.76	−1.338	−1.282	6.518	0.221	3.90	0.04
		−1.519	−1.147	6.946	0.200	4.28	0.32

Source: Howard et al. (1995).

short symmetric hydrogen bond has an O—H length of only 1.221 Å. Combined X-ray and neutron diffraction studies, followed by topological analysis, give a $\nabla^2\rho$ value at the O—H critical point of $-6.8(10)$ eÅ^{-5}, and a considerable contraction in the plane perpendicular to the bond path (Hessian eigenvalues $-12.0(5)$, $-11.8(5)$ and $+17.0(2)$ eÅ^{-5}) (Flensburg et al. 1995). A second study of the short hydrogen bonds in methylammonium hydrogen maleate confirms these results with Laplacian values at the bond critical points of the two independent, symmetric short O—H bonds of $-5.9(9)$ and $-7.1(9)$ eÅ^{-5} (Madsen et al., to be published).

12.3.3.2 Molecular Electrostatic Moments in the Solid State

We noted in chapter 7 that there is both experimental and theoretical evidence that the electrostatic moments of molecules in crystals are enhanced by the interactions between molecules. Considerable progress in quantum-mechanical (Cramer and Truhlar 1992a) and combined quantum-mechanical/molecular mechanics methods has made it possible to calculate the effect of solvent on dipole moments of molecules in solution (Gao 1996). Such calculations indicate that the dipole moment of the water molecule in the liquid is enhanced from 1.86 D to 2.15 D, an increase of 0.29 D (Gao and Xia 1992). Not surprisingly, the effects are larger for more extended molecules containing aromatic rings. For 4-nitroaniline, the theoretical values are 7.64 D (isolated molecule) and 10.8 D (aqueous solution), respectively, an increase of 50%. For nucleotide bases, the increase in dipole moment in aqueous solution over the gas-phase values is found to be as large as 39–75% with, for example, adenine showing an increase from 2.17 to 3.81 D (Cramer and Truhlar 1992b, 1993; Gao 1994).

The induced polarization is important in the calculation of molecular properties, such as the hyperpolarizability discussed earlier in this chapter, and for the prediction of molecular packing and macromolecular folding. The diffraction method has the unique capability of yielding experimental values of the molecular electrostatic moments in a variety of environments. That the effects in solids can be of the magnitude indicated by the theoretical results on solutions is confirmed by the charge density analysis of 2-methyl-4-nitroaniline (Howard et al. 1992), a prototype nonlinear optical solid in which molecular chains, formed by "head-to-tail" hydrogen bonding between the NO_2 and NH_2 substituents, all point in one direction parallel to the glide plane of the space group Cc. The molecular dipole moment is found to be enhanced by the intermolecular interactions from an isolated molecule theoretical value of 8.2–8.8 D to about 22–24 D. Though the experimental standard deviation is large (≈ 8 D), a significant increase of μ appears beyond doubt, comparable to the increase calculated theoretically for water-solvated 4-nitroaniline. A calculation of the crystal-field effect, using the field induced by the surrounding 8.8 D molecular dipoles across the central molecule, predicts an increase from 8.8 D to 14.5 D. Since this calculation is not self-consistent, that is, no second calculation was performed with the higher dipole moment, it must underestimate the polarization. It is in qualitative agreement with the experimentally observed dipole-moment enhancement.

12.4 Concluding Remarks

A very large, and rapidly increasing, body of charge density information on molecular crystals is now available. For the many high-quality studies of molecular crystals not discussed above, the reader is referred to the original papers and to literature reviews on the subject.

Appendix A

Tensor Notation

A.1 Variant and Covariant Quantities

In the tensor notation (Patterson 1959, Sands 1982) the basis vectors of the direct lattice are written as a_i ($i = 1, 2, 3$), and the coordinates of a direct space vector as x^i. Thus, for a vector $\mathbf{v}$, we can write

$$\mathbf{v} = \sum x^i a_i = x^i a_i \tag{A.1}$$

where, following the *Einstein summation convention*, summation over repeated indices is implicitly assumed.

The terms variant and covariant refer to the transformation properties of the quantities. A transformation may be defined by the transformation matrix $\mathbf{T}$ operating on the direct space basis a_i such that

$$\mathbf{a}' = \mathbf{Ta} \quad \text{or} \quad a'_j = T^i_j a_i \tag{A.2}$$

As the vector $\mathbf{v}$ must be invariant under the transformation, the coordinates x^j must transform as

$$\mathbf{x}' = \mathbf{xT}^{-1} \quad \text{or} \quad x^{j'} = F^k_j x^j \quad \text{(with } T^i_k F^k_j = \delta^i_j) \tag{A.3}$$

as can be verified by multiplication of Eqs. (A.1) and (A.2). Quantities that transform as a_i are called *covariant*; those that transform as x^i are called *contravariant*.

Since $\mathbf{a}_i \cdot \mathbf{a}^*_j = \delta_{ij}$, or in tensor notation, $a_i \cdot a^j = \delta^j_i$, the reciprocal axes are contravariant and are written as a^i. As the Miller indices are the coordinates in the reciprocal base system, they must be covariant and are written as h_i. Thus, the Miller indices transform like the direct axes, both being covariant.

A.2 The Relation Between the Contravariant and Covariant Bases a^i and a_i

We may write

$$a^i = A^{ij} a_j \tag{A.4}$$

By taking the scalar product of both sides of this equation with a^k, we obtain

$$a^i \cdot a^k = A^{ij} a_j \cdot a^k = A^{ik} \tag{A.5}$$

since $a_j \cdot a^k = \delta_j^k$.

The matrix A^{ik} with elements $a^i \cdot a^k$ is called the reciprocal metric g^{ik}. Combining Eqs. (A.4) and (A.5) gives

$$a^i = g^{ij} a_j \tag{A.6}$$

The scalar product of both sides of Eq. (A.6) with a_k gives

$$a^i \cdot a_k = g^{ij} a_j \cdot a_k \qquad \text{or} \qquad \delta_k^i = g^{ij} g_{jk} \tag{A.7}$$

that is, the real-space metric and the reciprocal-space metric are each other's inverse; in matrix notation:

$$\mathbf{g}^* = \mathbf{g}^{-1} \tag{A.8}$$

A.3 The Length of a Vector and the Angle Between Two Vectors

In analogy to $a^i = g^{ij} a_j$, we can write

$$x^i = g^{ij} x_j \qquad \text{and} \qquad x_i = g_{ij} x^j \tag{A.9}$$

The square of the length l of a vector $\mathbf{v} = x^i a_i$ is given by

$$l^2 = x^i g_{ij} x^j = x_i g^{ij} x_j$$

and the angle between two vectors $\mathbf{v} = x^i a_i$ and $\mathbf{w} = y^i a_i$

$$\cos (v, w) = \frac{x^i g_{ij} y^j}{(x^i g_{ij} x^j)^{1/2} (y^i g_{ij} y^j)^{1/2}} \tag{A.10}$$

Appendix B

Symmetry and Symmetry Restrictions

B.1 Symmetry Operations

The 230 three-dimensional space groups are combinations of rotational and translational symmetry elements. A symmetry operation S transforms a vector $\mathbf{r}$ into $\mathbf{r}'$

$$S(\mathbf{r}) = \mathbf{r}' \tag{B.1}$$

such that $\rho(\mathbf{r}') = \rho(\mathbf{r})$, where ρ is the electron density.

A symmetry operation can have both rotational and translational components, and is described in the Seitz notation as $\{\mathbf{R} \,|\, \mathbf{s}\}$. The terms $\mathbf{R}$ and $\mathbf{s}$ are the rotational and translational parts of the **3d** symmetry element, respectively, such that

$$\mathbf{r}' = \mathbf{R}\mathbf{r} + \mathbf{s} \tag{B.2}$$

An example is $\{\sigma_x \,|\, 0\ 1/2\ 1/2\}$ for an n-glide $\perp a$.

Several types of symmetry operations can be distinguished in a crystalline substance. Purely translational operations, such as the translations defining the crystal lattice, are represented by $\{\mathbf{I} \,|\, n_1, n_2, n_3\}$, with n_1, n_2, n_3 being integers. Proper rotational operations are represented by the n-fold rotation axes $\{\mathbf{n} \,|\, 000\}$ ($n = 2, 3, 4, 6$). Rotation–inversion axes such as the $\bar{2}$ axis are improper rotation operations, while screw axes and glide planes are combined rotation–translation operations.

In a rectangular coordinate system, the rotation matrix $\mathbf{n}$ given by

$$\mathbf{n} = \begin{pmatrix} \cos 2\pi/n & -\sin 2\pi/n & 0 \\ \sin 2\pi/n & \cos 2\pi/n & 0 \\ 0 & 0 & 1 \end{pmatrix} \tag{B.3}$$

290

and the rotation–inversion matrix by

$$
\bar{n} = \begin{pmatrix} -\cos 2\pi/n & \sin 2\pi/n & 0 \\ -\sin 2\pi/n & -\cos 2\pi/n & 0 \\ 0 & 0 & -1 \end{pmatrix} \tag{B.4}
$$

A product of two symmetry elements of a group must also be an element of that group, or

$$
\{R\,|\,s\}\{T\,|\,u\}r = \{R(Tr + u) + s\}r = \{RT\,|\,Ru + s\}r \tag{B.5}
$$

Thus, $\{RT\,|\,Ru + s\}$ must be an element of the group. Similarly, the inverse of a symmetry element must also be an element of the group. If $\{T\,|\,u\}$ is the inverse of $\{R\,|\,s\}$, then $\{RT\,|\,Ru + s\} = \{I\,|\,0\}$, or $T = R^{-1}$. Therefore, $Ru + s = 0$, or $u = -R^{-1}s$. As the translational component of a screw axis or glide plane is always parallel to that symmetry element, it is not affected by the symmetry operation. Thus, in that case, $R^{-1}s = s$, and $u = -s$.

A *special position* in the crystal is repeated in itself by at least one symmetry element, that is, $r = r'$. According to Eq. (B.2), this means that $|s|$ must be zero if a symmetry element is to give rise to a special position. It follows that translations, screw operations, and glide planes do not generate special positions. On the other hand, positions located on proper rotation axes or centers of symmetry have lower multiplicity than general positions in the unit cell.

B.2 Symmetry in Reciprocal Space

If the contravariant components of a vector transform as $r' = Rr$, the covariant components (such as h, k, l) transform as $r^{*\prime} = (R^{-1})^T r^*$, or, in tensor notation, $x^j = R_i^j x^i$ and $x_j = R_j^i x_i$ (appendix A). Thus, a reflection H, where $H = h_i a^i$, is repeated by the symmetry element, S, at

$$
H' = R'H \tag{B.6}
$$

where $R' = (R^{-1})^T$. If we use a Cartesian coordinate system, R will be a unitary matrix and $R' = R$.

As noted above, if $\{R\,|\,0\}$ is an element of the space group of the crystal, $\{R^{-1}\,|\,0\}$ must also be a symmetry element. We may therefore write

$$
\begin{aligned}
F(H) &= \sum f_v \exp{(2\pi i H \cdot r_v)} \\
&= \sum_{1/2} f_v \exp{(2\pi i H \cdot r_v)} + \sum_{1/2} f_v \exp{(2\pi i H \cdot R^{-1} r_v)} \\
&= \sum_{1/2} f_v \exp{(2\pi i H \cdot r_v)} + \sum_{1/2} f_v \exp{(2\pi i RH \cdot r_v)}
\end{aligned} \tag{B.7}
$$

because if H is rotated in the opposite direction and R kept stationary, the dot product will remain unaffected.

For the reflection at $\mathbf{RH}$, we may write, similarly,

$$F(\mathbf{RH}) = \sum_{1/2} f_v \exp 2\pi i (\mathbf{RH} \cdot \mathbf{r}_v) + \sum_{1/2} f_v \exp 2\pi i (\mathbf{RH} \cdot \mathbf{Rr}_v)$$

$$= \sum_{1/2} f_v \exp 2\pi i (\mathbf{RH} \cdot \mathbf{r}_v) + \sum_{1/2} f_v \exp 2\pi i (\mathbf{H} \cdot \mathbf{r}_v) \qquad (\text{B.8})$$

Since Eqs. (B.7) and (B.8) are equal, this implies that a rotational symmetry element of direct space is also a rotational symmetry element of reciprocal space. This result must be correct; since X-ray scattering is a physical property of the crystal, it must at least have the point-group symmetry of the crystal.

The effect of the translational component of the element $\{\mathbf{R} | \mathbf{s}\}$ can be considered in an analogous manner

$$F(\mathbf{H}) = \sum_{1/2} f_v \exp \{2\pi i \mathbf{H} \cdot \mathbf{r}_v\} + \sum_{1/2} f_v \exp \{2\pi i \mathbf{H} \cdot (\mathbf{R}^{-1} \mathbf{r}_v - \mathbf{s})\} \qquad (\text{B.9})$$

$$F(\mathbf{RH}) = \sum_{1/2} f_v \exp \{2\pi i \mathbf{RH} \cdot \mathbf{r}_v + \sum_{1/2} f_v \exp \{2\pi i \mathbf{RH} \cdot (\mathbf{Rr}_v + \mathbf{s})\}$$

$$= \sum_{1/2} f_v \exp \{2\pi i \mathbf{H} \cdot \mathbf{R}^{-1} \mathbf{r}_v\} + \sum_{1/2} f_v \exp \{2\pi i \mathbf{H} \cdot (\mathbf{r}_v + \mathbf{R}^{-1} \mathbf{s})\}$$

$$= \sum_{1/2} f_v \exp \{2\pi i \mathbf{H} \cdot \mathbf{R}^{-1} \mathbf{r}_v\} + \sum_{1/2} f_v \exp \{2\pi i \mathbf{H} \cdot (\mathbf{r}_v + \mathbf{s})\} \qquad (\text{B.10})$$

Comparison of Eqs. (B.9) and (B.10) shows that the structure factor amplitudes of reflections at $\mathbf{H}' = \mathbf{RH}$ and $\mathbf{H}$ are related by the expression

$$F(\mathbf{H}') = \exp (2\pi i \mathbf{H} \cdot \mathbf{s}) F(\mathbf{H}) \qquad (\text{B.11})$$

Equation (B.11) implies that $I(\mathbf{H}') = I(\mathbf{H})$, that is, the rotational symmetry of the space group, is repeated in the diffraction pattern. In addition, if the atomic scattering factors f are real, which is the case when resonance effects are negligible, a center of symmetry is added to the diffraction pattern, that is, $I(\mathbf{H}) = F(\mathbf{H}) F^*(\mathbf{H}) = I(-\mathbf{H})$ even in the absence of an inversion center, which is Friedel's law.

B.3 Systematic Absences

The symmetry elements will leave certain classes of reflections invariant, or $F(\mathbf{H}') = F(\mathbf{H})$. Examination of Eq. (B.11) shows that unless $F(\mathbf{H}) = 0$, $\exp (-2\pi i \mathbf{H} \cdot \mathbf{s})$ must be equal to 1, or

$$h s_1 + k s_2 + l s_3 = 0 \ (\text{mod } 1) \qquad (\text{B.12a})$$

In tensor notation, this is expressed as

$$h_i s^i = 0 \ (\text{mod } 1) \qquad (\text{B.12b})$$

A reflection will be systematically absent if this condition is not fulfilled. As an example, we take the element $(\sigma_x | 0 \ 1/2 \ 1/2)$. The $0kl$ reflections are invariant. Therefore, only $0kl$ reflections with $k + l = 2n$ will have nonzero intensity.

B.4 Symmetry Restrictions of Tensor Elements

Second-rank tensors transform according to the expression

$$\sigma' = \mathbf{R}\sigma\mathbf{R}^T \tag{B.13a}$$

or, in tensor notation,

$$\sigma^{ij} = R^i_k \sigma^{kl} R^j_l \tag{B.13b}$$

Symmetry restrictions exist for tensors describing macroscopic physical properties of all but triclinic crystals, and for tensors describing the local properties of atoms at sites with point-group symmetries higher than $\bar{1}$.

As an example, we consider an atom on a site of four-fold symmetry. The matrix $\mathbf{R}$ in this case is given by

$$\mathbf{R}(C_4) = \begin{pmatrix} 0 & -1 & 0 \\ 1 & 0 & 0 \\ 0 & 0 & 1 \end{pmatrix} \tag{B.14}$$

Applying the transformation (B.13) to the symmetric tensor β_{ij} leads to the equation

$$\begin{pmatrix} \beta_{11} & \beta_{12} & \beta_{13} \\ \beta_{12} & \beta_{22} & \beta_{23} \\ \beta_{13} & \beta_{23} & \beta_{33} \end{pmatrix} = \begin{pmatrix} \beta_{22} & -\beta_{12} & -\beta_{23} \\ -\beta_{12} & \beta_{11} & \beta_{13} \\ -\beta_{23} & \beta_{13} & \beta_{33} \end{pmatrix} \tag{B.15}$$

which implies that $\beta_{11} = \beta_{22}$, and $\beta_{12} = \beta_{13} = \beta_{23} = 0$; in other words, the representation quadric of the tensor is an ellipsoid of revolution, oriented along the four-fold axis.

Symmetry restrictions for a number of crystal systems are summarized in Table B.1. The local symmetry restrictions for a site on a symmetry axis are the same as those for the crystal system defined by such an axis, and may thus be higher than those of the site. This is a result of the implicit *mmm* symmetry of a symmetric second-rank tensor property. For instance, for a site located on a mirror plane, the symmetry restrictions are those of the monoclinic crystal system.

The third-rank tensors, as occur in the expression for the anharmonic temperature factor (chapter 2), the restrictions may be derived by use of the transformation law:

$$\sigma^{ijk} = R^i_m R^j_n R^k_p \sigma^{mnp} \tag{B.16}$$

Symmetry restrictions for third- and fourth-order anharmonic temperature parameters are lised in the *International Tables for X-ray Crystallography* Vol. IV (1974). A more complete list for elements up to rank eight has been derived by Kuhs (1984).

Symmetry restrictions for spherical harmonic functions are given in appendix D.

TABLE B.1 Symmetry Restrictions for the Components of a Symmetric Second-Rank Tensor Referred to the Crystal Axes

Crystal System	σ^{11}	σ^{12}	σ^{13}	σ^{22}	σ^{23}	σ^{33}
Triclinic ($\bar{1}$)	σ^{11}	σ^{12}	σ^{13}	σ^{22}	σ^{23}	σ^{33}
Monoclinic $\left(\dfrac{2}{m}\right)$	σ^{11}	0	σ^{13}	σ^{22}	0	σ^{33}
Orthorhombic (*mmm*)	σ^{11}	0	0	σ^{22}	0	σ^{33}
Tetragonal $\left(\dfrac{4}{m}mm\right)$	σ^{11}	0	0	σ^{11}	0	σ^{33}
Hexagonal, trigonal $\left(\dfrac{6}{m}mm, \dfrac{3}{m}mm\right)$	σ^{11}	$\frac{1}{2}\sigma^{11}$	0	σ^{11}	0	σ^{33}
Cubic $\left(\dfrac{4}{m}3m\right)$	σ^{11}	0	0	σ^{11}	0	σ^{11}

Source: Sands (1982).

Appendix C

The 50% Probability Ellipsoid

The three-dimensional Gaussian distribution function is, in tensor notation, given by

$$P(\mathbf{u}) = \frac{|\sigma^{-1}|^{1/2}}{(2\pi)^{3/2}} \exp\left\{-\tfrac{1}{2}\mathbf{u}^T\sigma^{-1}\mathbf{u}\right\} \qquad (2.21b)$$

with $\sigma^{ij} = \langle u^i u^j \rangle$.

For an ellipsoidal surface defined by $\mathbf{u}^T\sigma^{-1}\mathbf{u} = c^2$, the probability of the atom being inside the surface is a function of c.

Using the principal axes system, we may write

$$\mathbf{u}^T\sigma^{-1}\mathbf{u} = \frac{u_1^2}{\sigma_1^2} + \frac{u_2^2}{\sigma_2^2} + \frac{u_3^2}{\sigma_3^2} = c^2 \qquad (C.1)$$

Without loss of generality, we may change the metric of space such that $\sigma_1^2 = \sigma_2^2 = \sigma_3^2$, which reduces the equations to those of the equivalent isotropic distribution with $\sigma^2 = \sigma_1^2 = \sigma_2^2 = \sigma_3^2$, or, with Eq. (C.1):

$$\frac{3u_1^2}{\sigma^2} = c^2 \qquad (C.2)$$

Thus, on the surface of the sphere defined by Eq. (C.1), $r = u_1\sqrt{3} = c\sigma$. The probability P_c that the atom is within the sphere is given by the integral over the isotropic Gaussian distribution:

$$P_c = \int_{r=0}^{c\sigma} \int_{\vartheta=0}^{\pi} \int_{\varphi=0}^{2\pi} \frac{|\sigma^{-1}|^{1/2}}{(2\pi)^{3/2}} \exp\left\{-\frac{r^2}{2\sigma^2}\right\} d\tau \qquad (C.3)$$

or, with $d\tau = r^2 \sin\theta \, dr \, d\theta \, d\varphi$, after integration over θ and ϕ

$$P_c = \frac{4\pi}{(2\pi\sigma^2)^{3/2}} \int_0^{c\sigma} r^2 \exp\left(-\frac{r^2}{2\sigma^2}\right) dr \tag{C.4}$$

and with the substitution $r' = r/\sigma \rightarrow$

$$P_c = \frac{4\pi}{(2\pi)^{3/2}} \int_0^c r'^2 \exp\left(-\frac{r'^2}{2}\right) dr' = \left(\frac{2}{\pi}\right)^{1/2} \int_0^c r^2 \exp -\frac{r^2}{2} dr \tag{C.5}$$

which is the expression reported by Johnson and Levy (1974).

The integral can be related to the error function by partial integration, which gives

$$\int_0^K x \exp(-x) \, dx = \int_0^K \exp(-x) \, dx - K \exp(-K) \tag{C.6}$$

Use of tabulated values of the error function shows that for $c = 1.5382$, $P_c = 0.5$. Alternatively, a tabulation of the integral in Eq. (C.5) (Owen 1962) may be used directly.

Appendix D

Spherical Harmonic Functions

D.1 Real Spherical Harmonic Functions and Associated Normalization Constants (x, y, and z are Direction Cosines)

l	Symbol	C_{lm} [a]	Angular function, c_{lmp} [b]	Normalization for Wave Functions, M_{lm} [c] Expression	Numerical Value	Normalization for Density Functions, L_{lm} [d] Expression	Numerical Value
0	00	1	1	$(1/4\pi)^{1/2}$	0.28209	$1/4\pi$	0.07958
1	11+ 11− 10	1	$\left.\begin{array}{c} x \\ y \\ z \end{array}\right\}$	$(3/4\pi)^{1/2}$	0.48860	$1/\pi$	0.31831
2	20	1/2	$3z^2 - 1$	$(5/16\pi)^{1/2}$	0.31539	$\dfrac{3\sqrt{3}}{8\pi}$	0.20675
	21+ 21− 22+ 22−	3 6	$\left.\begin{array}{c} xz \\ yz \\ (x^2 - y^2)/2 \\ xy \end{array}\right\}$	$(15/4\pi)^{1/2}$	1.09255	$3/4$	0.75
3	30	1/2	$5z^3 - 3z$	$(7/16\pi)^{1/2}$	0.37318	$\dfrac{10}{13\pi}$	0.24485
	31+ 31−	3/2	$\left.\begin{array}{c} x[5z^2 - 1] \\ y[5z^2 - 1] \end{array}\right\}$	$(21/32\pi)^{1/2}$	0.45705	$\left(ar^f + \dfrac{14}{5} - \dfrac{\pi}{4}\right)^{-1}$	0.32033
	32+ 32−	15	$\left.\begin{array}{c} (x^2 - y^2)z \\ 2xyz \end{array}\right\}$	$(105/16\pi)^{1/2}$	1.44531	1	1
	33+ 33−	15	$\left.\begin{array}{c} x^3 - 3xy^2 \\ -y^3 + 3x^2y \end{array}\right\}$	$(35/32\pi)^{1/2}$	0.59004	$4/3\pi$	0.42441

(continued)

(continued)

l	Symbol	C_{lm} [a]	Angular function, c_{lmp} [b]	Normalization for Wave Functions, M_{lm} [c] — Expression	Numerical Value	Normalization for Density Functions, L_{lm} [d] — Expression	Numerical Value
4	40	1/8	$35z^4 - 30z^2 + 3$	$(9/256\pi)^{1/2}$	0.10579	[e]	0.06942
	41+ 41−	5/2	$x[7z^3 - 3z]$ $y[7z^3 - 3z]$	$(45/32\pi)^{1/2}$	0.66905	$\dfrac{735}{512\sqrt{7}+196}$	0.47400
	42+ 42−	15/2	$(x^2 - y^2)[7z^2 - 1]$ $2xy[7z^2 - 1]$	$(45/64\pi)^{1/2}$	0.47309	$\dfrac{105\sqrt{7}}{4(136 + 28\sqrt{7})}$	0.33059
	43+ 43−	105	$(x^3 - 3xy^2)z$ $(-y^3 + 3x^2y)z$	$(315/32\pi)^{1/2}$	1.77013	5/4	1.25
	44+ 44−	105	$x^4 - 6x^2y^2 + y^4$ $4x^3y - 4xy^3$	$(315/256\pi)^{1/2}$	0.62584	15/32	0.46875
5	50	1/8	$63z^5 - 70z^3 - 15z$	$(11/256\pi)^{1/2}$	0.11695		0.07674
	51+ 51−	15/8	$(21z^4 - 14z^2 + 1)x$ $(21z^4 - 14z^2 + 1)y$	$(165/256\pi)^{1/2}$	0.45295		0.32298
	52+ 52−	105/2	$(3z^3 - z)(x^2 - y^2)$ $2xy(3z^3 - z)$	$(1155/64\pi)^{1/2}$	2.39677		1.68750
	53+ 53−	105/2	$(9z^2 - 1)(x^3 - 3xy^2)$ $(9z^2 - 1)(3x^2y - y^3)$	$(385/512\pi)^{1/2}$	0.48924		0.34515
	54+ 54−	945	$z(x^4 - 6x^2y^2 + y^4)$ $z(4x^3y - 4xy^3)$	$(3465/256\pi)^{1/2}$	2.07566		1.50000
	55+ 55−	945	$x^5 - 10x^3y^2 + 5xy^4$ $5x^4y - 10x^2y^3 + y^5$	$(693/512\pi)^{1/2}$	0.65638		0.50930
6	60	1/16	$231z^6 - 315z^4 + 105z^2 - 5$	$(13/1024\pi)^{1/2}$	0.06357		0.04171
	61+ 61−	21/8	$(33z^5 - 30z^3 + 5z)x$ $(33z^5 - 30z^3 + 5z)y$	$(273/256\pi)^{1/2}$	0.58262		0.41721
	62+ 62−	105/8	$(33z^4 - 18z^2 + 1)(x^2 - y^2)$ $2xy(33z^4 - 18z^2 + 1)$	$(1365/2048\pi)^{1/2}$	0.46060		0.32611
	63+ 63−	315/2	$(11z^3 - 3z)(x^3 - 3xy^2)$ $(11z^3 - 3z)(3x^2y - y^3)$	$(1365/512\pi)^{1/2}$	0.92121		0.65132
	64+ 64−	945/2	$(11z^2 - 1)(x^4 - 6x^2y^2 + y^4)$ $(11z^2 - 1)(4x^3y - 4xy^3)$	$(819/1024\pi)^{1/2}$	0.50457		0.36104
	65+ 65−	10395	$z(x^5 - 10x^3y^2 + 5xy^4)$ $z(5x^4y - 10x^2y^3 + y^5)$	$(9009/512\pi)^{1/2}$	2.36662		1.75000
	66+ 66−	10395	$x^6 - 15x^4y^2 + 15x^2y^4 - y^6$ $6x^5y - 20x^3y^3 + 6xy^5$	$(3003/2048\pi)^{1/2}$	0.68318		0.54687
7	70	1/16	$429z^7 - 693z^5 + 315z^3 - 35z$	$(15/1024\pi)^{1/2}$	0.06828		0.04480
	71+ 71−	7/16	$(429z^6 - 495z^4 + 135z^2 - 5)x$ $(429z^6 - 495z^4 + 135z^2 - 5)y$	$(105/4096\pi)^{1/2}$	0.09033		0.06488
	72+ 72−	63/8	$(143z^5 - 110z^3 + 15z)(x^2 - y^2)$ $2xy(143z^5 - 110z^3 + 15z)$	$(315/2048\pi)^{1/2}$	0.22127		0.15732
	73+ 73−	315/8	$(143z^4 - 66z^2 + 3)(x^3 - 3xy^2)$ $(143z^4 - 66z^2 + 3)(3x^2y - y^3)$	$(315/4096\pi)^{1/2}$	0.15646		0.11092
	74+ 74−	3465/2	$(13z^3 - 3z)(x^4 - 6x^2y^2 + y^4)$ $(13z^3 - 3z)(4x^3y - 4xy^3)$	$(3465/1024\pi)^{1/2}$	1.03783		0.74044

(continued)

(*continued*)

			Angular function,	Normalization for Wave Functions, M_{lm}[c]		Normalization for Density Functions, L_{lm}[v]	
l	Symbol	C_{lm}[a]	c_{lmp}[b]	Expression	Numerical Value	Expression	Numerical Value
7	75+ 75−	10395/2	$(13z^3 - 1)(x^5 - 10x^3y^2 + 5xy^4)$ $(13z^3 - 1)(5x^4y - 10x^2y^3 + y^5)$	$(3465/4096\pi)^{1/2}$	0.51892		0.37723
	76+ 76−	135135	$z(x^6 - 15x^4y^2 + 15x^2y^4 - y^6)$ $z(6x^5y + 20x^3y^3 - 6xy^5)$	$(45045/2048\pi)^{1/2}$	2.6460		2.00000
	77+ 77−	135135	$x^7 - 21x^5y^2 + 35x^3y^4 - 7xy^6$ $7x^6y - 35x^4y^3 + 21x^2y^5 - y^7$	$(6435/4096\pi)^{1/2}$	0.70716		0.58205

[a] Common factor such that $C_{lm}c_{lmp} = P_l^m(\cos\theta)_{\sin m\varphi}^{\cos m\varphi}$.

[b] $x = \sin\theta\cos\varphi$, $y = \sin\theta\sin\varphi$, $z = \cos\theta$.

[c] As defined by $y_{lmp} = M_{lmp}c_{lmp}$ where c_{lmp} are Cartesian functions.

[d] Paturle and Coppens (1988), as defined by $d_{lmp} = L_{lmp}c_{lmp}$ where c_{lmp} are Cartesian functions.

[e] $N_{ang} = \{14A_-^5 - 144A_+^5 - 20A_-^3 + 6A_- - 6A_+)2\pi\}^{-1}$ where: $A\pm = [(30 \pm \sqrt{480})/70]^{1/2}$.

[f] ar = arctan (2).

D.2 Kubic Harmonic Functions

(a) Wave function-normalized Kubic harmonics as linear combinations of wave function-normalized spherical harmonic functions. Coefficients in the expression

$$K_{lj} = \sum_{mp} k_{mpj}^l y_{lmp}.$$

Wave function-type normalization defined as

$$\int_0^\pi \int_0^{2\pi} |K_{lj}|^2 \sin\theta \, d\theta \, d\varphi = 1.^{a,b}$$

Even l				mp			
l	j	0+	2+	4+	6+	8+	10+
0	1	1					
4	1	$\frac{1}{2}(\frac{7}{3})^{1/2}$ 0.76376		$\frac{1}{2}(\frac{5}{3})^{1/2}$ 0.64550			
6	1	$\frac{1}{2}(\frac{1}{2})^{1/2}$ 0.35355		$-\frac{1}{2}(\frac{7}{2})^{1/2}$ −0.93541			
6	2		$\frac{1}{4}(11)^{1/2}$ 0.82916		$-\frac{1}{4}5^{1/2}$ −0.55902		
8	1	$\frac{1}{8}(33)^{1/2}$ 0.71807		$\frac{1}{4}(\frac{7}{3})^{1/2}$ 0.38188		$\frac{1}{8}(\frac{65}{3})^{1/2}$ 0.58184	

(*continued*)

(*continued*)

Even *l*				mp		
l *j*	0+	2+	4+	6+	8+	10+
10 1	$\frac{1}{8}(\frac{65}{6})^{1/2}$ 0.41143		$-\frac{1}{4}(\frac{11}{2})^{1/2}$ −0.58630		$-\frac{1}{8}(\frac{187}{6})^{1/2}$ −0.69784	
10 2		$\frac{1}{8}(\frac{247}{6})^{1/2}$ 0.80202		$\frac{1}{16}(\frac{19}{3})^{1/2}$ 0.15729		$\frac{1}{16}85^{1/2}$ 0.57622

l *j*	2−	4−	6−	8−
3 1	1			
7 1	$\frac{1}{2}(\frac{13}{6})^{1/2}$ 0.73598		$\frac{1}{2}(\frac{11}{16})^{1/2}$ 0.41458	
9 1	$\frac{1}{4}3^{1/2}$ 0.43301		$-\frac{1}{4}(13)^{1/2}$ −0.90139	
9 2	$\frac{1}{2}(\frac{17}{6})^{1/2}$ 0.84163		$-\frac{1}{2}(\frac{7}{6})^{1/2}$ −0.54006	

(b) Density-normalized Kubic harmonics as linear combinations of unnormalized spherical harmonic functions. Coefficients in the expression

$$K_{lj} = \sum_{mp} k'^{l}_{mpj} u_{lmp}$$

where

$$u_{lm\pm} = P_l^m \cos(\theta)^{\cos m\varphi}_{\sin m\varphi}.$$

Density type normalization is defined as

$$N_{lj} \int_0^{\pi} \int_0^{2\pi} |K_{lj}| \sin\theta \, d\theta \, d\varphi = 2 - \delta_{l0}.^{b}$$

Even *l*					mp			
l *j*	N_{lj}	0+	2+	4+	6+	8+	10+
0 1	$1/4\pi = 0.07958$	1					
4 1	0.43454	1		+1/168			
6 1	0.25220	1		−1/360			
6 2	0.02083		1		−1/792		

(*continued*)

(continued)

Even l			mp					
l j	N_{lj}		0+	2+	4+	6+	8+	10+
8 0	0.56292		1		1/5940		$\dfrac{1}{672}\cdot\dfrac{1}{5940}$	
10 1	0.36490		1		1/5460		$\dfrac{1}{4320}\cdot\dfrac{1}{5460}$	
10 2	0.00952		1			1/43680		$-\dfrac{1}{456}\cdot\dfrac{1}{43680}$

l j			2−	4−	6−	8−
3 1	0.06667		1			
7 1	0.01461		1		1/1560	
9 1	0.00596		1		1/2520	
9 2	0.00015			1		−1/4080

(c) Density-normalized Kubic harmonics as linear combinations of density-normalized spherical harmonic functions. Coefficients in the expression

$$K_{lj} = \sum_{mp} k''^{l}_{mpj} d_{lmp}.$$

Density-type normalization is defined as

$$\int_0^\pi \int_0^{2\pi} |K_{lj}| \sin\theta\, d\theta\, d\varphi = 2 - \delta_{l0}.$$

Even l		mp				
l j	0+	2+	4+	6+	8+	10+
0 1	1					
4 1	0.78245		0.57939			
6 1	0.37790		−0.91682			
6 2		0.83848		−0.50000		

(continued)

(*continued*)

Even l		mp			
l j		$2-$	$4-$	$6-$	$8-$
3 1		1			
7 1		0.73145		0.63290	

[a] *Source*: Paturle and Coppens (1988).
[b] *Source*: Su and Coppens (1994b).

D.3 Symmetry-Restrictions for Spherical Harmonic Functions

(a) Index-picking rules of site-symmetric spherical harmonics with $l \leq 6$ (l, m, and j are integers).[a]

Symmetry	Choice of Coordinate Axes	Indices of Allowed d_{lmp}
1	Any	All $(l, m, \pm)$
$\bar{1}$	Any	$(2\lambda, m, \pm)$
2	$2 \parallel z$	$(l, 2\mu, \pm)$
m	$m \perp z$	$(l, l - 2j, \pm)$
$2/m$	$2 \parallel z, m \perp z$	$(2\lambda, 2\mu, \pm)$
222	$2 \parallel z, 2 \parallel y$	$(2\lambda, 2\mu, +), (2\lambda + 1, 2\mu, -)$
$mm2$	$2 \parallel z, m \perp y$	$(l, 2\mu, +)$
mmm	$m \perp z, m \perp y, m \perp x$	$(2\lambda, 2\mu, +)$
4	$4 \parallel z$	$(l, 4\mu, \pm)$
$\bar{4}$	$\bar{4} \parallel z$	$(2\lambda, 4\mu, \pm), (2\lambda + 1, 4\mu + 2, \pm)$
$4/m$	$4 \parallel z, m \perp z$	$(2\lambda, 4\mu, \pm)$
422	$4 \parallel z, 2 \parallel y$	$(2\lambda, 4\mu, +), (2\lambda + 1, 4\mu, -)$
$4mm$	$4 \parallel z, m \perp y$	$(l, 4\mu, +)$
$\bar{4}2m$	$\bar{4} \parallel z, 2 \parallel x$	$(2\lambda, 4\mu, +), (2\lambda + 1, 4\mu + 2, -)$
	$m \perp y$	$(2\lambda, 4\mu, +), (2\lambda + 1, 4\mu + 2, +)$
$4/mmm$	$4 \parallel z, m \perp z, m \perp x$	$(2\lambda, 4\mu, +)$
3	$3 \parallel z$	$(l, 3\mu, \pm)$
$\bar{3}$	$\bar{3} \parallel z$	$(2\lambda, 3\mu, \pm)$
32	$3 \parallel z, 2 \parallel y$	$(2\lambda, 3\mu, +), (2\lambda + 1, 3\mu, -)$
	$2 \parallel x$	$(3\mu + 2j, 3\mu, +),$ $(3\mu + 2j + 1, 3\mu, -)$
$3m$	$3 \parallel z, m \perp y$	$(l, 3\mu, +)$
	$m \perp x$	$(l, 6\mu, +), (l, 6\mu + 3, -)$
$\bar{3}m$	$\bar{3} \parallel z, m \perp y$	$(2\lambda, 3\mu, +)$
	$m \perp x$	$(2\lambda, 6\mu, +), (2\lambda, 6\mu + 3, -)$

(*continued*)

(*continued*)

Symmetry	Choice of Coordinate Axes	Indices of Allowed d_{lmp}
6	$6 \parallel z$	$(l, 6\mu, \pm)$
$\bar{6}$	$\bar{6} \parallel z$	$(2\lambda, 6\mu, \pm), (2\lambda + 1, 6\mu + 3, \pm)$
6/m	$6 \parallel z, m \perp z$	$(2\lambda, 6\mu, \pm)$
622	$6 \parallel z, 2 \parallel y$	$(2\lambda, 6\mu, +), (2\lambda + 1, 6\mu, -)$
6mm	$6 \parallel z, m \parallel y$	$(l, 6\mu, +)$
$\bar{6}m2$	$\bar{6} \parallel z, m \perp y$	$(2\lambda, 6\mu, +), (2\lambda + 1, 6\mu + 3, +)$
	$m \perp x$	$(2\lambda, 6\mu, +), (2\lambda + 1, 6\mu + 3, -)$
6/mmm	$6 \parallel z, m \perp z, m \perp y$	$(2\lambda, 6\mu, +)$

(b) Symmetry-allowed Kubic harmonic functions.

		point group				
		23	$m\bar{3}$	432	$\bar{4}3m$	$m\bar{3}m$
l	j	T	T_h	O	T_d	O_h
0	1	×	×	×	×	×
3	1	×			×	
4	1	×	×	×	×	×
6	1	×	×	×	×	×
6	2	×	×			
7	1	×			×	
8	1	×	×	×	×	×
9	1	×			×	
9	2	×		×		
10	1	×	×	×	×	×
10	2	×	×			

[a] *Source*: Kara and Kurki-Suonio (1981).

D.4 Transformation of Real Spherical Harmonic Density Functions on Rotation of the Coordinate System

The following treatment follows derivations given by Su (1993) and Su and Coppens (1994a).

D.4.1 Rotation of the Coordinate System

Let (r, θ, ϕ) and (r, θ', ϕ') be the spherical coordinates of a vector,

$$\mathbf{X} = (x_1, x_2, x_3) \begin{pmatrix} \mathbf{e}_1 \\ \mathbf{e}_2 \\ \mathbf{e}_3 \end{pmatrix} = (x'_1, x'_2, x'_3) \begin{pmatrix} \mathbf{e}'_1 \\ \mathbf{e}'_2 \\ \mathbf{e}'_3 \end{pmatrix}$$

The unitary matrix which transforms the two right-handed Cartesian bases **e** and **e'** can be written in terms of Eulerian angles α, β, and γ, (Arfken 1970, Steinborn and Ruedenberg 1973, Edmonds 1974) such that

$$
\begin{pmatrix} e'_1 \\ e'_2 \\ e'_3 \end{pmatrix} = \begin{pmatrix} \cos\alpha\cos\beta\cos\gamma - \sin\alpha\sin\gamma & \cos\alpha\sin\gamma + \sin\alpha\cos\beta\cos\gamma & -\sin\beta\cos\gamma \\ -\cos\alpha\cos\beta\sin\gamma - \sin\alpha\cos\gamma & \cos\alpha\cos\gamma - \sin\alpha\cos\beta\sin\gamma & \sin\beta\sin\gamma \\ \cos\alpha\sin\beta & \sin\alpha\sin\beta & \cos\beta \end{pmatrix}
$$

$$
\times \begin{pmatrix} e_1 \\ e_2 \\ e_3 \end{pmatrix}
$$

$$
= \mathbf{R} \begin{pmatrix} e_1 \\ e_2 \\ e_3 \end{pmatrix} \tag{D.1}
$$

That is, the transformation is represented as successive rotations of γ, β, α about the e_3, e_2, and e_1 axes. A positive rotation is a counterclockwise rotation.[1] Since **R** is unitary, it follows that the Cartesian coordinates transform as

$$
\begin{pmatrix} x'_1 \\ x'_2 \\ x'_3 \end{pmatrix} = \mathbf{R} \begin{pmatrix} x_1 \\ x_2 \\ x_3 \end{pmatrix} \tag{D.2}
$$

The Eulerian angles have a domain of definition $0 \leq \alpha \leq 2\pi$, $0 \leq \beta \leq \pi$, and $0 \leq \gamma \leq 2\pi$. From Eqs. (D.1) and (D.2), α, β, and γ can be expressed in terms of the elements of **R**:

$$
\beta = \arccos(R_{33}) \tag{D.3}
$$

$$
\alpha = \begin{cases} \arccos(R_{31}/\sin\beta) & \text{if } R_{32}/\sin\beta \geq 0 \\ 2\pi - \arccos(R_{31}/\sin\beta) & \text{if } R_{32}/\sin\beta < 0 \end{cases} \tag{D.4}
$$

$$
\gamma = \begin{cases} \arccos(-R_{13}/\sin\beta) & \text{if } R_{23}/\sin\beta \geq 0 \\ 2\pi - \arccos(R_{13}/\sin\beta) & \text{if } R_{23}/\sin\beta < 0 \end{cases} \tag{D.5}
$$

Equations (D.4) and (D.5) are valid if $R_{33} \neq \pm 1$.

If $R_{33} = \pm 1$, then $\beta = \begin{cases} 0 \\ \pi \end{cases}$. We set $\gamma = 0$, and find α from

$$
\alpha = \begin{cases} \arccos(R_{11}/R_{33}) & \text{if } R_{12}/R_{33} \geq 0 \\ 2\pi - \arccos(R_{11}/R_{33}) & \text{if } R_{12}/R_{33} < 0 \end{cases} \tag{D.6}
$$

---

[1] *Note:* α, β, and γ are related to the diffractometer angles ω, χ, and φ, except that, in the conventional definition, the rotation β is around the y axis, rather than the x axis.

Equations (D.3)–(D.6) ensure that the angles are within the domain of definition and unambiguously defined.

D.4.2 Rotation of the Complex Spherical Harmonic Functions

The complex spherical harmonic functions, defined by Eq. (3.22), transform under rotation according to (Rose 1957, Arfken 1970)

$$Y_l^m(\theta', \phi') = \sum_{m'=-l}^{l} Y_l^{m'}(\theta, \phi) D_{m'm}^{(l)}(\alpha, \beta, \gamma) \tag{D.7}$$

where the $D^{(l)}$ terms are $(2l + 1) \times (2l + 1)$ matrices, which form the $(2l + 1)$-dimensional irreducible representation of the rotation group. We may write

$$D_{m'm}^{(l)}(\alpha, \beta, \gamma) = e^{-im'\alpha} \delta_{m'm}^{(l)}(\beta) e^{-im\gamma} \tag{D.8}$$

The elements $\delta_{m'm}^{(l)}(\beta)$ are given by (Rose 1957, Steinborn and Ruedenberg 1973)

$$\delta_{m'm}^{(l)}(\beta) = \left(\frac{(l + m')! (l - m')!}{(l + m)! (l - m)!} \right)^{1/2} (-1)^{m'-m} \sum_k (-1)^k \binom{l+m}{k} \binom{l-m}{l-m'-k}$$

$$\times \left[\cos \frac{\beta}{2} \right]^{2l-m'+m-2k} \left[\sin \frac{\beta}{2} \right]^{2k-m+m'} \tag{D.9}$$

with the range of integer k defined by $\max (0, m - m') \le k \le \min (l - m', l + m)$, and

$$\binom{a}{b} = \frac{a!}{(a - b)! \, b!}.$$

An alternative way of evaluating the functions $\delta_{m'm}^{(l)}(\beta)$ is described by Su and Coppens (1994a).

D.4.3 Rotation of the Real Spherical Harmonic Density Functions

As the real spherical harmonic density functions $d_{lmp}(\theta, \phi)$ are directly related to the $Y_l^m(\theta, \phi)$ terms, their rotation follows from Eq. (D.7) (Arfken 1970). The results are, for $d_{lm+}(\theta, \phi)$,

$$d_{lm+}(\theta', \phi') = \frac{L_{lm}}{M_{lm}} (-1)^m \delta_{0m}^{(l)}(\beta) \cos (m\gamma) \sqrt{2} \, \frac{M_{l0}}{L_{l0}} d_{l0}(\theta, \phi) + \frac{L_{lm}}{M_{lm}} \sum_{m'=1}^{l}$$

$$\times \left\{ \left[(-1)^{m+m'} \delta_{m'm}^{(l)}(\beta) \cos (m\gamma + m'\alpha) + (-1)^m \delta_{-m'm}^{(l)}(\beta) \right. \right.$$

$$\left. \left. \times \cos (m\gamma - m'\alpha) \right] \frac{M_{lm'}}{L_{lm'}} d_{lm'+} \right.$$

$$+ [(-1)^{m+m'} \delta^{(l)}_{m'm}(\beta) \sin (m\gamma + m'\alpha) - (-1)^m \delta^{(l)}_{-m'm}(\beta)$$

$$\times \sin (m\gamma - m'\alpha)] \frac{M_{lm'}}{L_{lm'}} d_{lm'-}(\theta, \phi)\} \tag{D.10a}$$

and for $d_{lm}(\theta', \phi')$

$$d_{lm-}(\theta', \phi') = \frac{L_{lm}}{M_{lm}} (-1)^{m+1} \delta^{(l)}_{0m}(\beta) \sin (m\gamma) \sqrt{2} \frac{M_{l0}}{L_{l0}} d_{l0}(\theta, \phi) + \frac{L_{lm}}{M_{lm}} \sum_{m'=1}^{l}$$

$$\times \left\{ \left[(-1)^{m+m'+1} \delta^{(l)}_{m'm}(\beta) \sin (m\gamma + m'\alpha) + (-1)^{m+1} \delta^{(l)}_{-m'm}(\beta) \right. \right.$$

$$\left. \times \sin (m\gamma - m'\alpha) \right] \frac{M_{lm'}}{L_{lm'}} d_{lm'+}$$

$$+ [(-1)^{m+m'} \delta^{(l)}_{m'm}(\beta) \cos (m\gamma + m'\alpha) - (-1)^m \delta^{(l)}_{-m'm}(\beta)$$

$$\left. \times \cos (m\gamma - m'\alpha)] \frac{M_{lm'}}{L_{lm'}} d_{lm'-}(\theta, \phi) \right\} \tag{D.10b}$$

and for $m = 0$

$$d_{l0}(\theta', \phi') = \delta^{(l)}_{00}(\beta) d_{l0}(\theta, \phi) + \frac{1}{\sqrt{2}} \frac{L_{l0}}{M_{l0}} \sum_{m'=1}^{l}$$

$$\times \left\{ [(-1)^{m'} \delta^{(l)}_{m'0}(\beta) + \delta^{(l)}_{-m'0}(\beta)] \cos (m'\alpha) d_{lm'+}(\theta, \phi) \frac{M_{lm'}}{L_{lm'}} \right.$$

$$\left. + [(-1)^{m'} \delta^{(l)}_{m'0}(\beta) + \delta^{(l)}_{-m'0}(\beta)] \sin (m'\alpha) d_{lm'-}(\theta, \phi) \frac{M_{lm'}}{L_{lm'}} \right\} \tag{D.10c}$$

D.4.4 Transformation of Population Parameters

Let $\mathbf{f}$, $\mathbf{P}$ and $\mathbf{f}'$, $\mathbf{P}'$ be $(2l + 1) \times 1$ matrices representing the density-function normalized spherical harmonics and their population parameters, before and after rotation, respectively. Then, by using Eq. (D.10), we construct a $(2l + 1) \times (2l + 1)$ matrix $\mathbf{M}$ such that

$$\mathbf{f}' = \mathbf{M}\mathbf{f} \tag{D.11}$$

The population parameters transform according to

$$\mathbf{P}' = (\mathbf{M}^{-1})^T \mathbf{P} \tag{D.12}$$

For the dipolar terms, $\mathbf{M}$ is unitary, and $(\mathbf{M}^{-1})^T = \mathbf{M}$, but this is not the case for the higher moments. For the dipolar populations, the expressions are particularly simple, $(\mathbf{M} = \mathbf{R})$:

$$\begin{pmatrix} P'_{11} \\ P'_{11} \\ P'_{10} \end{pmatrix} = \mathbf{R} \begin{pmatrix} P_{11} \\ P_{11} \\ P_{10} \end{pmatrix} \tag{D.13}$$

where $\mathbf{R}$ is defined in Eq. (D.1).

Appendix E

Products of Spherical Harmonic Functions

E.1 Expressions for the Integrals over Products of Three Real Spherical Harmonic Functions

The integral over the product of three real spherical harmonic functions (Su 1993) is defined as

$$
C' \begin{pmatrix} m_1 & m_2 & m_3 \\ l_1 & l_2 & l_3 \\ p_1 & p_2 & p_3 \end{pmatrix} = \int_0^{2\pi} \int_0^{\pi} y_{l_1 m_1 p_1}(\theta, \phi) y_{l_2 m_2 p_2}(\theta, \phi) \, y_{l_3 m_3 p_3}(\theta, \phi) \sin \theta \, d\theta \, d\phi
$$

(E.1)

This integral will be zero unless $|l_2 - l_3| \leq l_1 \leq l_2 + l_3$, $l_1 + l_2 + l_3 = $ even, $p_1 = p_2 \cdot p_3$, $m_1 = |m_2 - m_3|$, or $m_1 = m_2 + m_3$. When these conditions are fulfilled, the integrals C' can be written as

$$
C' \begin{pmatrix} m & m' & m + m' \\ m & m' & l'' \\ p & p' & p \cdot p' \end{pmatrix}
$$

(E.2)

The integrals C' can be expressed in terms of the integrals of the product of three complex spherical harmonic functions:

$$
C \begin{pmatrix} m & m' & m + m' \\ l & l' & l'' \end{pmatrix} = \int_0^{2\pi} \int_0^{\pi} Y_l^m(\theta, \phi) \, Y_{l'}^{m'}(\theta, \phi) \, Y_{l''}^{m+m'}(\theta, \phi)^* \sin \theta \, d\theta \, d\phi
$$

(E.3)

The result is

$$
C'\begin{pmatrix} m & m' & m+m' \\ l & l' & l'' \\ p & p' & p \cdot p' \end{pmatrix} = \pm \frac{\sqrt{2}}{2} C\begin{pmatrix} m & m' & m+m' \\ l & l' & l'' \end{pmatrix} \tag{E.4a}
$$

if at least one of m, m', or $m + m'$ equal zero, and

$$
C'\begin{pmatrix} m & m' & m+m' \\ l & l' & l'' \\ p & p' & p \cdot p' \end{pmatrix} = \pm C\begin{pmatrix} m & m' & m+m' \\ l & l' & l'' \end{pmatrix} \tag{E.4b}
$$

if none of m, m', or $m + m'$ equal zero. The minus sign in Eq. (E.4a) and (E.4b) applies only when both p and p' are negative. In all other cases, the sign is $+$. Note that, by definition, $y_{l0} = y_{l0+}$.

The integral of the product of three complex spherical harmonic functions [Eq. (E.3)] is given by

$$
C\begin{pmatrix} m & m' & m+m' \\ l & l' & l'' \end{pmatrix} = (-1)^{(l-l')} \sqrt{\frac{(2l+1)(2l'+1)}{4\pi}} \begin{pmatrix} l & l' & l'' \\ 0 & 0 & 0 \end{pmatrix} (lml'm' \mid l''m + m') \tag{E.5}
$$

where $(lml'm' \mid l''m'')$ is the Clebsch–Gordan coefficient the numerical value of which can be calculated readily (Abramowitz and Stegun 1964, Edmonds 1974), and

$$
\begin{pmatrix} l_1 & l_2 & l_3 \\ 0 & 0 & 0 \end{pmatrix} = \begin{cases} 0 \quad \text{if } J = l_1 + l_2 + l_3 = \text{odd} \\[2mm] (-1)^{J/2} \sqrt{\dfrac{(J - 2l_1)!(J - 2l_2)!(J - 2l_3)!}{(J+1)!}} \\[4mm] \times \dfrac{(J/2)!}{(J/2 - l_1)!(J/2 - l_2)!(J/2 - l_3)!} \quad \text{if } J = l_1 + l_2 + l_3 = \text{even} \end{cases} \tag{E.6}
$$

For example, using Eqs. (E.5) and (E.6),

$$
(1011 \mid 21) = \frac{1}{\sqrt{2}},
$$

and

$$
\begin{pmatrix} 1 & 1 & 2 \\ 0 & 0 & 0 \end{pmatrix} = \sqrt{\frac{2}{15}},
$$

which gives

$$
C'\begin{pmatrix} 0 & 1 & 1 \\ 1 & 1 & 2 \\ + & - & - \end{pmatrix} = \frac{1}{10}\sqrt{\frac{15}{\pi}} = 0.218\,509\,686\,118\,416
$$

Expressions for the products of two spherical harmonic functions are given in Tables E.1 and E.2. Multiplication of both sides of the expressions by a spherical harmonic function appearing on the right-hand side, and subsequent integration, leads to equations of the type of Eq. (E.1). Thus, coefficients in Tables E.1 and E.2 are identical to the integrals C and C'.

Table E.3 lists the products of the real spherical harmonic functions in terms of the density-normalized spherical harmonic functions d_{lmp}.

TABLE E.1 Products of Two Normalized Complex Spherical Harmonic Functions

$$Y_{00}Y_{00} = 0.28209479\,Y_{00}$$
$$Y_{10}Y_{00} = 0.28209479\,Y_{10}$$
$$Y_{10}Y_{10} = 0.25231325\,Y_{20} + 0.28209479\,Y_{00}$$
$$Y_{11}Y_{00} = 0.28209479\,Y_{11}$$
$$Y_{11}Y_{10} = 0.21850969\,Y_{21}$$
$$Y_{11}Y_{11} = 0.30901936\,Y_{22}$$
$$Y_{11}Y_{11-} = -0.12615663\,Y_{20} + 0.28209479\,Y_{00}$$
$$Y_{20}Y_{00} = 0.28209479\,Y_{20}$$
$$Y_{20}Y_{10} = 0.24776669\,Y_{30} + 0.25231325\,Y_{10}$$
$$Y_{20}Y_{11} = 0.20230066\,Y_{31} - 0.12615663\,Y_{11}$$
$$Y_{20}Y_{20} = 0.24179554\,Y_{40} + 0.18022375\,Y_{20} + 0.28209479\,Y_{00}$$
$$Y_{21}Y_{00} = 0.28209479\,Y_{21}$$
$$Y_{21}Y_{10} = 0.23359668\,Y_{31} + 0.21850969\,Y_{11}$$
$$Y_{21}Y_{11} = 0.26116903\,Y_{32}$$
$$Y_{21}Y_{11-} = -0.14304817\,Y_{30} + 0.21850969\,Y_{10}$$
$$Y_{21}Y_{20} = 0.22072812\,Y_{41} + 0.09011188\,Y_{21}$$
$$Y_{21}Y_{21} = 0.25489487\,Y_{42} + 0.22072812\,Y_{22}$$
$$Y_{21}Y_{21-} = -0.16119702\,Y_{40} + 0.09011188\,Y_{20} + 0.28209479\,Y_{00}$$
$$Y_{22}Y_{00} = 0.28209479\,Y_{22}$$
$$Y_{22}Y_{10} = 0.18467439\,Y_{32}$$
$$Y_{22}Y_{11} = 0.31986543\,Y_{33}$$
$$Y_{22}Y_{11-} = -0.08258890\,Y_{31} + 0.30901936\,Y_{11}$$
$$Y_{22}Y_{20} = 0.15607835\,Y_{42} - 0.18022375\,Y_{22}$$
$$Y_{22}Y_{21} = 0.23841361\,Y_{43}$$
$$Y_{22}Y_{21-} = -0.09011188\,Y_{41} + 0.22072812\,Y_{21}$$
$$Y_{22}Y_{22} = 0.33716777\,Y_{44}$$
$$Y_{22}Y_{22-} = 0.04029926\,Y_{40} - 0.18022375\,Y_{20} + 0.28209479\,Y_{00}$$

TABLE E.2 Products of Two Real Spherical Harmonic Functions y_{lmp}, with Normalization Defined in Appendix D

$$y_{00}y_{00} = 0.28209479\,y_{00}$$
$$y_{10}y_{00} = 0.28209479\,y_{10}$$
$$y_{10}y_{10} = 0.25231325\,y_{20} + 0.28209479\,y_{00}$$
$$y_{11\pm}y_{00} = 0.28209479\,y_{11\pm}$$
$$y_{11\pm}y_{10} = 0.21850969\,y_{21\pm}$$
$$y_{11+}y_{11\pm} = 0.21850969\,y_{22+} - 0.12615663\,y_{20} + 0.28209479\,y_{00}$$
$$y_{11+}y_{11-} = 0.21850969\,y_{22-}$$
$$y_{20}y_{00} = 0.28209479\,y_{20}$$

(continued)

(*continued*)

$$y_{20}y_{10} = 0.24776669y_{30} + 0.25231325y_{10}$$
$$y_{20}y_{11\pm} = 0.20230066y_{31\pm} - 0.12615663y_{11\pm}$$
$$y_{20}y_{20} = 0.24179554y_{40} + 0.18022375y_{20} + 0.28209479y_{00}$$
$$y_{21\pm}y_{00} = 0.28209479y_{21\pm}$$
$$y_{21\pm}y_{10} = 0.23359668y_{31\pm} + 0.21850969y_{11\pm}$$
$$y_{21\pm}y_{11\pm} = 0.18467439y_{32+} - 0.14304817y_{30} + 0.21850969y_{10}$$
$$y_{21\pm}y_{11\mp} = -0.18467439y_{32-}$$
$$y_{21\pm}y_{20} = 0.22072812y_{41\pm} + 0.09011188y_{21\pm}$$
$$y_{21\pm}y_{21\pm} = 0.18022375y_{42\pm} \pm 0.15607835y_{22+} - 0.16119702y_{40} + 0.09011188y_{20} + 0.28209479y_{00}$$
$$y_{21+}y_{21-} = -0.18022375y_{42-} + 0.15607835y_{22-}$$
$$y_{22\pm}y_{00} = 0.28209479y_{22\pm}$$
$$y_{22\pm}y_{10} = 0.18467439y_{32\pm}$$
$$y_{22\pm}y_{11\pm} = \pm0.22617901y_{33+} - 0.05839917y_{31+} + 0.21850969y_{11+}$$
$$y_{22\pm}y_{11\mp} = 0.22617901y_{33-} \pm 0.05839917y_{31-} \mp 0.21850969y_{11-}$$
$$y_{22\pm}y_{20} = 0.15607835y_{42\pm} - 0.18022375y_{22\pm}$$
$$y_{22\pm}y_{21\pm} = \pm0.16858388y_{43+} - 0.06371872y_{41+} + 0.15607835y_{21+}$$
$$y_{22\pm}y_{21\mp} = 0.16858388y_{43-} \pm 0.06371872y_{41-} \mp 0.15607835y_{21-}$$
$$y_{22\pm}y_{22\pm} = \pm0.23841361y_{44+} + 0.04029926y_{40} - 0.18022375y_{20} + 0.28209479y_{00}$$
$$y_{22+}y_{22-} = 0.23841361y_{44-}$$

TABLE E.3 Products of Two Real Spherical Harmonic Functions y_{lmp}, with Normalization Defined in Appendix D

$$y_{00}y_{00} = 1.0000d_{00}$$
$$y_{10}y_{00} = 0.43301d_{10}$$
$$y_{10}y_{10} = 0.38490d_{20} + 1.0d_{00}$$
$$y_{11\pm}y_{00} = 0.43302d_{11\pm}$$
$$y_{11\pm}y_{10} = 0.31831d_{21\pm}$$
$$y_{11\pm}y_{11\pm} = 0.31831d_{22+} - 0.19245d_{20} + 1.0d_{00}$$
$$y_{11+}y_{11-} = 0.31831d_{22-}$$
$$y_{20}y_{00} = 0.43033d_{20}$$
$$y_{20}y_{10} = 0.37762d_{30} + 0.38730d_{10}$$
$$y_{20}y_{11\pm} = 0.28864d_{31\pm} - 0.19365d_{11\pm}$$
$$y_{20}y_{20} = 0.36848d_{40} + 0.27493d_{20} + 1.0d_{00}$$
$$y_{21\pm}y_{00} = 0.41094d_{21\pm}$$
$$y_{21\pm}y_{10} = 0.33329d_{31\pm} + 0.33541d_{11\pm}$$
$$y_{21\pm}y_{11\pm} = 0.26691d_{32+} - 0.21802d_{30} + 0.33541d_{10}$$
$$y_{21\pm}y_{11\mp} = -0.26691d_{32-}$$
$$y_{21\pm}y_{20} = 0.31155d_{41\pm} + 0.13127d_{21\pm}$$
$$y_{21\pm}y_{21\pm} = 0.25791d_{42+} + 0.22736d_{22+} - 0.24565d_{40} + 0.13747d_{20} + 1.0d_{00}$$
$$y_{21+}y_{21-} = 0.25790d_{42-} + 0.22736d_{22-}$$
$$y_{22\pm}y_{00} = 0.41094d_{22\pm}$$
$$y_{22\pm}y_{10} = 0.26691d_{32\pm}$$
$$y_{22\pm}y_{11\pm} = \pm0.31445d_{33+} - 0.083323d_{31+} + 0.33541d_{11+}$$
$$y_{22\pm}y_{11\mp} = 0.31445d_{33-} \pm 0.083323d_{31-} \mp 0.33541d_{11-}$$
$$y_{22\pm}y_{20} = 0.22335d_{42\pm} - 0.26254d_{22\pm}$$
$$y_{22\pm}y_{21\pm} = \pm0.23873d_{43+} - 0.089938d_{41+} + 0.22736d_{21+}$$
$$y_{22\pm}y_{21\mp} = 0.23873d_{43-} \pm 0.089938d_{41-} \mp 0.22736d_{21-}$$
$$y_{22\pm}y_{22\pm} = \pm0.31831d_{44+} + 0.061413d_{40} - 0.27493d_{20} + 1.0d_{00}$$
$$y_{22+}y_{22-} = 0.31831d_{44-}$$

Appendix F

Energy-Optimized Single-ζ Slater Values for Subshells of Isolated Atoms

TABLE F.1 Best Values of ζ (au^{-1}) for the Ground States of Neutral Atoms[1]

Element	1s	2s	2p	3s	3p	4s	3d	4p
He	1.6875							
Li	2.6906	0.6396						
Be	3.6848	0.9560						
B	4.6795	1.2881	1.2107					
C	5.6727	1.6083	1.5679					
N	6.6651	1.9237	1.9170					
O	7.6579	2.2458	2.2266					
F	8.6501	2.5638	2.5500					
Ne	9.6421	2.8792	2.8792					
Na	10.6259	3.2857	3.4009	0.8358				
Mg	11.6089	3.6960	3.9129	1.1025				
Al	12.5910	4.1068	4.4817	1.3724	1.3552			
Si	13.5745	4.5100	4.9725	1.6344	1.4284			
P	14.5578	4.9125	5.4806	1.8806	1.6288			
S	15.5409	5.3144	5.9885	2.1223	1.8273			
Cl	16.5239	5.7152	6.4966	2.3561	2.0387			
Ar	17.5075	6.1152	7.0041	2.5856	2.2547			
K	18.4895	6.5031	7.5136	2.8933	2.5752	0.8738		
Ca	19.4730	6.8882	8.0207	3.2005	2.8861	1.0995		
Sc	20.4566	7.2868	8.5273	3.4466	3.1354	1.1581	2.3733	
Ti	21.4409	7.6883	9.0324	3.6777	3.3679	1.2042	2.7138	
V	22.4256	8.0907	9.5364	3.9031	3.5950	1.2453	2.9943	
Cr	23.4138	8.4919	10.0376	4.1226	3.8220	1.2833	3.2522	
Mn	24.3957	8.8969	10.5420	4.3393	4.0364	1.3208	3.5094	

(continued)

[1] As defined by $R(r) = [\zeta^{n+3}/(n+2)!]\, r^n \exp(-\zeta r)$, where n is the principal quantum number.

(*continued*)

Element	1s	2s	2p	3s	3p	4s	3d	4p
Fe	25.3810	9.2995	11.0444	4.5587	4.2593	1.3585	3.7266	
Co	26.3668	9.7025	11.5462	4.7741	4.4782	1.3941	3.9518	
Ni	27.3526	10.1063	12.0476	4.9870	4.6950	1.4277	4.1765	
Cu	28.3386	10.5099	12.5485	5.1981	4.9102	1.4606	4.4002	
Zn	29.3245	10.9140	13.0490	5.4064	5.1231	1.4913	4.6261	
Ga	30.3094	11.2995	13.5454	5.6654	5.4012	1.7667	5.0311	1.5554
Ge	31.2937	11.6824	14.0411	5.9299	5.6712	2.0109	5.4171	1.6951
As	32.2783	12.0635	14.5368	6.1985	5.9499	2.2360	5.7928	1.8623
Se	33.2622	12.4442	15.0326	6.4678	6.2350	2.4394	6.1590	2.0718
Br	34.2471	12.8217	15.5282	6.7395	6.5236	2.6382	6.5197	2.2570
Kr	35.2316	13.1990	16.0235	7.0109	6.8114	2.8289	6.8753	2.4423

Source: Clementi and Raimondi (1963). For double zeta functions, see Clementi (1965).

Appendix G

Fourier–Bessel Transforms

TABLE G.1 Closed-Form Expressions for Fourier Transform of Slater-Type Functions

$$G_{N,k}(K, Z) \equiv \int_0^\infty r^N \exp(-Zr) j_k(Kr)\, dr, \quad \text{with} \quad K = 4\pi \sin\theta/\lambda$$

	N				
k	1	2	3	4	5
0	$\dfrac{1}{K^2 + Z^2}$	$\dfrac{2Z}{(K^2 + Z^2)^2}$	$\dfrac{2(3Z^2 - K^2)}{(K^2 + Z^2)^3}$	$\dfrac{24Z(Z^2 - K^2)}{(K^2 + Z^2)^4}$	$\dfrac{24(5Z^2 - 10K^2Z^2 + K^4)}{(K^2 + Z^2)^5}$
1		$\dfrac{2K}{(K^2 + Z^2)^2}$	$\dfrac{8KZ}{(K^2 + Z^2)^3}$	$\dfrac{8K(5Z^2 - K^2)}{(K^2 + Z^2)^4}$	$\dfrac{48KZ(5Z^2 - 3K^2)}{(K^2 + Z^2)^5}$
2			$\dfrac{8K^2}{(K^2 + Z^2)^3}$	$\dfrac{48K^2Z}{(K^2 + Z^2)^4}$	$\dfrac{48K^2(7Z^2 - K^2)}{(K^2 + Z^2)^5}$
3				$\dfrac{48K^3}{(K^2 + Z^2)^4}$	$\dfrac{384K^3Z}{(K^2 + Z^2)^5}$
4					$\dfrac{384K^4}{(K^2 + Z^2)^5}$
5					
6					
7					

(continued)

(*continued*)

k	N 6	7	8
0	$\dfrac{240Z(K^2-3Z^2)(3K^2-Z^2)}{(K^2+Z^2)^6}$	$\dfrac{720(7Z^6-35K^2Z^4+21K^4Z^2-K^6)}{(K^2+Z^2)^7}$	$\dfrac{40320(Z^7-7K^2Z^5+7K^4Z^3-K^6Z)}{(K^2+Z^2)^8}$
1	$\dfrac{48K(35Z^4-42K^2Z^2+3K^4)}{(K^2+Z^2)^6}$	$\dfrac{1920KZ(7Z^4-14K^2Z^2+3K^4)}{(K^2+Z^2)^7}$	$\dfrac{5760K(21Z^6-63K^2Z^4+27K^4Z^2-K^6)}{(K^2+Z^2)^8}$
2	$\dfrac{384K^2Z(7Z^2-3K^2)}{(K^2+Z^2)^6}$	$\dfrac{1152K^2(21Z^4-18K^2Z^2+K^4)}{(K^2+Z^2)^7}$	$\dfrac{11520K^2Z(21Z^4-30K^2Z^2+5K^4)}{(K^2+Z^2)^8}$
3	$\dfrac{384K^3(9Z^2-K^2)}{(K^2+Z^2)^6}$	$\dfrac{11520K^3Z(3Z^2-K^2)}{(K^2+Z^2)^7}$	$\dfrac{11520K^3(33Z^4-22K^2Z^2+K^4)}{(K^2+Z^2)^8}$
4	$\dfrac{3840K^4Z}{(K^2+Z^2)^6}$	$\dfrac{3840K^4(11Z^2-K^2)}{(K^2+Z^2)^7}$	$\dfrac{46080K^4Z(11Z^2-3K^2)}{(K^2+Z^2)^8}$
5	$\dfrac{3840K^5}{(K^2+Z^2)^6}$	$\dfrac{46080K^5Z}{(K^2+Z^2)^7}$	$\dfrac{40680K^5(13Z^2-K^2)}{(K^2+Z^2)^8}$
6		$\dfrac{46080K^6}{(K^2+Z^2)^7}$	$\dfrac{645120K^6Z}{(K^2+Z^2)^8}$
7			$\dfrac{645120K^7}{(K^2+Z^2)^8}$

Source: Avery and Watson (1977), Su and Coppens (1990).

Appendix H

Evaluation of the Integrals $A_{N,l_1,l_2,k}(Z,R)$ Occurring in the Expression for the Peripheral Contribution to the Electrostatic Properties

Expression (8.47) for the peripheral contributions to the electrostatic potential contains the integrals

$$A_{N,l_1,l_2,k}(Z, R) = \int_0^\infty G_{N+2,l_1}(Z, S) j_{l_2}(SR) S^k \, dS \tag{H.1}$$

They can be evaluated by substitution of the expressions for $G_{N+2,l_1}(Z, S)$ (Avery and Watson 1977, Su and Coppens 1990, Su 1993; see Appendix G) and $j_l(x)$ (Arfken 1970; see also Table 3.7), and subsequent use of the following integrals (Gradsteyn and Ryzhik 1965)

$$\int_0^\infty \frac{\sin (ax) \, dx}{x(x^2 + \beta^2)^{(n+1)}} = \frac{\pi}{2\beta^{(2n+2)}} \left[1 - \frac{e^{-a\beta}}{2^n n!} F_n(a\beta) \right]$$

$$[a>0, \, \mathrm{Re} \, \beta>0, \, F_0(z)=1, \, F_1(z)=z+2, \ldots, F_n(z)=(z+2n)F_{n-1}(z)-zF'_{n-1}(z)] \tag{H.2}$$

$$\int_0^\infty \frac{x^{(2m+1)} \sin (ax) \, dx}{(x^2 + z)^{(n+1)}} = \frac{(-1)^{(m+n)}}{n!} \frac{\pi}{2} \frac{d^n}{dz^n} (z^m e^{-a\sqrt{z}})$$

$$[a > 0, \, 0 \le m \le n, \, |\arg z| < \pi] \tag{H.3}$$

$$\int_0^\infty \frac{x^{(2m)} \cos (ax) \, dx}{(x^2 + z)^{(n+1)}} = \frac{(-1)^{(m+n)}}{n!} \frac{\pi}{2} \frac{d^n}{dz^n} (z^{(m-1/2)} e^{-a\sqrt{z}})$$

$$[a > 0, \, 0 \le m < n + 1, \, |\arg z| < \pi] \tag{H.4}$$

Some results are

$$A_{0,0,0,0}(Z, R) = \frac{\pi(2 - e^{-RZ}(RZ + 2))}{2RZ^3} \tag{H.5}$$

$$A_{0,0,1,1}(Z, R) = \frac{\pi(2 - e^{-RZ}[(RZ)^2 + 2RZ + 2]\}}{2R^2 Z^3} \tag{H.6}$$

$$A_{0,0,2,2}(Z, R) = \frac{\pi[6 - e^{-RZ}[(RZ)^3 + 3(RZ)^2 + 6RZ + 6]\}}{2R^3 Z^3} \tag{H.7}$$

Alternatively, the integrals in Eq. (H.1) can be reduced to integrals of the following form (Gradshteyn and Ryzhik 1965)

$$\int_0^\infty \frac{x^{\rho - 1} J_\nu(ax)}{(x^2 + k^2)^{\mu + 1}}\, dx = \frac{a^\nu k^{\rho + \nu - 2\mu - 2} \Gamma(\frac{1}{2}\rho + \frac{1}{2}\nu) \Gamma(\mu + 1 - \frac{1}{2}\rho - \frac{1}{2}\nu)}{2^{\nu + 1} \Gamma(\mu + 1) \Gamma(\nu + 1)}$$

$$\times {}_1F_2\left(\tfrac{1}{2}\rho + \tfrac{1}{2}\nu; \tfrac{1}{2}\rho + \tfrac{1}{2}\nu - \mu, \nu + 1; \frac{a^2 k^2}{4}\right)$$

$$+ \frac{a^{2\mu + 2 - \rho} \Gamma(\frac{1}{2}\rho + \frac{1}{2}\nu - \mu - 1)}{2^{2\mu + 3 - \rho} \Gamma(\mu + 2 + (\nu - \rho)/2)}$$

$$\times {}_1F_2\left(\mu + 1; \mu + 2 + \frac{\nu - \rho}{2}, \mu + 2 - \frac{\nu + \rho}{2}; \frac{a^2 k^2}{4}\right)$$

$$[a > 0,\ -Re\,\nu < Re\,\rho < 2Re\,\mu + \tfrac{7}{2}] \tag{H.8}$$

where the hypergeometric function ${}_1F_2(a; b, c; x)$ is defined as

$${}_1F_2(a; b,c; x) = \sum_{n=0}^{\infty} \frac{(a)_n}{(b)_n (c)_n n!} x^n \tag{H.9}$$

and $(a)_n = a(a + 1)\ldots(a + n - 1)$, with $(a)_0 = 1$.

Appendix I

The Matrix $\mathbf{M}^{-1}$ Relating d-Orbital Occupancies P_{ij} to Multipole Populations P_{lmp} (Eq. 10.9)

TABLE I.1 Matrix Relating d-Orbital Occupancies P_{ij} to Multipole Populations P_{lmp} (Eq. 10.9)

d-Orbital Populations	Multipole Populations					
	P_{00}	P_{20}	P_{22+}	P_{40}	P_{42+}	P_{44+}
P_{z}	0.200	1.039	0.00	1.396	0.00	0.00
P_{xz}	0.200	0.520	0.942	−0.931	1.108	0.00
P_{yz}	0.200	0.520	−0.942	−0.931	−1.108	0.00
$P_{x^2-y^2}$	0.200	−1.039	0.00	0.233	0.00	1.571
P_{xy}	0.200	−1.039	0.00	0.233	0.00	−1.571

Mixing Terms	P_{21}	P_{21-}	P_{22+}	P_{22-}	P_{41+}	P_{41-}
$P_{z^2/xz}$	1.088	0.00	0.00	0.00	2.751	0.00
$P_{z^2/yz}$	0.00	1.088	0.00	0.00	0.00	2.751
P_{z^2/x^2-y^2}	0.00	0.00	−2.177	0.00	0.00	0.00
$P_{z^2/xy}$	0.00	0.00	0.00	−2.177	0.00	0.00
$P_{xz/yz}$	0.00	0.00	0.00	1.885	0.00	0.00
P_{xz/x^2-y^2}	1.885	0.00	0.00	0.00	−0.794	0.00
$P_{xz/xy}$	0.00	1.885	0.00	0.00	0.00	−0.794
P_{yz/z^2-y^2}	0.00	−1.885	0.00	0.00	0.00	0.794
$P_{yz/xy}$	1.885	0.00	0.00	0.00	−0.794	0.00
$P_{x^2-y^2/xy}$	0.00	0.00	0.00	0.00	0.00	0.00

Mixing Terms	P_{42+}	P_{42-}	P_{43+}	P_{43-}	P_{44+}
$P_{z^2/xz}$	0.00	0.00	0.00	0.0	0.00
$P_{z^2/yz}$	0.00	0.00	0.00	0.0	0.00
P_{z^2/x^2-y^2}	1.919	0.00	0.00	0.0	0.00
$P_{z^2/xy}$	0.00	1.919	0.00	0.0	0.00
$P_{xz/yz}$	0.00	2.216	0.00	0.0	0.00
P_{xz/x^2-y^2}	0.00	0.00	2.094	0.0	0.00
$P_{xz/xy}$	0.00	0.00	0.00	2.094	0.00
P_{yz/z^2-y^2}	0.00	0.00	0.00	2.094	0.00
$P_{yz/xy}$	0.00	0.00	−2.094	0.00	0.00
$P_{x^2-y^2/xy}$	0.00	0.00	0.00	0.00	3.142

Source: Holladay et al. (1983), Coppens and Becker (1992).

Appendix J

The Interaction Between Two Nonoverlapping Charge Distributions

Let $T_{\alpha\beta\gamma\ldots\nu} = (4\pi\varepsilon_0)^{-1}\nabla_\alpha\nabla_\beta\nabla_\gamma\ldots\nabla_\nu R^{-1}$ be defined as a tensor which is proportional to $R^{-(l+1)}$, symmetric with respect to interchange of any pair of suffixes, and reduced to zero on contraction, that is, when at least two indices are equal. The first five T-tensors are

$$T = (4\pi\varepsilon_0)^{-1}R^{-1} \tag{J.1}$$

$$T_\alpha = (4\pi\varepsilon_0)^{-1}\nabla_\alpha R^{-1} = -(4\pi\varepsilon_0)^{-1}R_\alpha R^{-3} \tag{J.2}$$

$$T_{\alpha\beta} = (4\pi\varepsilon_0)^{-1}(3R_\alpha R_\beta - R^2\delta_{\alpha\beta})R^{-5} \tag{J.3}$$

$$T_{\alpha\beta\gamma} = (4\pi\varepsilon_0)^{-1}(-3)[5R_\alpha R_\beta R_\gamma - R^2(R_\alpha\delta_{\beta\gamma} + R_\beta\delta_{\gamma\alpha} + R_\gamma\delta_{\alpha\beta})]R^{-7} \tag{J.4}$$

$$\begin{aligned}
T_{\alpha\beta\gamma\delta} = (4\pi\varepsilon_0)^{-1}3[&35R_\alpha R_\beta R_\gamma R_\delta - 5R^2(R_\alpha R_\beta\delta_{\gamma\delta} + R_\alpha R_\gamma\delta_{\beta\delta}\\
&+ R_\alpha R_\delta\delta_{\beta\gamma} + R_\beta R_\gamma\delta_{\alpha\delta} + R_\beta R_\delta\delta_{\alpha\gamma} + R_\gamma R_\delta\delta_{\alpha\beta})\\
&+ R^4(\delta_{\alpha\beta}\delta_{\gamma\delta} + \delta_{\alpha\gamma}\delta_{\beta\delta} + \delta_{\alpha\delta}\delta_{\beta\gamma})]R^{-9}
\end{aligned} \tag{J.5}$$

The electrostatic interaction between two nonoverlapping charge distributions A and B, consisting of N_A and N_B atoms, respectively, and each represented by their atom-centered multipole moments, is given by (using the Einstein summation convention for the indices α, β, γ) (Buckingham 1978)

$$\begin{aligned}
E_{es} = \sum_i^{N_A}\sum_j^{N_B} &Tq_iq_j + T_\alpha(q_i\mu_{\alpha,j} - q_j\mu_{\alpha,i}) + T_{\alpha\beta}(\tfrac{1}{3}q_i\Theta_{\alpha\beta,j} + \tfrac{1}{3}q_j\Theta_{\alpha\beta,i} - \mu_{\alpha,i}\mu_{\beta,j})\\
&+ T_{\alpha\beta\gamma}(\tfrac{1}{15}q_i\Omega_{\alpha\beta\gamma,j} - \tfrac{1}{15}q_j\Omega_{\alpha\beta\gamma,i} - \tfrac{1}{3}\mu_{\alpha,i}\Theta_{\beta\gamma,i} + \tfrac{1}{3}\mu_{\alpha,j}\Theta_{\beta\gamma,i}\\
&+ T_{\alpha\beta\gamma\delta}(\tfrac{1}{9}\Theta_{\alpha\beta,i}\Theta_{\gamma\delta,j} + \cdots) + \cdots
\end{aligned}$$

$$= \sum_i \sum_j \sum_{l=0}^{\infty} \sum_{l'=0}^{\infty} (-1)^{l'} \frac{1}{1.3.5 \ldots (2l-1)} \frac{1}{1.3.5 \ldots (2l'-1)}$$

$$\times \, T_{\alpha\beta\gamma\ldots\nu\alpha'\beta'\gamma'\ldots\nu'} M^{(l)}_{\alpha\beta\gamma\ldots\nu,i} M^{(l')}_{\alpha'\beta'\gamma'\ldots\nu',j} \tag{J.6}$$

in which $M^{(l)}_{\alpha\beta\gamma\ldots\nu,i}$ is the lth multipole moment of atom i, as defined by expression (7.3).

Appendix K

Conversion Factors

TABLE K.1 Atomic Versus SI Units

Quantity	Symbol	Name or Value[a]	Expression	Value
Charge	e	Millikan	e	$1.6021773 \cdot 10^{-19}$ C
Mass	m_e	Thomson	m_e	$9.1093897 \cdot 10^{-31}$ kg
Action	$\hbar$	Planck	$\hbar$	$1.05457266 \cdot 10^{-34}$ Js
Length	a_0	Bohr	$\hbar^2 4\pi\varepsilon_0/m_e e^2$	$5.29177249 \cdot 10^{-11}$ m
Energy	E_h	Hartree	$e^2/4\pi\varepsilon_0 a_0$	$4.3597482 \cdot 10^{-18}$ J
Time	t_0	Jiffy	$4\pi\varepsilon_0 \hbar a_0/e^2$	$2.418884468 \cdot 10^{-17}$ s
Force			$e^2/4\pi\varepsilon_0 a_0^2$	$8.23872833 \cdot 10^{-8}$ N
Velocity	v_0		$a_0 t_0^{-1}$	$2.187691291 \cdot 10^6$ ms^{-1}
Momentum	p_0	Dumond	$m v_0 = \hbar a_0^{-1}$	$1.99285340 \cdot 10^{-24}$ kg ms^{-1}
Electron density			$e a_0^{-3}$	$1.08121026 \cdot 10^{12}$ C cm^{-3}
Electrostatic potential			$e/4\pi\varepsilon_0 a_0$	27.2113961 C m^{-1}
Velocity of light	c	137.036 Bohr jiffy^{-1}	c	$2.99792458 \cdot 10^8$ ms^{-1}
Electric field	E	Stark	$e/4\pi\varepsilon_0 a_0^2$	$5.1422082 \cdot 10^{11}$ C m^{-2}
Dipole moment	μ		$e a_0$	$8.47835792 \cdot 10^{-30}$ C m^2

[a] Value only given if different from 1.

Source: Smith (1982) (Modified).

TABLE K.2 Conversion Factors

$Energy$:   1 au energy $= 1\ E_h = 1$ hartree $= \hbar^2 m_e^{-1} a_0^{-2}$

$$= 27.211\ \text{eV} = 2\ \text{Rydbergs}$$

$$= 627.509\ \text{kcal mol}^{-1} = 2625.50\ \text{kJ mol}^{-1}$$

$$= 4.35975 \cdot 10^{-11}\ \text{ergs} = 4.35975 \cdot 10^{-18}\ \text{J}$$

$Length$:   1 au length $= 1$ bohr $= 1\ a_0 = 0.529177$ Å

$$= 0.529177 \cdot 10^{-10}\ \text{m}$$

$Electron\ density$:   1 au electron density $= e a_0^{-3} = 6.748315$ eÅ^{-3}

$$= 1.0812 \cdot 10^{12}\ \text{C m}^{-3}$$

$$1\ \text{eÅ}^{-3} = 0.148185\ e\ (\text{au})^{-3}$$

$Electrostatic\ potential$:   1 au electrostatic potential $= e/4\pi e_0 a_0 = 1.889726$ eÅ^{-1}

$$= 27.2114\ \text{V}$$

$Electric\ field\ gradient$:   1 au electric field gradient $= e a_0^{-3} = 6.74833$ eÅ^{-3}

$$= 9.7173646 \cdot 10^{21}\ \text{cm}^{-3}$$

TABLE K.3 Dipole and Quadrupole (or Second) Moment Conversion Factors[a]

	Dipole Moments				
	$e\,a_0$	e Å	C m	esu	D
1 $e\,a_0$	1.0	0.529177	$8.47836 \cdot 10^{-30}$	$2.54175 \cdot 10^{-18}$	2.54175
1 eÅ	1.88973	1.0	$1.60218 \cdot 10^{-29}$	$4.80321 \cdot 10^{-18}$	4.80321
1 C m	$1.17947 \cdot 10^{29}$	$6.24151 \cdot 10^{28}$	1.0	$2.99792 \cdot 10^{11}$	$2.99792 \cdot 10^{29}$
1 esu	$3.93430 \cdot 10^{17}$	$2.08194 \cdot 10^{17}$	$3.33564 \cdot 10^{-12}$	1.0	$1.0 \cdot 10^{18}$
1 D	0.393430	0.208194	$3.33564 \cdot 10^{-30}$	$1.0 \cdot 10^{-18}$	1.0

	Quadrupole and Second Moments				
	$e\,a_0^2$	e Å^2	C m^2	esu	B
1 $e\,a_0^2$	1.0	0.280029	$4.48655 \cdot 10^{-40}$	$1.34504 \cdot 10^{-26}$	1.34504
1 e Å^2	3.57106	1.0	$1.60218 \cdot 10^{-39}$	$4.80321 \cdot 10^{-26}$	4.80321
1 C m^2	$2.22888 \cdot 10^{39}$	$6.24151 \cdot 10^{38}$	1.0	$2.99792 \cdot 10^{13}$	$2.99792 \cdot 10^{39}$
1 esu	$7.43475 \cdot 10^{25}$	$2.08194 \cdot 10^{25}$	$3.33564 \cdot 10^{-14}$	1.0	$1.0 \cdot 10^{26}$
1 B	0.743475	0.208194	$3.33564 \cdot 10^{-40}$	$1.0 \cdot 10^{-26}$	1.0

[a] Fundamental constants are from Cohen and Taylor (1987) ($e = 1.60217733 \cdot 10^{-19}$ C, $a_0 = 0.529177249 \cdot 10^{-10}$ m, $4\pi\varepsilon_0 = 10^7\ c^{-2}$, $c = 2.99792458 \cdot 10^8$ m s^{-1}, esu of charge $= 3.33564095 \cdot 10^{-10}$ C).

$Source$: Spackman (1992).

Appendix L

Selected Exercises

Chapter 1

(1) A hydrogen atom is described by the Gaussian radial density function

$$\rho(r) = N \exp(-\kappa^2 \alpha r^2)$$

Derive the scattering factor expression for this atom.

Chapter 2

(1) Evaluate the bias in the isotropic temperature factor of the hydrogen atom of problem 1.1 as a function of the true κ value, if κ is arbitrarily set to 1 in a refinement.

Chapter 3

(1) A radial density function of an atom is given by

$$\frac{\zeta^{n+3}}{(n+2)!} r^n \exp(-\xi r)$$

where $n > 0$. Determine the position of the maximum of this function. For $n = 4$, which value of ξ causes the maximum to be at 0.5 Å from the nucleus?

(2) The valence scattering factor of an isolated sulfur atom, normalized to one electron, is given as a function of $\sin \theta / \lambda$ by the following values (interval in $\sin \theta / \lambda = 0.05 \, \text{Å}^{-1}$)

1.0; 0.920; 0.720; 0.484; 0.278; 0.131; 0.040; -0.006

Derive the valence scattering factor for $\kappa = 0.95$, using linear interpolation.

(3) Use the normalization factors described in chapter 3 to derive the coefficients in the expression

$$y_{10}y_{10} = Ad_{20} + Bd_{00}$$

from the corresponding expression in terms of the wave function normalized spherical harmonic functions.

(4) An experiment is performed to evaluate the deformation density in an O—O bond in a molecule. Two different reference states are used to calculate deformation densities. The first is the spherical-atom reference state (the promolecule density); the second is a prepared-atom reference state in which oxygen atoms have the configuration

$$(1s)^2(2s)^2(2p_x)^2(2p_y)^1(2p_z)^1$$

The O—O bond is 1.200 Å long, and directed along the z axis of the coordinate system of both oxygen atoms. The p-orbital radial functions of the oxygen atoms are given by [compare expression (3.34)]

$$R(r) = \kappa^{3/2}\frac{(2\zeta)^{5/2}}{4!^{1/2}}(\kappa r)\exp(-\kappa\zeta r) \qquad \text{with } \zeta = 2.227 \text{ au}^{-1} \text{ and } \kappa = 0.95$$

$$(3.34)$$

(a) Calculate the difference in *electron density* between the two deformation densities at the bond midpoint.

(b) Use the expression of problem 3 to express the difference between the two deformation densities in terms of the multipolar functions d_{lmp}.

(5a) Given the Slater-type radial function for the density:

$$R(r) = \kappa^3\zeta^{(k+3)}/(k+2)!(\kappa r)^k \exp(-\zeta\kappa r)$$

find the Fourier–Bessel transforms $\langle j_n \rangle$ for $n = 2, 3, 4$ and $k = 3, 4$.

(5b) Evaluate the scattering factors at $\sin\theta/\lambda = 0.4 \text{ Å}^{-1}$.

(6) Two Cartesian coordinate systems are related by the rotation

$$\begin{pmatrix} X' \\ Y' \\ Z' \end{pmatrix} = \begin{pmatrix} -0.6984 & -0.6604 & -0.3188 \\ -0.2511 & -0.1988 & 0.9473 \\ -0.6890 & 0.7241 & -0.0306 \end{pmatrix}\begin{pmatrix} X \\ Y \\ Z \end{pmatrix}$$

(a) Find the Eulerian angles describing the rotation.

(b) In the unprimed system, a pseudo atom has the following dipolar populations $P_{10} = 0.15$, $P_{11+} = 0.33$, $P_{11-} = 0.05$. What is the magnitude of the atomic dipole moment if $\kappa = 1$, $k = 2$ and $\zeta = 3.2 \text{ au}^{-1}$, and what are the dipolar populations in the primed system?

(c) In the unprimed system, a pseudoatom has the following nonzero quadrupolar populations

$$P_{20} = 0.14, \qquad P_{21+} = 0.03, \qquad P_{22+} = 0.12, \qquad \text{and } P_{22-} = 0.01$$

What are the quadrupolar populations in the primed system?

Chapter 4

(1) An experimental curve is fitted by a Gaussian function centered at x_0, that is:

$$y = \frac{N}{\sigma\sqrt{2\pi}} \exp\left[-\frac{(x - x_0)^2}{2\sigma^2}\right]$$

The variables are σ, x_0, and N. Derive the expressions for all unique elements of the least-squares matrix **B**.

(2) The cyclic molecule of tetrasulfurtetraimide $S_4(NH)_4$ has idealized 4 mm symmetry, with the mirror planes containing the NH groups. A picture of the molecule is shown in Fig. L.1. In the crystal, the symmetry of the isolated molecule is reduced to m due to extensive hydrogen bonding (Gregson et al. 1988). The charge-density basis set applied is the same as that used for tetrasulfur tetranitride, S_4N_4, discussed in chapter 4.

Derive the charge-density parameters required for molecular symmetries equal to 1, to m, and to 4 mm by

(a) applying chemical equivalence imposed by the molecular symmetry;
(b) applying both chemical equivalence and local symmetry at each of the atoms.

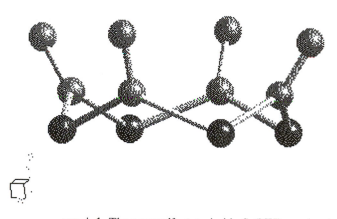

FIG. L.1 The tetrasulfurtetraimide $S_4(NH)_4$ molecule.

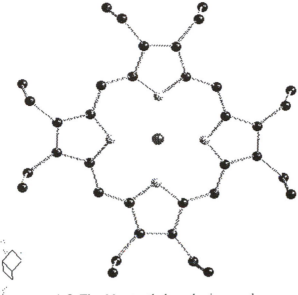

FIG. L.2 The M-octaethylporphyrin complex.

(3)[1] Bis-fluoro germanium octaethylporphyrin $C_{36}H_{44}N_4GeF_2$ (Ge-OEPF$_2$) crystallizes in space group $I4_1/a$ ($Z = 4$) (Fig. L.2) with Ge at a site with $\bar{4}$ symmetry. Figure L.2 is a view of M-OEP, where M is the cation; in Ge-OEPF$_2$ the F atoms are in axial positions coordinated to Ge. In the charge density refinement of the complex, the multipolar density functions are defined in local coordinate systems on each of the atoms.

(a) What are the constraints due to (i) crystallographic and (ii) chemical equivalence?
(b) Which local symmetry would you choose for the Ge, F, N, and C atoms?
(c) Choose local coordinate systems on each of the atoms such that the constraints of part (b) can be satisfied.
(d) Which are the nonzero elements of the U^{ij} tensor for site symmetries 2 and $\bar{4}$?
(e) Which are the nonvanishing multipolar populations for site symmetries 2, $\bar{4}$, $mm2$, and m? Choose the z axis along the main symmetry axis, noting that $m = \bar{2}$.

---

[1] Contributed by C. Lecomte and V. Pichon-Pesme, University of Nancy, France.

Chapter 7

(1) Derive the expression

$$\Theta_{yy} = \tfrac{1}{2}(-3\Theta_{22+} - \tfrac{1}{2}\Theta_{20})$$

from the definitions of the electrostatic moments.

(2) Hirshfeld's stockholder partitioning applied to a theoretical density of hydrogen cyanide, HCN, gives the following values for the atomic charges, dipole moments, and second moments μ_{zz}:

HCN	H	C	N
q (e)	+0.133	+0.066	−0.201
μ_z(eÅ)	−0.104	−0.161	−0.045
μ_{zz}(eÅ^2)	+0.089	+0.046	−0.134

The bond lengths in the molecule are: C≡N: 1.18 Å, C—H: 1.05 Å. The positive z axis is towards the nitrogen atom.

Derive the molecular dipole moment in D, and the molecular second moment μ_{zz} in units of 10^{-40} Cm^2.

Chapter 8

(1) Use the net charges for HCN as calculated by the stockholder concept from the theoretical density (problem 7.2) to evaluate:

(a) the electrostatic potential,
(b) the electric field,

at two points outside the molecule, located on the molecular axis at 2 Å from the H and the N atoms, respectively.

(2) Repeat (1) using all known atomic moments.

Chapter 10

(1) The population parameters on the iron atom (bispyridyl)iron(II) tetra-phenylporphyrin were determined from the data on two different crystals. Results are as follows:

	Crystal 1	Crystal 2
Fe symmetry	D_{4h}	D_{4h}
κ	1.00	1.00
κ'	0.89(1)	0.90(1)
$P_{00}(3d)$	6.82(4)	7.30(4)
P_{20}	0.02(2)	0.08(2)
P_{40}	−0.21(2)	−0.21(2)
P_{44+}	−0.20(2)	−0.27(2)

Derive the orbital populations in both cases.

(2) The population parameters of the iron atom in FeS_2 are given below. The symmetry of the iron site is $\bar{3}$. The z axis is along the $\bar{3}$ axis.

$$P_{00} = 5.85\ (15)$$

$$P_{20} = 0.05\ (4)$$

$$P_{40} = 0.30\ (3)$$

$$P_{43+} = -0.38\ (3)$$

$$P_{43-} = 0.01\ (3)$$

(a) Explain why P_{43-} is very small. What does this imply for the local symmetry of the Fe d-electron density?

(b) Use the results to calculate the diagonal elements of the electric field gradient at the iron nucleus. Express the results in SI units. What are the values of the off-diagonal elements of ∇E?

(c) Derive the d-orbital populations and the associated errors, given the following correlation coefficients (only the lower triangle of the matrix is given).

	P_{00}	P_{20}	P_{40}	P_{43+}	P_{43+}
P_{00}	1.0				
P_{20}	0.02	1.0			
P_{40}	0.11	0.03	1.0		
P_{43+}	0.04	0.04	0.04	1.0	
P_{43-}	0.01	0.00	0.03	0.03	1.0

(d) Repeat the previous calculation for the case that $\gamma(P_{00}, P_{20}) = 0.6$, while all other values are unaltered.

References

Chapter 1

Arfken, G., *Mathematical Methods for Physicists*, 2nd ed., Academic Press: New York, London (1970), p. 522.

Bacon, G. E., *Neutron Diffraction*, Clarendon Press: Oxford (1962).

Blume, M., *J. Appl. Phys.* **57**, 3615 (1985).

Blume, M., in *Resonance Anomalous X-ray Scattering. Theory and Applications*, G. Materlik, C. J. Sparks, and K. Fischer (eds.), North-Holland: Amsterdam (1994), p. 495.

Cohen-Tannoudji, C., Diu, B., and Laloe, F., *Quantum Mechanics*, John Wiley and Sons: New York (1977).

Cromer, D. T. and Liberman, D., *J. Chem. Phys.* **53**, 1891 (1970), FORTRAN program FPRIME.

Feil, D., *Isr. J. Chem.* **16**, 103 (1975).

Feil, D., *Lecture Notes on Light and Matter*, University of Twente, Enschede, The Netherlands (1992).

Finkelstein, K. D., Shen, Q., and Shastri, S., *Phys. Rev. Lett.* **69**, 1612 (1992).

Gerward, L., Thuesen, G., Stibius Jensen, M., and Alstrup, I., *Acta Cryst.* A**35**, 852 (1979).

Giacovazzo, C., in *Fundamentals of Crystallography*, C. Giacovazzo (ed.), Oxford University Press: Oxford (1992), p. 167.

Hendrickson, W. A., Smith, J. L., Phizackerley, R. P., and Merritt, E. A., *Proteins: Structure, Function, and Genetics* **4**, 77 (1988).

Hoyt, J. J., De Fontaine, D., and Warburton, W. K., *J. Appl. Cryst.* **17**, 344 (1984).

International Tables for Crystallography, Vol. C, Kluwer: Dordrecht (1992).

International Tables for X-ray Crystallography, Vol. IV, Kynoch Press: Birmingham, England (1974).

Jackson, J. D., *Classical Electrodynamics*, John Wiley and Sons: New York (1977).

James, R. W., *The Optical Principles of the Diffraction of X-rays*. OxBow Press: Woodbridge, CT (1982).

Kissel, L. and Pratt, R. H., *Acta Cryst.* A**46**, 170 (1990).

Kissel, L., Zhou, B., Roy, S. C., Sen Gupta, S. K., and Pratt, R. H., *Acta Cryst.* A**51**, 271 (1995).

Pickering, I. J., Sansone, M., Marsch, J. J., and George, G. N., *J. Am. Chem. Soc.* **115**, 6302 (1993).

Sorensen, L. B., Cross, J. O., Newville, M., Ravel, B., Rehr, J. J., Stragier, H., Bouldin, C. E., and Woicik, J.C., in *Resonance Anomalous X-ray Scattering. Theory and Applications*, G. Materlik, C. J. Sparks, and K. Fischer, (eds.), North-Holland: Amsterdam (1994), p. 389.

Squires, G. L., *Thermal Neutron Scattering*, Cambridge University Press: Cambridge (1978).

Templeton, D. H., in *Resonance Anomalous X-ray Scattering. Theory and Applications*, G. Materlik, C. J. Sparks, and K. Fischer, (eds.), North-Holland: Amsterdam (1994), p. 1.

Templeton , D. H., and Templeton, L. K., *ACA Abstracts*, Ser. 2 **21**, 72 (1993).

Warren, B. E., *X-ray Diffraction*, Addison-Wesley Publishing Company: Reading, MA, (1967) p. 253.

Chapter 2

Born, M. and Huang, K., *Dynamical Theory of the Crystal Lattice*, Clarendon: Oxford (1954).

Born, M. and von Kármán, T., *Phys. Zeitschr.* **13**, 297 (1912).

Born, M. and von Kármán, T., *Phys. Zeitschr.* **14**, 15 (1913).

Bu, X., Cisarova, I., and Coppens, P., *Acta Cryst.* C**48**, 1558 (1992).

Coppens, P., Thermal Smearing and Chemical Bonding, in *Electron and Magnetization Densities in Molecules and Solids*, P. J. Becker (ed.), Plenum Press: New York (1980), p. 521.

Cruickshank, D. W. J., *Acta Cryst.* **9**, 754 (1956).

Dawson, B., *Proc. R. Soc. Lond.*, Ser. A **298**, 235 (1967).

Dawson, B., Hurley, A. C., and Maslen, V. W., *Proc. R. Soc. Lond.*, Ser. A **298**, 289 (1967).

Dunitz, J. D., *X-Ray Analysis and the Structure of Organic Molecules*, Cornell University Press: Ithaca, London (1979).

Harel, M. and Hirshfeld, F. L., *Acta Cryst.* B**31**, 162 (1975).

Hirshfeld, F. L., *Acta Cryst.* A**32**, 239 (1976).

Hummel, W., Rasseli, A., and Bürgi, H.-B., *Acta Cryst.* B**47**, 683 (1990).

Johnson, C. K., *Acta Cryst.* A**25**, 187 (1969).

Johnson, C. K., ACA Program and Abstracts, 1970 Winter Meeting, Tulane University (1970), p. 60.

Johnson, C. K. and Levy, H. A., in *International Tables for X-ray Crystallography*, Vol. IV, Kynoch Press: Birmingham, England (1974), p. 311.

Kendal, M. G. and Stuart, A., *The Advanced Theory of Statistics*, Griffin: London (1958).

Kontio, A. and Stevens, E. D., *Acta Cryst.* A**38**, 623 (1982).

Kuhs, W. F., *Acta Cryst.* A**39**, 148 (1983).

Kuhs, W. F., *Acta Cryst.* A**48**, 80 (1992).

Mair, S., *J. Phys. C.: Solid State Phys.* **13**, 2857 (1980).

Pawley, G. S., *Phys. Stat. Sol.* **20**, 347 (1967).

Prince, E., *Mathematical Techniques in Crystallography and Materials Science*, Springer-Verlag: New York, Heidelberg, Berlin (1982), p. 147.

Scheringer, C., *Acta Cryst.* A**41**, 73 (1985).

Scheringer, C., *Acta Cryst.* A**41**, 79 (1985).

Schomaker, V. and Trueblood, K. N., *Acta Cryst.* B**24**, 63 (1968).

Stewart, R. F. and Feil, D., *Acta Cryst.* A**36**, 503 (1980).

Takusagawa, F. and Koetzle, T. F., *Acta Cryst.* B**35**, 2126 (1979).

Tanaka, K. and Marumo, F., *Acta Cryst.* **A39**, 631 (1983).

Willis, B. T. M., *Acta Cryst.* **A25**, 277 (1969).

Willis, B. T. M. and Pryor, A. W., *Thermal Vibrations in Crystallography*, Cambridge University Press: Cambridge (1975), p. 66.

Zucker, U. H. and Schulz, H., *Acta Cryst.* **A38**, 563 (1982).

Chapter 3

Arfken, G., *Mathematical Methods for Physicists*, 2nd ed., Academic Press: New York, London (1970).

Avery, J. S. and Watson, K. J., *Acta Cryst.* **A33**, 679 (1977).

Baert, F., Coppens, P., Stevens, E. D., and DeVos, L., *Acta Cryst.* **A38**, 143 (1982).

Bats, J. W., Coppens, P., and Koetzle, T. F., *Acta Cryst.* **B33**, 37 (1977).

Becker, P., Coppens, P., and Ross, F. K., *J. Am. Chem. Soc.* **95**, 7604 (1973).

Bentley, J. and Stewart, R. F., *Acta Cryst.* **A30**, 60 (1974).

Bragg, W. H., *The Intensity of X-ray Reflection by Diamond*, Proceedings of the Physical Society of London **33**, 301 (1921).

Brown, G. M. and Levy, H. A., *Acta Cryst.* **B29**, 790 (1973).

Brown, A. S. and Spackman, M. A., *Acta Cryst.* **A47**, 21 (1991).

Clementi, E. and Raimondi, D. L., *J. Chem. Phys.* **38**, 2686 (1963).

Clementi, E. and Roetti, C., *At. Data Nucl. Data Tables* **14**, 177 (1974).

Cochran, W., *Acta Cryst.* **9**, 924 (1956).

Cohen-Tannoudji, C., Diu, B., and Laloë, F., *Quantum Mechanics*, John Wiley and Sons: New York, London, Sydney, Toronto, and Hermann: Paris (1977) pp. 948–949.

Condon, E. V. and Shortley, G. H., *The Theory of Atomic Spectra*, Cambridge University Press: London, New York (1957).

Coppens, P., in *Neutron Diffraction*, H. Dachs (ed.), Springer-Verlag: Berlin, *Top. Curr. Phys.* **6**, 71–111 (1978).

Coppens, P. and Vos, A., *Acta Cryst.* **B28**, 146 (1971).

Coppens, P., Sabine, T. M., Delaplane, R. G., and Ibers, J. A., *Acta Cryst.* **B25**, 2451 (1969).

Coppens, P., Guru Row, T. N., Leung, P., Stevens, E. D., Becker, P. J., and Yang, Y. W., *Acta Cryst.* **A35**, 63 (1979).

Coulson, C. A., *Valence*, Oxford University Press: Oxford (1961).

Dawson, B., *Proc. R. Soc. Lond.*, Ser. A **298**, 225, 264 (1967).

DeMarco, J. J. and Weiss, R. J., *Phys. Rev.* **A137**, 1869 (1965).

Deutsch, M., *Phys. Rev.* **B45**, 646 (1992).

Drück, U. and Kotuglu, A., *Zeitschr. für Kristallographie*, **166**, 233 (1984).

Elkaim, E., Tanaka, K., Coppens, P., and Scheidt, W. R., *Acta Cryst.* **B43**, 457 (1987).

Freeman, A. J., *Acta Cryst.* **12**, 261 (1959).

Frishberg, C. and Massa, L. J., *Phys. Rev.* **B24**, 7018 (1981).

Hansen, N. K., *Study of the Electron Density Distribution in Molecular Crystals by Analysis of X-ray Diffraction Data Using Non-Spherically Symmetric Scattering Functions*, Thesis, University of Århus, Denmark (1978).

Hansen, N. K. and Coppens, P., *Acta Cryst.* **A34**, 909 (1978).

Hanson, J. C., Sieker, L. C., and Jensen, L. H., *Acta Cryst.* **B29**, 797 (1973).

Hellner, E., *Acta Cryst.* **B33**, 239 (1977).

Hirshfeld, F. L., *Acta Cryst.* **B27**, 769 (1971).

Hirshfeld, F. L., *Isr. J. Chem.* **16**, 226 (1977).

Hirshfeld, F. L. and Rzotkiewicz, S., *Mol. Physics* **27**, 1319 (1974).

International Tables for Crystallography, Vol. C, Kluwer: Dordrecht (1992).

International Tables for X-ray Crystallography, Vol. IV, Kynoch Press: Birmingham, England (1974).

James, R. W., *The Optical Principles of the Diffraction of X-rays*, OxBow Press: Woodbridge, CT (1982).

Jeffrey, G. A. and Cruickshank, D. W. J., *Quart. Rev. Chem. Soc. Lond.* **7**, 335 (1953).

Jones, D. S., Pautler, D., and Coppens, P., *Acta Cryst.* A**28**, 635 (1972).

Kurki-Suonio, K., *Acta Cryst.* A**24**, 379 (1968).

Kurki-Suonio, K., *Isr. J. Chem.* **16**, 115 (1977a).

Kurki-Suonio, K., *Isr. J. Chem.* **16**, 132 (1977b).

Lecomte, C., in *The Application of Charge Density Research to Chemistry and Drug Design*, NATO ASI Series, Series B, Vol. 250, A. Jeffrey and J. F. Piniella (eds.), Plenum Press: New York, p. 121.

Massa, L. J. and Clinton, W. L., *Trans. Am. Cryst. Assoc.* **8**, 149 (1972).

Massa, L. J., Goldberg, M., Frishberg, C., Boehme, R. F., and LaPlaca, S. L., *Phys. Rev. Lett.* **55**, 622 (1985).

Matthews, D. A. and Stucky, G. D., *J. Am. Chem. Soc.* **93**, 5954 (1971).

Matthews, D. A., Swanson, J., and Stucky, G. D., *J. Am. Chem. Soc.* **93**, 5945 (1971).

Moss, G. R., Souhassou, M., Blessing, R. H., Espinosa, E., and Lecomte, C., *Acta Cryst.* B**51**, 650 (1995).

Pearlman, D. A. and Kim, S.-H., *J. Mol. Biol.* **24**, 327 (1985).

Ransil, B. J., *Rev. Mod. Phys.* **32**, 245 (1960).

Scheringer, C. and Kotuglu, A., *Acta Cryst.* A**39**, 899 (1983).

Slater, J. C., *Phys. Rev.* **42**, 33 (1932).

Stewart, R. F., *J. Chem. Phys.* **51**, 4569 (1969).

Stewart, R. F., *Acta Cryst.* A**32**, 565 (1976).

Stewart, R. F., *Isr. J. Chem.* **16**, 124 (1977).

Stewart, R. F., in *Electron and Magnetization Densities of Molecules and Solids*, P. Becker (ed.), Plenum Press: New York (1980), p. 439.

Stewart, R. F., Davidson, E. R., and Simpson, W. T., *J. Chem. Phys.* **42**, 3175 (1965).

Su, Z. and Coppens, P., *J. Appl. Cryst.* **23**, 71 (1990).

Szabo, A. and Ostlund, N. S., *Modern Quantum Chemistry, Introduction to Advanced Electronic Structure Theory*, McGraw-Hill: New York (1989).

Taylor, J. C. and Sabine, T. M., *Acta Cryst.* B**28**, 3340 (1972).

Tomii, Y., *Proc. Phys. Soc. Japan* **13**, 1030 (1958).

Verschoor, G. C. and Keulen, E., *Acta Cryst.* B**27**, 134 (1971).

Von der Lage, F. C. and Bethe, H. A., *Phys. Rev.* **71**, 612 (1947).

Chapter 4

Becker, P. J. and Coppens, P., *Acta Cryst.* A**30**, 129 (1974a).

Becker, P. J. and Coppens, P., *Acta Cryst.* A**30**, 148 (1974b).

Becker, P. J. and Coppens, P., *Acta Cryst.* A**31**, 417 (1975).

Blessing, R. H., *Acta Cryst.* B**51**, 816 (1995).

Coppens, P., *Acta. Cryst.* B**24**, 1272 (1968).

Coppens, P., Project Reporter, *Acta Cryst.* A**40**, 184 (1984).

Coppens, P., Boehme, R., Price, P. F., and Stevens, E. D., *Acta Cryst.* A**37**, 857 (1981).

Craven, B. M. and McMullan, R. K., *Acta Cryst.* B**35**, 934 (1979).

DeLucia, M. L., *Selected Studies in X-ray Diffraction*, Thesis, State University of New York at Buffalo (1977), p. 62.

Diamond, R., *Acta Cryst.* **21**, 253 (1966).

Feynman, R. P., *Phys. Rev.* **56**, 340 (1939).

Hamilton, W. C., *Statistical Methods in the Physical Sciences*, Ronald Press: New York (1964).

Hirshfeld, F. L., *Isr. J. Chem.* **16**, 1226 (1977).

Hirshfeld, F. L., *Acta Cryst.* B**40**, 613 (1984).

Hirshfeld, F. L. and Rzotkiewicz, S., *Mol. Phys.* **27**, 1319 (1974).

Hughes, E., *J. Am. Chem. Soc.* **63**, 1737 (1941).

Jones, D. S., Pautler, D., and Coppens, P., *Acta Cryst.* A**28**, 635 (1972).

Levine, I. N., *Quantum Chemistry*, 3rd ed., Allyn and Bacon: Boston (1983).

McCandlish, L. E., Stout, G. H., and Andrews, L. C., *Acta Cryst.* A**31**, 245 (1975).

Pawley, G. S., in *Advances in Structure Research by Diffraction Methods*, Vol. 4, W. Hoppe and R. Mason (eds.), Pergamon: Braunschweig (1972), p. 1.

Raymond, K. N., *Acta Cryst.* A**28**, 163 (1972).

Scheringer, C., Kotuglu, A., and Mullen, D., *Acta Cryst.* A**34**, 481 (1978).

Schwarzenbach, D. and Lewis, J., in *Electron Distributions and the Chemical Bond*, P. Coppens and M. B. Hall (eds.), Plenum Press: New York (1982), p. 413ff.

Stevens, E. D., in *Electron and Magnetization Densities in Molecules and Crystals*, P. J. Becker (ed.), Plenum Press: New York (1980), p. 476.

Stevens, E. D. and Coppens, P., *Acta Cryst.* A**31**, 612 (1975).

Stewart, R. F., *Acta Cryst.* A**32**, 565 (1976).

Watkin, D., *Acta Cryst.* A**40**, 411 (1994).

Whittaker, E. and Robinson, G., *The Calculus of Observations*, 4th ed., Dover Press: New York (1967).

Chapter 5

Alkire, R. W., Yelon, W. B., and Schneider, J. R., *Phys. Rev.* B**26**, 3097 (1982).

Arfken, G., *Mathematical Methods for Physicists*, 2nd ed., Academic Press: New York, London (1970).

Bader, R. F. W. and Bandrauk, A. D., *J. Chem. Phys.* **49**, 1653 (1968).

Bader, R. F. W., Keaveny, I., and Cade, P. E., *J. Chem. Phys.* **47**, 3381 (1967).

Cade, P. E., Bader, R. F. W., Henneker, W. H., and Keaveny, I., *J. Chem. Phys.* **50**, 5313 (1969).

Collins, D. M., *Nature* **298**, 49 (1982).

Coppens, P., *Science* **158**, 1577 (1967).

Coppens, P., *Acta Cryst.* B**30**, 255 (1974).

Coppens, P., *Coord. Chem. Rev.* **65**, 285 (1985).

Coppens, P. and Lehmann, M. S., *Acta Cryst.* B**32**, 1777 (1976).

Cruickshank, D. W., *Acta Cryst.* **2**, 65 (1949).

Dawson, B., *Acta Cryst.* **17**, 990 (1964).

De Vries, R. Y., Briels, W. J., and Feil, D., *Acta Cryst.* A**50**, 383 (1994).

Dunitz, J. D. and Seiler, P., *Acta Cryst.* B**29**, 589 (1973).

Dunitz, J. D. and Seiler, P., *J. Am. Chem. Soc.* **105**, 7056 (1983).

Feil, D., *Isr. J. Chem.* **16**, 103 (1977).

Glusker, J. P. and Trueblood, K. N., *Crystal Structure Analysis—A Primer*, 2nd ed., Oxford University Press: New York, Oxford (1985), p. 134.

Gull, S. F. and Daniell, G. J., *Nature* **272**, 686 (1978).

Hall, M. B., in *Electron Distributions and the Chemical Bond*, P. Coppens and M. B. Hall (eds.), Plenum Press: New York (1982), p. 205.

Hall, M. B., *Chem. Scripta* **26**, 389 (1986).

Hanson, J. C., Sieker, L. C., and Jensen, L. H., *Acta Cryst.* **B29**, 797 (1973).

Heijser, W., Baerends, E. J., and Ros, P., *J. Mol. Struct.* **63**, 109 (1980).

Hermansson, K., *Acta Cryst.* **B41**, 161 (1985).

Hirshfeld, F. L., *Acta Cryst.* **B40**, 613 (1984).

Hirshfeld, F. L., in *Accurate Molecular Structures*, A. Domenicano and I. Hargittai (eds.), Oxford University Press: Oxford, New York (1992), p. 237.

Hirshfeld, F. L. and Rzotkiewicz, S., *Mol. Phys.* **27**, 1319 (1974).

Jauch, W., *Acta Cryst.* **A50**, 650 (1994).

Jauch W. and Palmer, A., *Acta Cryst.* **A49**, 590 (1993).

Jaynes, E. T., *IEEE Trans. Syst. Sci. Cybern.* **SSC-4**, 227 (1968).

Jeffrey, G. A. and Cruickshank, D. W. J., *Quart. Rev. Chem. Soc. Lond.* **7**, 335 (1953).

Kunze, K. and Hall, M. B., *J. Am. Chem. Soc.* **108**, 5122 (1986).

Kunze, K. and Hall, M. B., *J. Am. Chem. Soc.* **109**, 7617 (1987).

Lehmann, M. S. and Coppens, P., *Acta Chem. Scand.* **A31**, 530 (1977).

Low, A. A. and Hall, M. B., *J. Phys. Chem.* **94**, 628 (1990).

Mensching, L., Von Niessen, W., Valtazanos, P., Ruedenberg, K., and Schwarz, W. H. E., *J. Am. Chem. Soc.* **111**, 6933 (1989).

Morooka, M., Ohba, S., and Toriumi, K., *Acta Cryst.* **B48**, 459 (1992).

O'Connell, A. M., Rae, A. I. M., and Maslen, E. N., *Acta Cryst.* **21**, 208 (1966).

Papoular, R. J. and Gillon, B., *Europhys. Lett.* **13**, 429 (1990a).

Papoular, R. J. and Gillon, B., in *Neutron Scattering Data Analysis*, M. W. Johnson (ed.), *Inst. Phys. Conf. Ser.* **107**, Bristol (1990b), p. 101.

Papoular, R. J., Prandl, W., and Schiebel, P., in *Maximum Entropy and Bayesian Methods*, C. Ray Smith, G. J. Erickson, and P. O. Neudorfer (eds.), Dordrecht: Kluwer (1992) p. 359ff.

Papoular, R. J., Vekhter, Y., and Coppens, P., *Acta Cryst.* **A52**, 397 (1996).

Rees, B., *Acta Cryst.* **A32**, 483 (1976).

Rees, B., *Acta Cryst.* **A34**, 254 (1978).

Rees, B. and Mitschler, A., *J. Am. Chem. Soc.* **98**, 7918 (1976).

Roux, M. and Daudel, R., *Compt. Rend. Acad. Sci.* **240**, 90 (1955).

Ruedenberg, K. and Schwarz, W. H. E, *J. Chem. Phys.* **92**, 4596 (1990).

Ruysink, A. F. J. and Vos, A., *Acta Cryst.* **A30**, 503 (1974).

Saka, T. and Kato, N., *Acta Cryst.* **A42**, 469 (1968).

Sakata, M. and Sato, M., *Acta Cryst.* **A46**, 263 (1990).

Sakata, M., Uno, T., Takata, M., and Howard, C. J., *J. Appl. Cryst.* **26**, 159 (1993).

Savariault, J. M. and Lehmann, M. S., *J. Am. Chem. Soc.* **102**, 1298 (1980).

Scheringer, C., Kotuglu, A., and Mullen, D., *Acta Cryst.* **A34**, 481 (1978).

Schwarz, W. H. E., Ruedenberg, K., and Mensching, L., *J. Am. Chem. Soc.* **111**, 6926 (1989).

Souhassou, M., Lecomte, C., Blessing, R. H., Aubry, A., Rohmer, M.-M., Wiest, R., Benard, M., and Marraud, M., *Acta Cryst.* **B47**, 253 (1991).

Stevens, E. D. and Coppens, P., *Acta Cryst.* **A32**, 915 (1976).

Tischler, J. Z. and Batterman, B. W., *Phys. Rev.* **B30**, 7060 (1984).

Yamabe, S. and Morokuma, K., *J. Am. Chem. Soc.* **97**, 4458 (1975).

Zobel, D., Luger, P., Dreissig, W., and Koritsanszky, T., *Acta Cryst.* **B48**, 837 (1992).

Chapter 6

Bachrach, S. M. and Streitweiser, A., *J. Comp. Chem.* **10**, 514 (1989).

Bader, R. F. W., *Atoms in Molecules: A Quantum Theory*, Clarendon Press, Oxford Science Publications: Oxford, (1990).

Bader, R. F. W. and Essén, H., *J. Chem. Phys.* **80**, 1943 (1984).

Bader, R. F. W. and Laidig, K. E., in *The Application of Charge Density Research to Chemistry and Drug Design*, NATO ASI Series, Series B, Vol. 250, G. A. Jeffrey and J. F. Piniella (eds.), Plenum Press: New York (1991), p. 34.

Bader, R. F. W., Nguyen-Dang, T. T., and Tal, Y., *Rep. Prog. Phys.* **44**, 893 (1981).

Bader, R. F. W., Slee, T. S., Cremer, D., and Kraka, E., *J. Am. Chem. Soc.* **105**, 5061 (1983).

Coppens, P., *Phys. Rev. Lett.* **35**, 98 (1975).

Coppens, P. and Hamilton, W. C., *Acta Cryst.* **B24**, 925 (1968).

Coppens, P., Petricek, V., Levendis, D., Larsen, F. K., Paturle, A., Gao, Y., and LeGrand, A. D., *Phys. Rev. Lett.* **59**, 1695 (1987).

Gatti, C., Bianchi, R., Destro, R., and Merati, F., *J. Mol. Struc. (Theochem.)* **255**, 409 (1992).

Hirshfeld, F. L., *Isr. J. Chem.* **16**, 198 (1977a).

Hirshfeld, F. L., *Theoret. Chim. Acta* **44**, 129 (1977b).

Howard, S. T., Hursthouse, M. B., Lehmann, C. W., and Poyner, E. A., *Acta Cryst.* **B51**, 328 (1995).

Johnson, C. K., *ACA Abstracts*, Ser. 2 **29**, 105 (1992).

Kurki-Suonio, K., *Acta Cryst.* **A24**, 379 (1968).

Kurki-Suonio, K., *Analysis of Crystal Atoms on the Basis of X-ray Diffraction*, Proceedings Italian Crystallographic Association Meeting, Bari, Italy (1971).

Kurki-Suonio, K. and Salmo, P., *Ann. Acad. Sci. Fenn.*, Ser. A **VI**, 369 (1971).

McLean, A. D. and Yoshimine, M., *Tables of Linear Molecule Wave Functions*, I.B.M. Corp., San Jose, CA (1967).

Moss, G. and Coppens, P., *Chem. Phys. Lett.* **75**, 298 (1980).

Politzer, P., *Theoret. Chim. Acta* **23**, 203 (1971).

Politzer, P. and Reggio, P. H., *J. Am. Chem. Soc.* **94**, 8308 (1972).

Pouget, J. P., Khanna, S. K., Denoyer, F., Comes, R., Garito, A. F., and Heeger, A. J., *Phys. Rev. Lett.* **37**, 437 (1976).

Snyder, L. C. and Basch, H., *Molecular Wavefunctions and Properties*, John Wiley and Sons: New York (1972).

Stevens, E. D. and Coppens, P., *Acta Cryst.* **A31**, 512 (1975).

Stewart, R. F., in *The Application of Charge Density Research to Chemistry and Drug Design*, NATO ASI Series, Series B, Vol. 250, G. A. Jeffrey and J. F. Piniella (eds.), Plenum Press: New York (1991), p. 63ff.

Vahvaselkä, A. and Kurki-Suonio, K., *Physica Fennica* **10**, 87 (1975).

Velders, G. J. M., *The Electron Density of Molecules: A Tool for Understanding Molecular Response*, PhD Thesis, University of Twente, The Netherlands (1992), p. 49.

Weiss, R. J., *X-ray Determination of Electron Distributions*, John Wiley and Sons: New York (1966), p. 77.

Chapter 7

Aroney, M. J., Le Fevre, R. J. W., and Singh, A. N., *J. Chem. Soc.* 3179 (1965).

Baert, F., Coppens, P., Stevens, E. D., and DeVos, L., *Acta Cryst.* **A38**, 143 (1982).

Baert, F. Schweiss, P., Heger, G., and More, M., *J. Mol. Struct.* **178**, 29 (1988).

Bats, J. W. and Fuess, H., *Acta Cryst.* **B42**, 26 (1986).

Bats, J. W., Fuess, H., and Elerman, Y., *Acta Cryst.* **B42**, 552 (1986).

Battaglia, M. R., Buckingham, A. D., and Williams, J. H., *Chem. Phys. Lett.* **78**, 421 (1981).

Bergmann, E. D., Weiler-Feilchenfeld, H., and Neiman, Z., *J. Chem. Soc.* **B**, 1334 (1970).

Berkovitch-Yellin, Z. and Leiserowitz, L., *J. Am. Chem. Soc.* **102**, 7677 (1980).

Boyd, R. J., *Can. J. Phys.* **55**, 452 (1977).

Brown, R. D., Godfrey, P. D., and Storey, J., *J. Mol. Spectrosc.* **58**, 445 (1975).

Buckingham, A. D., in *Physical Chemistry. An Advanced Treatise*, Vol. IV (*Molecular Properties*), D. Henderson (ed.), Academic Press: New York (1970), pp. 349–386.

Buckingham, A. D., Graham, C., and Williams, J. H., *Mol. Phys.* **49**, 703 (1983).

Calderbank, K. E., Calvert, R. L., Lukins, P. B., and Ritchie, G. L. D., *Aust. J. Chem.* **34**, 1835 (1981).

Christen, D., Griffiths, J. H., and Sheridan, J., *Z. Naturforsch.* **37a**, 1378 (1982).

Coonan, M. H. and Ritchie, G. L. D., *Chem. Phys. Lett.* **202**, 237 (1993).

Coppens, P., Guru Row, T. N., Leung, P., Stevens, E. D., Becker, P. J., and Yang, Y. W., *Acta Cryst.* **A35**, 63 (1979).

Coppens, P., Moss, G., and Hansen, N. K., in *Computing in Crystallography*, R. Diamond, S. Ramaseshan, and K. Venkatesan (eds.), Indian Academy of Sciences: Bangalore (1980), pp. 16.01–16.18.

Corfield, P. W. R. and Shore, S. G., *J. Am. Chem. Soc.* **95**, 1480 (1973).

Craven, B. M. and Benci, P., *Acta Cryst.* **B37**, 1584 (1981).

Craven, B. M. and McMullan, R. K., *Acta Cryst.* **B35**, 934 (1979).

Craven, B. M. and Weber, H.-P., *Acta Cryst.* **B39**, 743 (1983).

Craven, B. M., Fox, R. O. Jr., and Weber, H.-P, *Acta Cryst.* **B38**, 1942 (1982).

Cromer, D. T., Larson, A. C., and Stewart, R. F., *J. Chem. Phys.* **65**, 336 (1976).

Dagg, I. R., Read, L. A. A., and Smith, W., *Can. J. Phys.* **60**, 1431 (1982).

Delaplane, R. G., Tellgren, R., and Olovsson, I., *Acta Cryst.* **B46**, 361 (1990).

Dennis, G. R., PhD Thesis, University of Sydney, Australia (1986).

Dennis, G. R. and Ritchie, G. L. D., *J. Phys. Chem.* **95**, 656 (1991).

Destro, R., Bianchi, R., and Morosi, G., *J. Phys. Chem.* **93**, 4447 (1989).

Dyke, T. R. and Muenter, J. S., *J. Chem. Phys.* **59**, 3125 (1973).

Edward, J. T., Farrell, P. G., and Lob, J. L., *J. Phys. Chem.* **77**, 2192 (1973).

Eisenstein, M., *Acta Cryst.* **B44**, 412 (1988).

Emrich, R. J. and Steele, W., *Mol. Phys.* **40**, 469 (1980).

Epstein, J., Ruble, J. R., and Craven, B. M., *Acta Cryst.* **B38**, 140 (1982).

Fabius, B., Cohen-Addad, C., Larsen, F. K., Lehmann, M. S., and Becker, P., *J. Am. Chem. Soc.* **111**, 5728 (1989).

Gatti, C., Saunders, V. R., and Roetti, C., *J. Chem. Phys.* **101**, 10686 (1994).

He, X. M., Swaminathan, S., Craven, B. M., and McMullan, R. K., *Acta Cryst.* **B44**, 271 (1988).

Hirshfeld, F. L., *Acta Cryst.* **B40**, 484 (1984).

Hirshfeld, F. L. and Hope, H., *Acta Cryst.* **B36**, 406 (1980).

Katritzky, A. R., Randall, E. W., and Sutton, L. E., *J. Chem. Soc.* 1769 (1957).

Khanarian, G. and Moore, W. J., *Aust. J. Chem.* **33**, 1727 (1980).

Klooster, W. T. and Craven, B. M., unpublished results.

Klooster, W. T., Swaminathan, S., Nanni, R., and Craven, B. M., *Acta Cryst.* **B48**, 217 (1992).

Kojima, T., Yano, E., Nakagawa, K., and Tsunekawa, S., *J. Mol. Spectrosc.* **122**, 408 (1987).

Koritsanszky, T., Buschmann, J., Denner, L., Luger, P., Knochel, A., Haarich, M., and Patz, M., *J. Am. Chem. Soc.* **113**, 8388 (1991).

Krijn, M. P. C. M., *Electron Density Distributions and the Hydrogen Bond*, Thesis, University of Twente, Enschede, The Netherlands (1988).

Krijn, M. P. C. M. and Feil, D., *J. Chem. Phys.* **89**, 4188 (1988).

Kukolich, S. G., Aldrich, P. D., Read, W. G., and Campbell, E. J., *J. Chem. Phys.* **79**, 1105 (1983).

Kulakowska, I., Geller, M., Lesyng, B., and Wierzchowski, K. L., *Biochim. Biophys. Acta* **361**, 119 (1974).

Kumler, W. D. and Fohlen, G. M., *J. Am. Chem. Soc.* **64**, 1944 (1942).

Kurki-Suonio, K., *Analysis of Crystal Atoms on the Basis of X-ray Diffraction*, Proceedings Italian Crystallographic Association Meeting, Bari, Italy (1971).

Kurki-Suonio, K. and Salmo, P., *Ann. Acad. Sci. Fenn.*, Ser. A **VI**, 369 (1971).

Kurland, R. J. and Wilson, E. B., *J. Chem. Phys.* **27**, 585 (1957).

Larsen, N. W., Nygaard, L., Pedersen, T., Pedersen, C. T., and Davy, H., *J. Mol. Struct.* **118**, 89 (1984).

Leavers, D. R. and Taylor, W. T., *J. Phys. Chem.* **81**, 2257 (1977).

Lefebvre, J., *Solid State Commun.* **13**, 1873 (1973).

Lumbroso, H., Segard, C., and Roques, B., *J. Organometal. Chem.* **61**, 249 (1973).

Moss, G. and Coppens, P., *Chem. Phys. Lett.* **75**, 298 (1980).

Moss, G. and Coppens, P., in *Chemical Applications of Atomic and Molecular Electrostatic Potentials*, P. Politzer and D. G. Truhlar (eds.), Plenum Press: New York, (1981).

Moss, G. and Feil, D., *Acta Cryst.* **A37**, 414 (1981).

Palmer, M. H., Wheeler, J. R., Kwiatkowski, J. S., and Lesyng, B., *J. Mol. Struct.* **92**, 283 (1983).

Pearlman, D. A. and Kim, S.-H., *J. Mol. Biol.* **211**, 171 (1990).

Pottel, R., Adolph, D., and Kaatze, U., *Ber. Bunsenges Phys. Chem.* **79**, 278 (1975).

Price, P. F., Maslen, E. N., and Delaney, W. T., *Acta Cryst.* **A34**, 194 (1978).

Rees, B. and Coppens, P., *Acta Cryst.* **B29**, 2516 (1973).

Sakellaridis, P. U. and Karageorgopolous, E. K., *Z. Naturforsch.* **29**a, 1834 (1974).

Schneider, W. C., *J. Am. Chem. Soc.* **72**, 761 (1950).

Sears, P. G., Fortune, W. H., and Blumenshine, R. L., *J. Chem. Eng. Data* **11**, 406 (1966).

Soundararajan, S., *Trans. Faraday Soc.* **54**, 1147 (1958).

Spackman, M. A., unpublished results (1991).

Spackman, M. A., *Chem. Rev.* **92**, 1769 (1992).

Stevens, E. D., *Acta Cryst.* **B34**, 544 (1978).

Stevens, E. D., *Mol. Phys.* **37**, 27 (1979).

Stevens, E. D. and Coppens, P., *Acta Cryst.* **B36**, 1864 (1980).

Stewart, R. F., *J. Chem. Phys.* **53**, 205 (1970).

Stewart, R. F., in *Critical Evaluation of Chemical and Physical Structural Information*; D. R. Lide and M. A. Paul (eds.), National Academy of Sciences: Washington, DC (1974), pp. 540–561.

Stewart, R. F., in *Electron and Magnetization Densities in Molecules and Solids*, P. Becker (ed.), Plenum Press: New York (1980), pp. 405–425.

Stolze, M. and Sutter, D. H., *Z. Naturforsch.* **42**a, 49 (1987).

Su, Z., *On the Evaluation of Coulomb Potential and Related Properties from the Multipole Description of the Charge Density*, Thesis, State University of New York at Buffalo (1993).

Su, Z. and Coppens, P., *Acta Cryst.* **A50**, 636 (1994); see also erratum: *Acta Cryst.* **A51**, 198 (1995).

Swaminathan, S. and Craven, B. M., *Acta Cryst.* **B40**, 511 (1984).

Swaminathan, S., Craven, B. M., Spackman, M. A., and Stewart, R. F., *Acta Cryst.* **B40**, 398 (1984).

Swaminathan, S., Craven, B. M., and McMullan, R. K., *Acta Cryst.* **B41**, 113 (1985).

Tigelaar, H. L. and Flygare, W. H., *J. Am. Chem. Soc.* **94**, 343 (1972).

Treiner, C., Skinner, J. F., and Fuoss, R. M., *J. Phys. Chem.* **68**, 3406 (1964). van Nes, G. J. H. and van Bolhuis, F., *Acta Cryst.* **B35**, 2580 (1979).

Verhoeven, J. and Dymanus, A., *J. Chem. Phys.* **52**, 3222 (1970).

Weber, H.-P., and Craven, B. M., *Acta Cryst.* **B43**, 202 (1987).

Weber, H.-P. and Craven, B. M., *Acta Cryst.* **B46**, 532 (1990).

Weber, H.-P., Craven, B. M., and McMullan, R. K., *Acta Cryst.* **B36**, 645 (1980).

Yanez, M. and Stewart, R. F., *Acta Cryst.* **A34**, 648 (1978).

Chapter 8

Arfken, G., *Mathematical Methods for Physicists*, 2nd ed., Academic Press: New York, London (1970).

Avery, J. S., Sommer-Larsen, P., and Grodzicki, M., in *Local Density Approximations in Quantum Chemistry, Chemistry and Solid State Physics*, P. Dahl and J. S. Avery (eds.), Plenum Press: New York, London (1984).

Baert, F., Coppens, P., Stevens, E. D., and Devos, L., *Acta Cryst.* **A38**, 143 (1982).

Basch, H., *Chem. Phys. Lett.* **6**, 337 (1970).

Becker, P. and Coppens, P., *Acta Cryst.* **A46**, 254 (1990).

Bentley, J., in *Chemical Applications of Atomic and Molecular Electrostatic Potentials*, P. Politzer and D. G. Truhlar (eds.), Plenum Press: New York, London (1981).

Bertaut, E. F., *J. Phys. Radium*, **13**, 499 (1952).

Buckingham, A. D., in *Intermolecular Interactions from Diatomics to Biopolymers*, B. Pullmann (ed.), John Wiley: Chichester, New York (1978), p. 21.

Chirlian, L. E. and Francl, M. M., *J. Comp. Chem.* **8**, 894 (1987).

Clementi, E. and Roetti, C., *Atomic Data and Nuclear Data Tables* **14**, 177 (1974).

Cohen-Tannoudji, C., Diu, B., and Laloë, F., *Quantum Mechanics*, John Wiley and Sons: New York, London, Sydney, Toronto, and Hermann: Paris (1977).

Destro, R., Bianchi, R., and Morosi, G., *J. Phys. Chem.* **93**, 4447 (1989).

Epstein, J. and Swanton, D. J., *J. Chem. Phys.* **77**(2), 1048 (1982).

Francl, M. M., Carey, C., Chirlian, L. E., and Gange, D. M., *J. Comp. Chem.* **17**, 367 (1996).

Frost-Jensen, A., Su, Z., Hansen, N. K., and Larsen, F. K., *Inorg. Chem.* **34**, 4244 (1995).

Gao, Y., Frost-Jensen, A., Pressprich, M. R., Coppens, P., Marquez, A., and Dupuis, M., *J. Am. Chem. Soc.* **114**, 9214 (1992).

Ghermani, N.-E., Bouhmaida, N., and Lecomte, C. *Acta Cryst.* **A49**, 781 (1993).

Hanse, N. K., *Z. für Naturforsch.* **48a**, 81 (1993).

He, H. M., Swaminathan, S., Craven, B. M., and McMullan, R. K., *Acta Cryst.* **B44**, 271 (1988).

Hirshfelder, J. O., Curtiss, C. F., and Bird, R. B., *Molecular Theory of Gases and Liquids*, John Wiley: New York (1954).

Howards, S. T., Hursthouse, M. B., Lehmann, C. W., Mallinson, P. R., and Frampton, C. S., *J. Chem. Phys.* **97**, 5616 (1992).

Jackson, J. D., *Classical Electrodynamics*, Wiley and Sons: New York (1975).

Jeffrey, G. A., Ruble, J. R., McMullan, R. K., and Pople, J. A., *Proc. R. Soc. Lond.*, Ser. A **414**, 47 (1987).

Khanarian, G. and Moore, W. J., *Aust. J. Chem.* **33**, 1727 (1980).

Koopmans, T., *Physica* **1**, 104 (1934).

Li, N., PhD Thesis, State University of New York at Buffalo (1989).

McClellan, A. L., *Tables of Experimental Dipole Moments*, Vol. 1, W. H. Freeman & Co.: San Francisco (1963).

McClellan, A. L., *Tables of Experimental Dipole Moments*, Vol. 2, Rahara Enterprises: El Cerrito, CA (1974).

Millet, F. S. and Dailey, B. P., *J. Chem. Phys.* **56**, 3249 (1972).

Momany, F. A., *J. Phys. Chem.* **82**, 592 (1978).

O'Keefe, M. and Spence, J. C. H., *Acta Cryst.* A**50**, 33 (1994).

Politzer, P., in *Chemical Applications of Atomic and Molecular Electrostatic Potentials*, P. Politzer and D. G. Truhlar (eds.), Plenum Press: New York, London (1981), pp. 7–28.

Politzer, P. and Truhlar, D. G. (eds.), *Chemical Applications of Atomic and Molecular Electrostatic Potentials*, Plenum Press: New York, London (1981), pp. 407–425.

Prosser, F. P. and Blanchard, C. H., *J. Chem. Phys.* **36**, 1112 (1962).

Saethre, L. J., Thomas, T. D., and Gropen, O., *J. Am. Chem. Soc.* **107**, 2581 (1985).

Saethre, L. J., Siggel, M. R. F., and Thomas, T. D., *J. Am. Chem. Soc.* **113**, 5224 (1991).

Schwartz, E., *Chem. Phys. Lett.* **6**, 631 (1970).

Siegbahn, K., Nordling, C., Fahlman, A., Nordberg, R., Hamrin, K., Hedman, J., Johansson, G., Bergmark, T., Karlson, S.-E., Lindgren, I., and Lindberg, B., *ESCA, Atomic and Molecular Structure Studied by Means of Electron Spectroscopy*, Almqvist and Wiksell: Uppsala (1967).

Spackman, M. A., *J. Comp. Chem.* **17**, 1 (1996).

Spackman, M. A. and Stewart, R. F., in *Chemical Applications of Atomic and Molecular Electrostatic Potentials*, P. Politzer and D. G. Truhlar (eds.), Plenum Press: New York, London (1981), pp. 407–425.

Spackman, M. A. and Stewart, R. F., in *Methods and Applications in Crystallographic Computing*, S. R. Hall and T. Ashida (eds.), Oxford University Press: Oxford (1984), p. 302.

Stewart, R. F., *Chem. Phys. Lett.* **65**, 335 (1979).

Stewart, R. F., in *Applications of Charge Density Research to Chemistry and Drug Design*, NATO ASI Series, Series B: Physics Vol. 250, G. A. Jeffrey and J. F. Piniella (eds.), Plenum Press: New York (1991).

Su, Z., *J. Comp. Chem.* **14**, 1036 (1993).

Su, Z., Thesis, State University of New York at Buffalo (1993), p. 116.

Su, Z. and Coppens, P., *Acta Cryst.* A**48**, 188 (1992).

Swaminathan, S. and Craven, B. M., *Acta Cryst.* B**40**, 511 (1984).

Swaminathan, S. and Craven, B. M., *Acta Cryst.* B**41**, 113 (1985).

Tegenfeldt, J. and Hermansson, K., *Chem. Phys. Lett.* **118**, 293 (1985).

Treiner, C., Skinner, J. F., and Fuoss, R. M., *J. Phys. Chem.* **68**, 3406 (1964).

Woods, R. J., Khalil, M., Pell, W., Moffat, S. H., and Smith, V. H. Jr., *J. Comp. Chem.* **11**, 297 (1990).

Chapter 9

Avery, J., Sommer-Larsen, P., and Grodzicki, M., in *Local Density Approximations in Quantum Chemistry and Solid State Physics*, J. P. Dahl and J. Avery (eds.), Plenum Press: New York, London (1984), p. 733.

Bentley, J., *J. Chem. Phys.* **70**, 159 (1979).

Bertaut, F., *J. Phys. Radium* **13**, 499 (1952).

Born, M. and Huang, K., *Dynamical Theory of Crystal Lattices*, International Series of Monographs on Physics, Clarendon Press: Oxford (1954), p. 25.

Buckingham, A. D., *J. Chem. Phys.* **30**, 1580 (1959).

Buckingham, A. D., in *Physical Chemistry. An Advanced Treatise*, Vol. 4 (*Molecular Properties*), D. Henderson (ed.), Academic Press: London (1970), p. 360.

Buckingham, A. D., in *Intermolecular Interactions from Diatomics to Biopolymers*, B. Pullman (ed.), John Wiley: Chichester, New York (1978), p. 16ff.

Coombes, D. S. and Price, S. L., Presented at BCA Charge Density Meeting, Durham, UK, December 1995.

Coombes, D. S., Price, S. L., Willock, D. J., and Leslie, M., *J. Phys. Chem.* **100**, 7352 (1996).

Cornell, W. D., Cieplak, P., Bayly, C. I., Gould, I. R., Merz, K. M. Jr., Ferguson, D. M., Spellmeyer, D. C., Fox, T., Caldwell, J. W., and Kollman, P. A., *J. Am. Chem. Soc.* **117**, 5179 (1995) and references therein.

Cox, S. R., Hsu, L.-Y., and Williams, D. E., *Acta Cryst.* A**37**, 293 (1981).

Dahl, J. P. and Avery, J. (eds.), *Local Density Approximations in Quantum Chemistry and Solid State Physics*, Plenum Press: New York, London (1984).

Delley, B., *Chem. Phys.* **110**, 329 (1986).

Desiraju, G. R., *Crystal Engineering: The Design of Organic Solids*, Elsevier: Amsterdam (1989), p. 56.

Ewald, P. P., *Annalen der Physik* **64**, 253 (1921).

Glasser, M. L. and Zucker, I. J., in *Theoretical Chemistry: Advances and Perspectives*, Vol. 5, Academic Press: New York, San Francisco, London (1980).

Gordon, R. G. and Kim, Y. S., *J. Chem. Phys.* **56**, 3122 (1972).

Hirshfeld, F. L., in *Accurate Molecular Structures*, A. Domenicano and I. Hargittai (eds.), Oxford University Press: Oxford, New York (1992), p. 237.

Hirshfeld, F. L. and Mirsky, K., *Acta Cryst.* A**35**, 366 (1979).

Hirshfeld, F. L. and Rzotkiewicz, S., *Mol. Phys.* **27**, 1319 (1974).

Hohenberg , P. and Kohn, W., *Phys. Rev.* B**136**, 864 (1964).

Howard, C. J. and Jones, R. D. G., *Acta Cryst.* A**33**, 776 (1977).

Hurst, G. J. B., Fowler, P. W., Stone, A. J., and Buckingham, A. D., *Int. J. Quantum Chem.* **29**, 1223 (1986).

Israelachvili, J., *Intermolecular and Surface Forces*, 2nd ed., Academic Press: London (1992).

Jackson, J. D., *Classical Electrodynamics*, 2nd ed., John Wiley and Sons: New York (1974), p. 142.

Kim, Y. S. and Gordon, R. G., *J. Chem. Phys.* **60**, 1842 (1974).

Kitaigorodskii, A. I., *Organic Chemical Crystallography*, Consultants Bureau: New York (1961).

Kittel, C., *Introduction to Solid State Physics*, 3rd ed., John Wiley and Sons: New York (1966).

Lide, D. R. (ed.), *CRC Handbook of Chemistry and Physics*, 74th ed., CRC Press: Boca Raton, FL (1993).

London, F., *Trans. Faraday Soc.* **33**, 8 (1937).

March, N. H., *Adv. Phys.* **6**, 1 (1957).

Matsuzawa, N. and Dixon, D. A., *J. Phys. Chem.* **98**, 2545 (1994).

Nijboer, B. R. A. and De Wette, F. W., *Physica* **23**, 309 (1957).

Parr, R. G. and Yang, W., *Density Functional Theory of Atoms and Molecules*, Oxford University Press: Oxford, New York (1989).

Politzer, P., *J. Chem. Phys.* **70**, 1067 (1979).

Politzer, P., in *Chemical Applications of Atomic and Molecular Electrostatic Potentials*, P. Politzer and D. G. Truhlar (eds.), Plenum Press: New York, London (1981), p. 7ff.

Press, W. H., Flannery, B. P., Teukolsky, S. A., and Vetterling, W. T., *Numerical Recipes, The Art of Scientific Computing*, Cambridge University Press: Cambridge, New York, (1986).

Sabin, J. R. and Trickey, S. B. in *Local Density Approximations in Quantum Chemistry and Solid State Physics*, J. P. Dahl and J. Avery (eds.), Plenum Press: New York, London (1984), p. 333.

Sangster, M. J. L. and Atwood, R. M., *J. Phys. C.: Solid State Phys.* **11**, 1541 (1978).

Sangster, M. J. L., Schröder, U., and Atwood, R. M., *J. Phys. C.: Solid State Phys.* **11**, 1523 (1978).

Spackman, M. A., *J. Chem. Phys.* **85**, 6579 (1986a).

Spackman, M. A., *J. Chem. Phys.* **85**, 6587 (1986b).

Spackman, M. A. and Maslen, E. N., *J. Phys. Chem.* **90**, 2020 (1986).

Spackman, M. A., Weber, H. P., and Craven, B. M., *J. Am. Chem. Soc.* **110**, 775 (1988).

Su, Z. and Coppens, P., *Z. Naturforsch.* **48a**, 85 (1993).

Su, Z. and Coppens, P., *Acta Cryst.* **A51**, 27 (1995).

Wampler, J. E., *Methods in Enzymology* **243**, 559 (1994).

Williams, D. E., *Acta Cryst.* **A27**, 452 (1971).

Williams, D. E., in *Crystal Cohesion and Conformational Energies*, R. M. Metzger (ed.), Springer-Verlag: Berlin, Heidelberg, New York (1981).

Ziegler, T., *Chem. Rev.* **91**, 651 (1991).

Chapter 10

Albright, T. A., Burdett, J. K., and Whangbo, M. H., *Orbital Interactions in Chemistry*, John Wiley and Sons: New York (1985).

Ballhausen, C., *Introduction to Ligand Field Theory*, McGraw-Hill: New York (1962).

Barraclough, C. G., Martin, R. L., Mitra, S., and Sherwood, R. C., *J. Chem. Phys.* **53**, 1643 (1970).

Bénard, M., *J. Am. Chem. Soc.* **100**, 2354 (1978a).

Bénard, M., *J. Am. Chem. Soc.* **100**, 7740 (1978b).

Bénard, M., Coppens, P., DeLucia, M. L., and Stevens, E. D., *Inorg. Chem.* **19**, 1924 (1980).

Berkovitch-Yellin, Z. and Ellis, D. E., *J. Am. Chem. Soc.* **103**, 6066 (1981).

Bethe, H., *Ann. Physik*, **3** 133 (1929).

Clementi, E. and Raimondi, D. L., *J. Chem. Phys.* **38**, 2686 (1963).

Clementi, E. and Roetti, C., Roothaan-Hartree-Fock Atomic Wavefunctions, *Atomic Data and Nuclear Data Tables* **14**, 177 (1974).

Clemente, D. A., Biagini, M. C., Rees, B., and Hermann, W. A., *Inorg. Chem.* **21**, 3741 (1982).

Coppens, P., *Trans. Am. Cryst. Ass.* **26**, 91 (1990).

Coppens, P. and Becker, P., in *International Tables for Crystallography*, Vol. C, Kluwer Academic Publishers: Dordrecht (1992), Chap. 8.7, pp. 627–652.

Coppens, P. and Li, L., *J. Chem. Phys.* **81**, 1983 (1984).

Cotton, F. A., *Acc. Chem. Res.* **11**, 225 (1978).

Cotton, F. A. and Stanley, T. T., *Inorg. Chem.* **16**, 2668 (1977).

Cotton, F. A., DeBoer, B. G., LaPrade, M. D., Pipal, J. R., and Ucko, D. A., *Acta Cryst.* **B27**, 1664 (1971).

Edwards, W. D., Weiner, B., and Zerner, M. C., *J. Am. Chem. Soc.* **108**, 2196 (1986).

Finklea, S. L., Cathey, L., and Amma, E. L., *Acta Cryst.* **A32**, 529 (1976).

Heijser, W., Baerends, E. J., and Ros, P., *Far. Disc. Roy. Soc.* **14**, 213 (1980).

Holladay, A., Leung, P. C., and Coppens, P., *Acta Cryst.* **A39**, 377 (1983).

Iwata, M., *Acta Cryst.* **B33**, 59 (1977).

Iwata, M. and Saito, Y., *Acta Cryst.* **B29**, 8222 (1973).

Jafri, J. A., Logan, J., and Newton, M. D., *Isr. J. Chem.* **19**, 340 (1980).

Kashiwagi, H., Takada, T., Obara, S., Miyoshi, E., and Ohno, K., *Int. J. Quant. Chem.* **XIV**, 13 (1978).

Kirfel, A. and Eichhorn, K., *Acta Cryst.* **A46**, 271 (1990).

Kissel, L. and Pratt, R. H., *Acta Cryst.* **A46**, 170 (1990).

Lecomte, C., Blessing, R. H., Coppens, P., and Tabard, A., *J. Am. Chem. Soc.* **108**, 6942 (1986).

Li, N., Landrum, J., and Coppens, P., *Inorg. Chem.* **27**, 482 (1988).

Li, N., Su, Z., Coppens, P., and Landrum, J., *J. Am. Chem. Soc.* **112**, 7294 (1990).

Litterst, F. J., Schichl, A., and Kalvius, K. M., *Chem. Phys.* **28**, 89 (1978).

Logan, J., Newton, M. D., and Noell, J. O., *Int. J. Quantum Chem. Symp.* **18**, 213 (1984).

Mallinson, P. R., Koritsanszky, T., Elkaim, E., Li, N., and Coppens, P., *Acta Cryst.* **A44**, 336 (1988).

Marathe, V. R. and Trautwein, A., in *Advances in Mössbauer Spectroscopy*, B. V. Thosar and P. K. Iyengar (eds.), Elsevier: Amsterdam (1983), Chap. 7.

Martin, M., Rees, B., and Mitschler, A., *Acta Cryst.* **B38**, 6 (1982).

Mispelter, J., Momenteau, M., and Lhoste, J. M., *J. Chem. Phys.* **72**, 1003 (1980).

Mitschler, A., Rees, B., and Lehmann, M. S., *J. Am. Chem. Soc.* **100**, 3390 (1978).

Nagel, S., *J. Phys. Chem. Solids* **46**, 905 (1985).

Nowack, E., Schwarzenbach, D., and Hahn, T., *Acta Cryst.* **B47**, 650 (1991).

Obara, S. and Kashiwagi, H., *J. Chem. Phys.* **77**, 3155 (1982).

Rawlings, D. C., Gouterman, M., Davidson, E. R., and Feller, D., *Int. J. Quantum Chem.* **28**, 797 (1985a).

Rawlings, D. C., Gouterman, M., Davidson, E. R., and Feller, D., *Int. J. Quantum Chem.* **28**, 733 (1985b).

Ray, S. N. and Das, T. P., *Phys. Rev.* **B16**, 4794 (1977).

Reed, C. A., Mashiko, T., Scheidt, W. R., Spartalian, K., and Lang, G., *J. Am. Chem. Soc.* **102**, 2302 (1980).

Rees, B. and Mitschler, A., *J. Am. Chem. Soc.* **98**, 7918 (1976).

Rohmer, M. M., *Chem. Phys. Lett.* **116**, 44 (1985).

Sontum, S. F., Case, D. A., and Karplus, M., *J. Chem. Phys.* **79**, 2881 (1983).

Sternheimer, R. M. and Foley, H. M., *Phys. Rev.* **102**, 731 (1956).

Stevens, E. D., DeLucia, M. L., and Coppens, P., *Inorg. Chem.* **19**, 813 (1980).

Su, Z. and Coppens, P. *Acta Cryst.* **A52**, 748 (1996).

Su, Z. and Coppens, P. (to be published).

Sugano, S., Tanabe, Y., and Kamimura, H., *Multiplets of Transition Metal Ions in Crystals*, Academic Press: New York, London (1970).

Tsirel'son, V. G., Strel'tsov, V. A., Makarov, E. F., and Ozerov, R. P., *Soviet Physics, JETP* **65**, 1065 (1987).

Van Vleck, J. H., *The Theory of Electric and Magnetic Susceptibilities*, Oxford University Press: Oxford (1932).

Wood, J. S., *Inorganica Chimica Acta* **229**, 407 (1995).

Zeng, Y. and Holzwarth, N. A. W., *Phys. Rev.* **B50**, 8214 (1994).

Zerner, M. and Gouterman, M., *Theor. Chim. Acta* **4**, 44 (1966).

Zerner, M., Gouterman, M., and Kobayashi, H., *Theor. Chim. Acta* **6**, 363 (1966).

Chapter 11

Adams, D. M., *Inorganic Solids, An Introduction to Concepts in Solid-State Structural Chemistry*, Wiley: London, New York (1974).

Aldred, P. J. E. and Hart, M., *Proc. R. Soc. Lond.*, Ser. A **332**, 223; 239 (1973).

Alkire, R. W., Yelon, W. B., and Schneider, J. R., *Phys. Rev.* **B26**, 3097 (1982).

Altmann, S. L., Coulson, C. A., and Hume-Rothery, W., *Proc. R.. Soc.*, Ser. A **240**, 145 (1957).

Borie, B., *Acta Cryst.* **A37**, 238 (1981).

Bragg, W. L., Claringbull, G. F., and Taylor, W. H., *Crystal Structure of Minerals*, Cornell University Press: Ithaca, NY (1965).

Brown, P. J., *Philos. Mag.* **26**, 1377 (1972).

Chou, M. Y., Lam, P. K., and Cohen, M. L., *Phys. Rev.* **B28**, 4179 (1983).

Cohen, R. E., *Rev. Mineral.* **29**, 369 (1994).

Cooper, M. J., *Rep. Prog. Phys.* **48**, 415 (1985).

Coppens, P. and Feil, D., in *Applications of Charge Density Research to Chemistry and Drug Design*, NATO ASI Series, Series B: Physics Vol. 250, G. A. Jeffrey and J. F. Piniella (eds.), Plenum Press: New York. (1991), p. 1.

Cummings, S. and Hart, M., *Aust. J. Phys.* **41**, 423 (1988).

Dawson, B., *Proc. R. Soc. Lond.*, Ser. A **298**, 255; 379 (1967).

Deutsch, M., *Phys. Rev.* **B45**, 646 (1992).

Deutsch, M., Hart, M., and Cummings, S., *Phys. Rev.* **B42**, 1248 (1990).

Diana, M. and Mazzone, G., *Phys. Rev.* **B5**, 3832 (1972).

Dovesi, R., Angonoa, G., and Causa, M., *Philos. Mag.* **45**, 601 (1982).

Downs, J. W., *J. Phys. Chem.* **99**, 6849 (1995).

Downs, J. W. and Swope, R. J., *J. Phys. Chem.* **96**, 4834 (1992).

Ekardt, H., Fritsche, L., and Noffke, J., *J. Phys. F.* **14**, 97 (1984).

Fox, A. G. and Tabernor, M. A., *Acta Metall. Mater.* **39**, 669 (1991).

Geisinger, K. L., Spackman, M. A., and Gibbs, G. V., *J. Phys. Chem.* **91**, 3237 (1987).

Ghermani, N. E., Lecomte, C., and Dusausoy, Y., *Phys. Rev.* **B53**, 5231 (1996).

Gibbs, J. V., Downs, J. W., and Boisen, J., *Rev. Mineral.* **29**, 331 (1994).

Göttlicher, S. and Wölfel, E., *Z. Elektrochem.* **63**, 891 (1959).

Hansen, N. K. and Coppens, P., *Acta Cryst.* **A34**, 909 (1978).

Hansen, N. K., Schneider, J. R., and Larsen, F. K., *Phys. Rev.* **B29**, 917 (1984).

Hansen, N. K., Schneider, J. R., Yellon, W. B., and Pearson, W. H., *Acta Cryst.* **A43**, 763 (1987).

Hattori, H., Kuriyama, H., Katagawa, T., and Kato, N., *J. Phys. Soc. Japan* **20**, 988 (1965).

Hill, R. J., Newton, M. D., and Gibbs, G. V., *Solid State Chem.* **47**, 185 (1983).

Inoue, S. T. and Yamashita, J., *J. Phys. Soc. Japan* **35**, 677 (1973).

International Tables for X-ray Crystallography, Vol IV, Kynoch Press: Birmingham, England (1974).

James, R. W., *The Optical Principles of the Diffraction of X-Rays*, G. Bells and Sons: London (1948).

Larsen, F. K. and Hansen, N. K., *Acta Cryst.* **B40**, 169 (1984).

Larsen, F. K., Lehmann, M. S., and Merisalo, M., *Acta Cryst.* **A36**, 159 (1980).

Lu, Z. W. and Zunger, A., *Acta Cryst.* **A48**, 545 (1992).

Lu, Z. W., Wei, S.-H., and Zunger, A., *Acta Metall. Mater.* **40**, 2155 (1992).

Lu, Z. W., Zunger, A., and Deutsch, M., *Phys. Rev.* **B47**, 9385 (1993).

Lu, Z. W., Zunger, A., and Fox, A. G., *Acta Metall. Mater.* **42**, 3929 (1994).

Mackenzie, J. K. and Mathieson, A. McL., *Acta Cryst.* **A48**, 231 (1992).

Mak, T. C. W. and Zhou, G.-D., *Crystallography in Modern Chemistry. A Resource Book of Crystal Structures*, John Wiley and Sons: New York (1992).

Matshushita, T. and Kohra, K., *Phys. Status Solidi* **24**, 531 (1974).

Merisalo, M., Järvinen, M., and Kurittu, J., *Phys. Scr.* **17**, 23 (1978).

Ohba, S., Sato, S., and Saito, Y., *Acta Cryst.* **A37**, 697 (1981).

Ohba, S., Saito, Y., and Wakoh, S., *Acta Cryst.* **A38**, 103 (1982).

Pauling, L., *The Nature of the Chemical Bond*, Cornell University Press: Ithaca, NY (1939).

Pindor, A. J., Vosko, S. H., and Umrigar, C. J., *J. Phys.* F**16**, 1207 (1986).

Rath, J. and Callaway, J., *Phys. Rev.* B**8**, 5398 (1973).

Roberto, J. B. and Batterman, B. W., *Phys. Rev.* B**2**, 3220 (1970).

Saka, T. and Kato, N., *Acta Cryst.* A**42**, 469 (1986).

Saka, T. and Kato, N., *Acta Cryst.* A**43**, 252 (1987).

Schneider, J. R., Hansen, N. K., and Kretschmer, H., *Acta Cryst.* A**37**, 711 (1981).

Seiler, P. and Dunitz, J. D., *Helvetica Chimica Acta* **69**, 1107 (1986).

Spackman, M. A., *Acta Cryst.* A**42**, 271 (1986).

Spackman, M. A. and Weber, H. P., *J. Phys. Chem.* **92**, 794 (1988).

Spence, J. C. H., *Acta Cryst.* A**49**, 231 (1993) and references therein.

Takama, T. and Sato, S., *Japan. J. Appl. Phys.* **20**, 1183 (1981).

Takama, T., Tsuchiya, K., Kobayashi, K., and Sato, S., *Acta Cryst.* A**46**, 514 (1990).

Tanemura, S. and Kato, N., *Acta Cryst.* A**28**, 69 (1972).

Teworte, R. and Bonse, U., *Phys. Rev.* B**29**, 2102 (1984).

Tischler, J. Z., *Temperature Dependence of Higher Order Forbidden Reflections in Silicon and Germanium using Synchrotron Radiation*, Thesis, Cornell University, Ithaca, NY (1983).

Tischler, J. Z. and Batterman, B. W., *Phys. Rev.* B**30**, 7060 (1984).

Trucano, P. and Batterman, B. W., *Phys. Rev.* B**6**, 3659 (1972).

Tsirelson, V. G., Evdokimova, O. A., Belokoneva, E. L., and Urusov, V. S., *Phys. Chem. Minerals* **17**, 275 (1990).

von Barth, U. and Pedroza, A. C., *Phys. Scr.* **32**, 353 (1985).

Williams, B. (ed.), *Compton Scattering*, McGraw-Hill: New York (1977).

Willis, B. T. M., *Acta Cryst.* A**25**, 277 (1969).

Willis, B. T. M. and Pryor, A. W., *Thermal Vibrations in Crystallography*, Cambridge University Press: Cambridge (1975), p. 150.

Yang, Y. W. and Coppens, P., *Acta Cryst.* A**34**, 61 (1978).

Chapter 12

Almlof, J., Kvick, A., and Thomas, J. O., *J. Chem. Phys.* **59**, 3901 (1973).

Bader, R. F. W., *Atoms in Molecules: A Quantum Theory*, International Series of Monographs on Chemistry, No. 22, Clarendon Press, Oxford Science Publications: Oxford, New York (1990).

Bats, J. W. and Fuess, H., *Acta Cryst.* B**42**, 26 (1986).

Bats, J. W., Fuess, H., and Ellerman, Y., *Acta Cryst.* B**42**, 552 (1986).

Berkovitch-Yellin, Z. and Leiserowitz, L., *Acta Cryst.* B**33**, 3657 (1976).

Boese, R., Miebach, T., and De Meijere, A., *J. Am. Chem. Soc.* **113**, 1743 (1991).

Breitenstein, M., Dannöhl, H., Meyer, H., Schweig, A., Seeger, R., Seeger, U., and Zittlau, W., *Int. Rev. Phys. Chem.* **3**, 335 (1983).

Brock, C. P., Dunitz, J. D., and Hirshfeld, F. L., *Acta Cryst.* B**47**, 789 (1991).

Cameron, T. S., Bozeka, B., and Kwiatkowski, W., *J. Am. Chem. Soc.* **116**, 1211 (1994).

Coppens, P., *Science* **158**, 1577 (1967).

Coppens, P., Project Reporter, *Acta Cryst.* A**40**, 184 (1984).

Coppens, P., Sabine, T. M., Delaplane, R. G., and Ibers, J. A., *Acta Cryst.* B**25**, 2451 (1969).

Coulson, C. A. and Moffit, W. E., *Phil. Mag.* **40**, 1 (1949).

Cramer, C. J. and Truhlar, D. G., *Science* **256**, 213 (1992a).

Cramer, C. J. and Truhlar, D. G., *Chem. Phys. Lett.* **198**, 74 (1992b).

Cramer, C. J. and Truhlar, D. G., *Chem. Phys. Lett.* **202**, 567 (1993).

Dam, J., Harkema, S., and Feil, D., *Acta Cryst.* **B39**, 760 (1983).

Dietrich, H. and Scheringer, C., *Acta Cryst.* **B34**, 54 (1978).

Eisenstein, M. and Hirshfeld, F. L., *Chem. Phys.* **42**, 465 (1979).

Eisenstein, M. and Hirshfeld, F. L., *Chem. Phys.* **54**, 159 (1981).

Eisenstein, M. and Hirshfeld, F. L., *Acta Cryst.* **B39**, 61 (1983).

Fernandes, N. G., Tellgren, R., and Olovsson, I., *Acta Cryst.* **B46**, 458 (1990).

Fkyerat, A., Guelzim, A., Baert, F., Paulus, W., Heger, G., Zyss, J., and Périgaud, A., *Acta Cryst.* **B51**, 197 (1995).

Flensburg, C., Larsen, S., and Stewart, R. F., *J. Chem. Phys.* **99**, 10130 (1995).

Fritchie, C. J. Jr., *Acta Cryst.* **20**, 27 (1966).

Gao, J., *Biophys. Chem.* **51**, 253 (1994).

Gao, J., in *Reviews in Computational Chemistry*, Vol. 7, K. B. Lipkowitz and D. B. Boyd (eds.), VCH: New York (1996), p. 119.

Gao, J. and Xia, X., *Science* **258**, 631 (1992).

Hamzaoui, F., Thesis, Université des Sciences et Technologies de Lille, France (1995).

Hartman, A. and Hirshfeld, F. L., *Acta Cryst.* **20**, 80 (1966).

Hermansson, K., *Acta Cryst.* **B41**, 161 (1985).

Hirshfeld, F. L., *Isr. J. Chem.* **2**, 87 (1964).

Hirshfeld, F. L., *Acta Cryst.* **B27**, 769 (1971).

Hirshfeld, F. L., in *Accurate Molecular Structure*, A. Domenicano and I. Hargittai (eds.), Oxford University Press: Oxford, New York (1992), p. 238.

Hirshfeld, H. and Hope, H., *Acta Cryst.* **B40**, 613 (1980).

Howard, S. T., Hursthouse, M. B., Lehmann, C. W., and Poyner, E. A., *Acta Cryst.* **B51**, 328 (1995).

Howard, S. T., Hursthouse, M. B., Lehmann, C. W., Mallinson, P. R., and Frampton, C. S., *J. Chem. Phys.* **97**, 5616 (1992).

Irngartinger, H. and Deuter, J., *Chem. Ber.* **123**, 341, (1990).

Irngartinger, H., Hase, H.-L., Schulte, K.-W., and Schweig, A., *Angew. Chem. Int. Ed.* **16**, 187 (1977).

Irngartinger, H., Deuter, J., Wingert, H., and Regitz, M., *Chem. Ber.* **123**, 345 (1990).

Ito, T. and Sakurai, T., *Acta Cryst.* **B29**, 1594 (1973).

Krijn, M. P. C. M. and Feil, D., *J. Chem. Phys.* **89**, 4199 (1988).

Krijn, M. P. C. M., Graafsma, H., and Feil, D., *Acta Cryst.* **B44**, 609 (1988).

Kwiatkowski, W., Bakshi, P. K., Cameron, T. S., and Liu, M. T. H., *J. Am. Chem. Soc.* **116**, 5747 (1994).

Lin, K.-J., Cheng, M.-C., and Wang, Y., *J. Phys. Chem.* **98**, 11685 (1994).

Madsen, D., Flensburg, C., and Larsen, S., to be published.

Matthews, D. A. and Stucky, G. D., *J. Am. Chem. Soc.* **93**, 5954 (1971).

Matthews, D. A., Swanson, J., Mueller, M. H., and Stucky, G. D., *J. Am. Chem. Soc.* **93**, 5945 (1971).

McIntyre, G. J., Ptasiewicz-Bak, H., and Olovsson, I., *Acta Cryst.* **B46**, 27 (1990).

Pichon-Pesme, V., Lecomte, C., and Lachekar, H., *J. Phys. Chem.* **99**, 6242 (1995a).

Pichon-Pesme, V., Lecomte, C., and Lachekar, H., Presented at BCA Charge Density Meeting, Durham, UK, December 1995b.

Robinson, F. N. H., *Bell System Technical Journal* 913 (1967).

Seiler, P., Belzner, J., Bunz, U., and Szeimis, G., *Helv. Chim. Acta* **71**, 2100 (1988).

Stein, A., Lehmann, C. W., and Luger, P., *J. Am. Chem. Soc.* **114**, 7684 (1992).

Stevens, E. D. and Coppens, P., *Acta Cryst.* **A32**, 915 (1976).

Stewart, R. F. and Jensen, L. H., *Acta Cryst.* **23**, 1102 (1967).

Walker, P. D. and Mezey, P. G., *J. Am. Chem. Soc.* **115**, 12423 (1993).

Walker, P. D. and Mezey, P. G., *J. Am. Chem. Soc.* **116**, 12022 (1994).

Appendix A

Patterson, A. L., *International Tables for X-ray Crystallography*, Vol. II, Kynoch Press: Birmingham, England (1959), p. 52.

Sands, D. E., *Vectors and Tensors in Crystallography*, Addison-Wesley Publishing Company, Reading, MA (1982).

Appendix B

International Tables for X-ray Crystallography, Vol. IV, Kynoch Press: Birmingham, England (1974).

Kuhs, W. F., *Acta Cryst.* **A40**, 133 (1984).

Sands, D. E., *Vectors and Tensors in Crystallography*, Addison-Wesley Publishing Company: Reading, MA (1982), p. 126.

Appendix C

Johnson, C. K. and Levy, H. A., in *International Tables for X-ray Crystallography*, Vol. IV, Kynoch Press: Birmingham, England (1974), p. 314.

Owen, D. B., *Handbook of Statistical Tables*, Addison-Wesley: Reading, MA (1962).

Appendix D

Arfken, G., *Mathematical Methods for Physicists*, 2nd ed., Academic Press: New York (1970).

Edmonds, A. R., *Angular Momentum in Quantum Mechanics*, 2nd ed., Princeton University Press: Princeton, NJ (1974).

Kara, M. and Kurki-Suonio, K., *Acta Cryst.* **A37**, 201 (1981).

Paturle, A. and Coppens, P., *Acta Cryst.* **A44**, 6 (1988).

Rose, M. E., *Elementary Theory of Angular Momentum*, John Wiley and Sons: New York (1957).

Steinborn, E. O. and Ruedenberg, K., in *Advances in Quantum Chemistry*, Vol. 7, P.-O. Löwdin (ed.), Academic Press: New York (1973).

Su, Z., PhD Thesis, State University of New York at Buffalo (1993), Appendix C, p. 108ff.

Su, Z. and Coppens, P., *Acta Cryst.* **A50**, 636 (1994a); see also erratum: **A51**, 198 (1995).

Su, Z. and Coppens, P., *Acta Cryst.* **A50**, 408 (1994b).

Appendix E

Abramowitz, M. and Stegun, I. A., *Handbook of Mathematical Functions*, National Bureau of Standards: Washington, DC (1964).

Edmonds, A. R., *Angular Momentum in Quantum Mechanics*, 2nd ed., Princeton University Press: Princeton, NJ (1974).

Su, Z., Thesis, State University of New York at Buffalo (1993), p. 128.

Appendix F

Clementi, E., *I.B.M. J. Res. Develop.* **9**, 2 (1965).
Clementi, E. and Raimondi, D. L., *J. Chem. Phys.* **38**, 2686 (1963).

Appendix G

Avery, J. and Watson, K. J., *Acta Cryst.* A**33**, 679 (1977).
Su, Z. and Coppens, P., *J. Appl. Cryst.* **23**, 71 (1990).

Appendix H

Arfken, G., *Mathematical Methods for Physicists*, 2nd ed., Academic Press: New York, London (1970).
Avery, J. and Watson, K. J., *Acta Cryst.* A**33**, 679 (1977).
Gradshteyn, I. S. and Ryzhik, I. M., *Table of Integrals Series and Products*, 4th ed., Academic Press: New York (1965).
Su, Z. and Coppens, P., *J. Appl. Cryst.* **23**, 71 (1990).
Su, Z., *On the Evaluation of Coulomb Potentials and Related Properties from the Multipole Description of the Charge Density*, Thesis, State University of New York at Buffalo, Buffalo, NY (1993).

Appendix I

Coppens, P. and Becker, P., in *International Tables for Crystallography*, Vol. C, Kluwer Academic Publishers: Dordrecht (1992), Chap. 8.7, p. 627.
Holladay, A., Leung, P. C., and Coppens, P., *Acta Cryst.* A**39**, 377 (1983).

Appendix J

Buckingham, A. D., in *Intermolecular Interactions from Diatomics to Biopolymers*, B. Pullman (ed.), Wiley and Sons: Chichester, New York (1978). p. 1.

Appendix K

Cohen, E. R and Taylor, B. N., *J. Res. Natl. Bur. Stand.* **92**, 85 (1987).
Smith, V. H., in *Electron Distributions and the Chemical Bond*, P. Coppens and M. B. Hall (eds.), Plenum Press: New York (1982).
Spackman, M. A., *Chem. Rev.* **92**, 1769 (1992).

Appendix L

Gregson, D., Klebe, G., and Fuess, H., *J. Am. Chem. Soc.* **110**, 8488 (1988).

Index

347